ISBN 978-1-334-66768-8
PIBN 10598637

FB &c Ltd, Dalton House, 60 Windsor Avenue, London, SW19 2RR.
Company number 08720141. Registered in England and Wales.

For support please visit www.forgottenbooks.com

Vol. III. December, 1902

University Studies

Published by the University of Nebraska

CONTENTS

Greek and Latin in Biological Nomenclature
Frederic E. Clements 1

LINCOLN, NEBRASKA

CONTENTS OF THE UNIVERSITY STUDIES, Vol. I

No. 1

1. *On the Transparency of the Ether*
By DeWITT B. BRACE

2. *On the Propriety of Retaining the Eighth Verb-Class in Sanscrit*
By A. H. EDGREN

3. *On the Auxiliary Verbs in the Romance Languages*
By JOSEPH A. FONTAINE

No. 2

1. *On the Conversion of Some of the Homologues of Benzol Phenol into Primary and Secondary Amines*
By RACHEL LLOYD

2. *Some Observations on the Sentence-Length in English Prose*
By L. A. SHERMAN

3. *On the Sounds and Inflections of the Cyprian Dialect*
By C. E. BENNETT

No. 3

1. *On the Determination of Specific Heat and of Latent Heat of Vaporization with the Vapor Calorimeter*
By HAROLD N. ALLEN

2. *On the Color Vocabulary of Children*
By HARRY K. WOLFE

3. *On the Development of the King's Peace and the English Local Peace Magistracy*
By GEORGE E. HOWARD

No. 4

1. *On a New Order of Gigantic Fossils*
By E. H. BARBOUR

2. *On Certain Facts and Principles in the Development of Form in Literature*
By L. A. SHERMAN

3. *On the* Dikanikos Logos *in Euripides*
By JAMES T. LEES

UNIVERSITY STUDIES

OF THE

UNIVERSITY OF NEBRASKA

VOLUME III

LINCOLN
PUBLISHED BY THE UNIVERSITY
1903

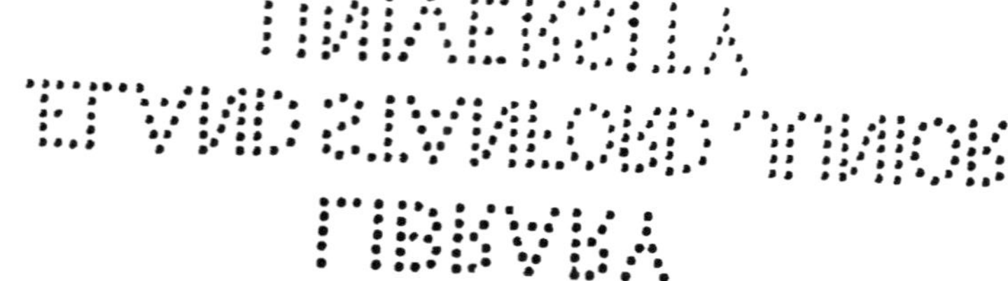

CONTENTS

CLEMENTS—Greek and Latin in Biological Nomenclature. 1

ENGBERG—The Degree of Accuracy of Statistical Data. 87

CARTMEL—The Anomalous Dispersion and Selective Absorption of Fuchsin 101

CARRIKER—Mallophaga from Birds of Costa Rica, Central America.. 123

FRYE—George Sand and her French Style 199

POUND—Notes on Certain Negative Verb Contractions in the Present. 223

MORITZ—On the Variation and Functional Relation of Certain Sentence-Constants in Standard Literature 229

BATES—On the Errors in the Methods of Measuring the Rotary Polarization of Absorbing Substances. 255

BATES—The Magnetic Rotary Dispersion of Solutions of Anomalous Dispersing Substances.......... 265

HARGITT—Regeneration in Hydromedusae........................... 275

WHITE—Some Peculiar Double Salts of Lead........................ 307

FLING—The Mémoires de Bailly......... 331

MORITZ—On the Representation of Numbers as Quotients of Sums and Differences of Perfect Squares........................ 355

UNIVERSITY STUDIES

VOL. III *DECEMBER, 1902* NO. 1

Greek and Latin in Biological Nomenclature

FREDERIC E. CLEMENTS

"Nomina Veterum Graecorum et Romanorum plantis imposita laudo, ad conspectum vero Recentiorum plurium horreo. Nec mirum factum! quis enim Tyro de nominibus fuit unquam instructus? quis unquam dedit circa denominationem plantarum praecepta, demonstrationes, exempla?" Linnaeus Critica Botanica I 1737.

The following treatise is intended to serve as a compendium of the principles of word-formation in Greek and Latin of sufficient thoroughness to enable the biologist to cônstruct in proper manner any derivative desired. Further than this, various unfortunate usages which have obtained in nomenclature and the many types of malformations will be considered in detail, and suggestions will be made for their correction or elimination. The treatment throughout is based upon the conviction that no biologist should be content with a nomenclature that is doubtful or crude in its philology. On the other hand, ultra-purism, together with the mooted questions pertaining solely to the classical philologist, will be avoided, since nomenclature for the sake of uniformity and stability must rest upon the assured. For these reasons, also, it is felt that, while he must conform to the best usage of the language, the nomenclator must go a step further, and, in the case of uncertain or various usage, establish a definiteness which the language itself did not know. Further warrant is found for this in the fact that the careless hand of

UNIVERSITY STUDIES, Vol. III, No. 1, December, 1902.

analogy is always busy throughout the life of a language, and, also, in the fact that the lexicon must take account of all usage, with the result that the cruder derivatives of formative and decadent periods of the language are found alongside of the purer, or at least more refined forms of the classical period.

While Kuntze's important contributions and the Rochester Code have been notable achievements on the way toward nomenclatural reform, it has been evident from the first that botanists had merely reached a temporary resting place, from which they must sooner or later go forward to the ultimate goal—a uniform and stable nomenclature and terminology of international recognition. The failure to deal with the matter of generic types and word-formation, both only less important than the cardinal principle of priority, made a reopening of the question inevitable, an event which is rapidly being brought about by the increasing frequency of papers upon nomenclature. The zoologists, while they have not gone so far in certain lines as the botanists, have greatly anticipated them by their action at the Zoological Congress of 1901, when they agreed to place zoological nomenclature upon a classical basis. Sooner or later, botanists must take the same action. When this time comes, biological nomenclature will be in a fair way to become a symmetrical, stable structure, based upon the two cardinal principles, priority and classicity. There can be little difference of opinion in regard to the repeated statement that nomenclature is merely an instrument in the hands of the biologist, and there should be just as little question that the instrument should be a worthy and ready one.

I.

Classical Greek and Latin are the basis of scientific nomenclature.

"Idiotae imposuere nomina absurda." Linnaeus Philosophia Botanica 158 1751.

There has never been any serious question concerning the necessity of a universal language for the natural sciences. The ancient and medieval development of biology, carried on first by Greek and Roman philosophers, and then perforce by men who had at least some knowledge of Greek and Latin, determined irrevocably that this scientific language should be Latin, immeasurably enriched by Greek derivatives. So natural and complete, indeed, was this linguistic heritage from the ancients and the herbalists that Linnaeus merely simplified the syntax, definitised the vocabulary, and modified the use of Latin, with its incorporated Greek, to obtain a great binomial system, without which taxonomy as it is to-day would have been impossible. Since Linnaeus, no botanist has questioned the right of Greek and Latin to constitute the language of science. DeCandolle did indeed point out the many advantages English would possess as an international means of communication between scientists, but it was hardly his thought that English would supplant Latin as the language of taxonomy. The realization of the suggestion, in view of the fact that biological publication is made in sixteen languages, among them Russian, Magyar, and Japanese, is anything but imminent. Yet, while biologists are agreed that Greek and Latin shall furnish the materials for nomenclature and terminology, their practice, unfortunately, is still very far from uniform. Personal and vernacular terms from all possible sources have increased to such an extent that nearly a sixth of our present generic and specific names are derived from vernacular tongues. The economy of time and intellectual effort obtained by the use of such names is so considerable that they will always appeal to the poorly prepared or indifferent descriptive biologist. But they offend all the canons of uniformity and taste, and the real taxonomist, whose work is thorough and painstaking from the first glimpse of a new organism to the final publication of its name and diagnosis, will avoid them.

The best Greek and the best Latin available are alone good enough for biological nomenclature. The Greek and Latin of Linnaeus were the work of no very certain hand, and should not constitute the standard, when a better standard is obtainable. Linné's knowledge of word-formation in Greek was often elusive, though his names are far superior as a rule to those of more recent coinage. Similarly, the formations of Byzantine

Greek and Late Latin, as well as those of many preclassic authors in both languages, have little value for the nomenclator. Classic Greek and Latin only can be fully satisfactory, since they are not merely the best Greek and Latin obtainable, but, also, because they present the best conditions for securing essential uniformity. Again, it should be clearly understood that classic Greek and Latin are not necessarily the Greek and Latin of the extreme purist.

II.

A name or term is invalid unless constructed according to the principles of word formation in classic Greek or Latin; alternatives are to be reduced to a uniform basis. Retroactively, all terms improperly constructed shall be corrected, except in the case of words of uncertain or unknown etymology, when no correction shall be made if any proper Greek or Latin construction will give such a word, with a possible meaning.

"Nomina generica ab uno vocabulo . . . fracto altera integra composita Botanicis indigna sunt." Critica Botanica 29 1737.

"Nomina generica ex duobus latinis vocabulis integris et conjunctis vix toleranda sunt." *Ibid.*, 26.

This rule finds its warrant in the fact that uniformity is a first requisite of nomenclature as purity is of linguistics. A malformation is not only unpleasant as well as incorrect philologically, but it is also extremely unfortunate by reason of the complications which it introduces into nomenclature. The philologist is satisfied only with most skilful handling of derivatives that is possible. He will no more be guilty of a malformation or a hybrid than the true scientist will be capable of a bit of superficial or bungling work. The latter must then learn to look upon linguistic matters with the same conscientiousness that he uses in scientific investigation. Ultimately, however, he must be prepared to go farther than the philologist even, for the sake of uniformity. The latter is chiefly concerned with the development of a language, or group of languages, and with him slightly different or alternative forms are of advantage rather than a source of difficulty. In science, where the form and application of each name or term should be absolutely fixed, alternative forms of words and alternative methods of composition lead inevitably to grave confusion. The nomenclator must in consequence outdo the philologist in his own field. When it is possible to obtain essentially the same derivative in several slightly different forms by varying the stem of the first term, the connecting vowel, or the form of the last term, or by proceeding from alternative forms of the same word or stem, then the nomenclator must make the most intelligent choice possible in the selection of the best form to use, or the best principle to govern. In so doing, he will often strengthen the hands of the philologist, since it is a well-known fact that many alternative forms are merely the bungling creations of the decadent period of a language.

In choosing a principle for guidance in dealing with alternative forms and methods of derivation, several courses have been considered. The first plan was to follow the usage in the case of each particular word, but it soon became evident that no one but a specialist in philology would be able to make derivatives at all, since the usage varied repeatedly in words of the same group. A similar attempt was made with regard to the best usage, but, while this led to somewhat greater uniformity, the results were not much more satisfactory, and the labor involved was enormous. From the first it was seen that, while an occasional word would deviate more or less regularly from the formation typical for its group, as in the case of the imparisyllabic neuter, *στόμα, στόματος* (mouth), which regularly enters into composition in its shortened stem form, the philologically correct stem, or the correct connective, was overwhelmingly predominant. Furthermore, since such usage includes the best usage in all cases, it was concluded that uniformity and purity could best be obtained by making this the invariable usage for all the stems of any group, as well as for all combinations of each stem.

The justification of such a rule may be readily found in a consideration of imparisyllabic stems, which have constituted the most fertile source of alternatives. The Greek neuters in *-μα*, gen. *-ματος*, furnish a large number of examples in which the shortened form of the nominative and the stem proper of the oblique cases alternate in word-formation. The Greek lexicon exhibits 1,782 neuters of this class, of which 231 appear in 969 derivatives as the first term. In the latter the proper stem appears in 781 words, while the shortened form appears in 188 words. The alternation of these stems in Greek has of necessity given rise to corresponding alternatives in nomenclature. Thus, there are found Grammonema Ag. 1832 and Grammatonema Kuetz. 1845, Lomaspora DC. 1821 and Lomatospora Reichenb. 1828, Spermodermia Tode 1790 and Spermatodermia Wallr. 1833, Stomotechium Lehm. 1818 and Stomatotechium Spach 1843. Unfortunate and confusing as the variants of the same generic name are, the case is very much worse when the variations of one stem furnish two otherwise valid generic names, as in the case of Dermatocarpus Eschw. 1823 and Dermocarpa Crouan 1858, Grammocarpus Seringe 1825 and Grammatocarpus Presl 1831, Haemospermum Reinw. 1825 and Haematospermum (Wallich) Lindl. 1836. In the former, we are concerned merely with uniformity, desirable as that may be, while in the latter the validity of a generic name is destroyed because of its essential identity with an earlier name, an identity of which the later author was probably unaware. Such fatal duplication of generic names can only be avoided by stringent rules for securing uniformity in methods of derivation. Myosurus L. 1737 (Myosuros Dill. 1719) and Myurus Endl. 1837 are again alternative forms of the same compound word, which have been applied to different genera. The former illustrates the rare and archaic type of syntactic composition, the latter follows the usual method of composition by stems. In the case of Coleosanthus Cassini April 1817 and Coleanthus Seidl July 1817, the latter, though correctly formed, falls by the working of priority before the former, which is a blunder, equally indefensible from the standpoint of syntactic or non-syntactic composition. Callitriche L. 1751 and Calothrix Ag. 1824 illustrate the confusion that arises from using alternative Greek words (καλλι-, καλός, beautiful) and from the variation of the termination of the last member of the compound. Either first term is correct, but their compounds are identical in meaning and essentially so in derivation. They are to be regarded merely as different forms of the same compound, and Calothrix becomes a homonym. The confusion wrought by alternative forms and blundering construction is nowhere better shown than in the following series of names, belonging to five different genera: Asterothrix Cassini 1827 (Asterotrix Brogn. 1843, Asterothria Gren. 1850), Asterotrichion Link 1840 (Asterostrichion "Klotsch" 1840, Asterotrichium Witts.), Asterotrichia Zanard. 1843 Astrotricha DC. 1829 (Astrotrichia Rchb. 1837), and Asterotrichum Bonord. 1851.

The retroactive application of this rule is imperative for the sake of uniformity and purity. By far the greater number of plant genera have already been recognized and named. The new names to be proposed for years to come will be relatively few, and a reform which affected even all of these would be barely worth while. Further than this, most new names are made after the pattern of names already in use, whether correctly or incorrectly formed, a practice certain to perpetuate the blunders of the past. Arguments from the standpoint of purity are equally cogent, but, as they would perhaps appeal to the philologist alone, they will not be insisted upon here. A rule of this sort to be at all worth while must be retroactive, for by retroaction alone can confusion be avoided and uniformity secured. The retroactive operation of the rule must be so safeguarded, however, that changes for reasons of uniformity or purity will be made upon real and not upon supposititious grounds. Framers of generic names have been extremely careless in the matter of indicating etymologies, but this is not sufficient warrant for reconstructing names upon the basis of supposed meanings. Many a genus has received a name of known or evident etymology, but of meaningless or mistaken application, a fact which should restrain us from correcting words of unknown derivation on the basis of an assumed etymology. In making changes to secure a more uniform and stable nomenclature, the greatest care must be taken to minimize the error arising from personal judgment. In many words of uncertain etymology, several derivations are equally plausible, or at least possible, and the exercise of personal choice would simply lead to greater confusion. For these reasons, changes in words where the etymology is not expressly indicated or clearly evident should not be made, unless the proper formation of such a word in Greek or Latin fails to give a name of any possible meaning. The correction of such words as fall under this rule can only be made upon the basis of greatest probability, which, unsatisfactory as it may be, will conduce to the ends sought.

WORD FORMATION IN GREEK

Greek words arise by derivation or by composition. In derivation, roots or stems acquire a new meaning through the addition of a suffix, a termination having no separate existence in the language, except in the rare case of certain words which have lost their real significance and are now found only as suffixes. In composition, two, rarely more, words are united according to certain rules to form a new term, or compound, in which the meaning of each may be traced. Formation by prefixes is really a sort of composition, except in the case of a few inseparable particles, which properly belong under the head of derivation. For the sake of convenience, however, all formation by prefixes will be considered under composition.

Greek has obtained its stems by derivation, i. e., by adding suffixes to roots, a process to which the origin of all simple words may be traced. Derivation belongs chiefly to the earlier development of the language, and, indeed, is very largely prehistoric, especially in the case of primary derivation. Composition, on the other hand, is a much later development, and must have attained its maximum in the classical period of Greek literature. Both derivation and composition afford the biologist the means of coining new words. For various reasons, among them convenience and usage, scientific terms have been taken directly from the Greek lexicon (sometimes, of course, they have been found already borrowed in Latin), or new words have been formed by composition. Formation by derivation is equally valid, and the fact that it almost invariably gives shorter words leads one to wonder that it should not have come into general use. The reason may be found in the fact that word-formation in biological nomenclature has been far from scholarly, and that derivation requires much greater care and knowledge than composition does. It is also true that the possibilities of derivation in Greek, though large, are necessarily limited by the relatively small number of suffixes, while the sources of composition are practically inexhaustible.

DERIVATION

Derivation consists in the addition of one or more suffixes to the primitive, irreducible portion of a word, which is termed a root. It may be distinguished as primary when one suffix is added to the root, making a stem, secondary when a second suffix is added to the stem, tertiary when a third suffix is attached, and so forth. For convenience, however, we may follow Henry,[1] and term those derivatives primary in which the root carries a single suffix, and secondary, all those in which the stem thus formed has been modified by one or more accretions. Furthermore, derivatives are classed as verbal when the suffix added permits of conjugation, and nominal when it permits of inflection. It is important that this be kept distinct from the fact that certain suffixes can be added only to verbal stems, while others can be attached only to nominal (denominative) ones. Nomenclature is not concerned with the construction of verbal stems, and the suffixes which follow are those which form nominal stems, i. e., nouns and adjectives.

Primary derivatives are formed by attaching the suffix immediately to the root, though rarely an adventicious -σ- intervenes. Secondary derivatives are made in similar manner by adding the suffix directly to the stem. In both cases, the groups of letters thus brought into contact conform to certain general phonetic principles of the language. For convenience in making the changes, which arise in this way in derivation and composition, a short summary of the phonetic mutations in Greek is given. Mutations peculiar to verbal stems are omitted. A more complete account of these phonetic laws may be found in any of the more comprehensive grammars.

[1] **Henry, Victor. A Short Comparative Grammar of Greek and Latin, 102. 1890.**

GENERAL PHONETIC PRINCIPLES

Aspirates

In composition, aspirates (χ (kh), φ (ph), θ (th)) arise when a surd (κ, π, τ), usually by elision of the final vowel of the stem of the first term, comes in contact with an initial aspirated vowel of the second term.

δέκ(α)-ἡμέρα – δεχήμερος, ten-day

ἐπ(ί)-ἕδρα – ἔφεδρος, seated upon

ἀντ(ί)-ὅρος – ἄνθορος, an opposite limit

Very rarely, this influence is exerted through an interposing consonant.

τέτρ(α)-ἵππος – τέθριππος, with four horses abreast

Accumulation of Consonants

As a rule, groups of consonants are modified to prevent harshness. Generally, three successive consonants, or a consonant and a double consonant, are avoided, or one letter is dropped, unless the first or last is a liquid (λ, μ, ν, ρ), or γ before a palatal (κ, γ, χ, ξ).

πέμπτος, fifth; σκληρός, hard; σάλπιγξ, trumpet

In composition, final κ or σ of the first term may stand before two other consonants.

ἐκστροφή, dislocation; ἐκφθείρω, to destroy utterly

The concurrence of two consonants, when it produces harshness, is avoided in several ways.

(1) When, by the transposition or loss of a letter, μ or ν stands immediately before λ or ρ, the corresponding sonant (β, δ) is inserted.

μές(ος)-ἡμέρα – μεσημ(ε)ρα – μεσημβρία, midday

ἀνήρ, genitive, *αν(ε)ρος – ἀνρός, – ἀνδρός, man

(2) A consonant is sometimes transposed to a more convenient position.

πυκνός, genitive, πνύξ, nominative, meeting place

Assimilation.

Two explosives can occur together only when the latter is a dental (τ, δ, θ). In such a group a palatal or labial must be of the same order, and another dental is changed to σ;

κ and π can alone stand before τ, γ and β before δ, and χ and φ before θ, while σ may occur before all three.

φλεκτικός, burning, from φλέγω, to burn; τριπτήρ, rubber, from τρίβω, to rub

πλέγδην, entwined, from πλέκω, to twist; γράβδην, grazing, from γράφω, to grave

σχιστός, cloven, from σχίζω, to cleave; πειστέον, persuaded, from πείθω, to persuade

ἐκ, from, always retains its final palatal in composition.

ἔκδημος, foreign; ἔκθυμα, pustule

Before σ, β and φ become π, γ and χ become κ, while τ, δ, and θ are dropped; κσ is then written ξ and πσ is written ψ.

χάλυβος gen., χαλυβ-ς nom. – χάλυψ, a Chalybean

γράφω, γραφ-σω – γράψω

μάστιγος gen., μαστιγ-ς – μάστιξ, whip

τριχός gen., θριχ-ς – θρίξ, hair

χάριτος gen., χαριτ-ς – χάρις, grace

λαμπάδος gen., λαμπαδ-ς – λαμπάς, torch

κόρυθος gen., κορυθ-ς – κόρυς, helmet

This rule applies to such groups as -κτ-, in which the τ is first dropped and the κ then passes into ξ.

νυκτός gen., νυκτ-ς – νύξ, night

Before μ, labials (π, β, φ) become μ, palatals (κ, χ) become γ, and dentals (τ, δ, θ, ζ) become σ.

βλεπ-μα (βλέπω) – βλέμμα, glance; τριβ-μα (τρίβω) – τρίμμα, anything rubbed; στεφ-μα (στέφω) – στέμμα, garland; πλεκ-μα (πλέκω) – πλέγμα, anything plaited; τευχ-μα (τεύχω) – τεῦγμα, a work; ᾀδ-μα (ᾄδω) – ᾆσμα, song; σχιδς-μα (σχίζω) – σχίσμα, cleft; πειθ-μα (πείθω) – πεῖσμα, cable.

ἐκ remains unchanged; ἔκμαγμα, a wax impression

The dentals (τ, δ, θ, ζ) are retained only before λ, ν, ρ. Before μ, they become σ (see above), as also before each other; before σ they are dropped.

πειθ-τικος (πείθω) – πειστικός, persuasive; ἡδ-θημα (ἥδομαι) – ἥσθημα, delight; σπερματ-σι, dat. (σπέρματα) – σπέρμασι; σχιδσ-σις (σχίζω) – σχίσις, cleaving.

Before another liquid (λ, μ, ρ), ν is assimilated to the liquid; before a labial (π, β, φ, ψ), it becomes μ; before a palatal (κ, γ, χ, ξ), it changes to γ; before σ, it is elided and the preceding vowel usually lengthens; before another ν, it is usually retained.

παλιμμήκης (πάλιν), very long; παλίλλυτος, loosed again; παλίρροια, back water; παλιμπλανής, wandering to and fro; σύμβλεμα (σύν), seam; σύμφυσις, a growing together; σύμψαλμα, harmony; παλίγκυρτος, fishing net; σύγγονος, congenital; σύγχροος, of like color; συγξέω, to smooth by scraping; μέλανος, μελαν-ς – μέλας; δαίμονος, δαιμον-σι – δαίμοσι; γίγαντος, γιγαντ-ς – γίγας.

Σύν drops its final before σ and a consonant, or before ζ, but the ν is simply assimilated before σ and a vowel.

σύστημα, system; σύζωμα, girdle

συσσεισμός, earthquake; σύσσωμος, united in one body

Πάλιν assimilates its ν before σ and a vowel, and usually retains it before σ and a consonant; before another ν, it is either dropped or retained.

παλίσσυτος, rushing back; παλίνσκιος or παλίσκιος, deeply shaded

παλίνζωος, living again; παλίννοστος or παλίνοστος, returning

Αγαν always drops ν, except where doubling or assimilation takes place.

ἀγάννιφος, ἀγάρροος

Εν does not change its final before ρ, σ, or ζ.

ἔνριζος, rooted; ἔνστασις, plan; ἐνζέννυμι, to boil in

Doubling of Consonants

(1) In word-formation, initial ρ is generally doubled when it follows a vowel, but remains single after a diphthong.

διαρρωγή, gap; γλυκύρριζα, sweet root; εὐρυρέων, broad-flowing; εὔριζος, well-rooted

(2) An aspirate is never doubled, but the corresponding surd takes the place of the first. Σαπφώ for Σαφφώ, etc.

(3) Doubling is a frequent phenomenon (mostly in verbs and comparatives) when the suffix *ya* (ι) follows the final consonant of root or stem; final κ, γ, χ, and, rarely, other explosives, absorbing ι, becoming σσ, δ becomes ζ, and λ becomes λλ.

φυλάσσω (φυλακ-ι-ω), μείζων (μεγ-ι-ων), ἄλλος (ἀλ-ι-ος).

Metathesis, or transposition of this ι takes place when it follows final ν or ρ.

θεράπαινα — * θεραπ-αν-γα — θεραπανια

σώτειρα — σωτηρ-ι-α

Syncope or elision of a vowel often occurs in the middle of a word.

πατρός for πατέρος

Contraction of vowels should be ignored, when an occasion for it might arise in forming scientific terms.

NOUN SUFFIXES

General

-μο- (-μος, m.) primary or secondary verbal oxytone: θυ-μός, heart; ἐρ-ισ-μός, strife

-μα- (-μη, f.), primary (or secondary?) verbal paroxytone: θέρ-μη, heat

-ο- (-ος, -ον, m. or n.) chiefly primary: νομ-ός, pasture; λύκ-ος, wolf

-α- (-η, f.) chiefly primary: φυγ-ή, flight; ῥο-ή, stream; λεύκ-η, white poplar

-ι- (-ις, m. or f.) chiefly primary, paroxytone: πόλ-ις, city

-εν-, -ον- (-ην, -ων, m. or f.) primary or secondary, mostly verbal: ἄρσ-ην, male; εἰκ-ών, image, αἰ-ών, age

-μεν- (-μην, m.) primary oxytone: λι-μήν, harbor

-μον- (-μων, m.) primary paroxytone: τέρ-μων, boundary

-μνο- (-μνον, n., -μνη, f.) primary, usually oxytone: στρω-μνή, bed

-ρο- (-ρος, m., -ρα, f., -ρον, n.) primary, mostly oxytone: ἕδ-ρα, seat, δῶ-ρον, gift

-λο- (-λος, m., -λη, f., -λον, n.) primary, mostly oxytone: φυ-λή, tribe; φῦ-λον, class. -ιλος, -ηλη, -ωλον, -ωλη, are widely extended false suffixes used after a consonant: they show the accretion of certain stem vowels.

-νο- (-νος, m., -νη, f., -νον, n.) primary, often oxytone: ὕπ-νος, sleep; ποι-νή, penalty; τέκ-νον, child. An adventicious α has given the suffix -ανο- (-ανος, -ανη, -ανον, seen in: στέφ-ανος, crown; μηχ-ανή, device; δρέπ-ανον, scythe.

-νι- (-νις, f.,) primary: μῆ-νις, wrath

-το- (-τος, m., -τη, f.) primary, usually paroxytone: χόρ-τος, yard; κοί-τη, bed

-ατ- (-αρ, -ωρ, n.) primary: ἧπ-αρ, liver; ὕδ-ωρ, water

-ακ- (-αξ, m.) primary paroxytone: ῥύ-αξ, torrent; ἅρπ-αξ, robber

-αδ- (-ας, f.) primary or secondary, verbal or denominative, oxytone: λαμπ-άς, torch; ἑβδ-ομ-άς, week

-ιδ-, -ιθ- (-ις, f., rarely m.) primary or secondary, mostly oxytone when feminine, and paroxytone when masculine: ὄρν-ις, bird; κλε-ίς, key; πα ῖς, child; ἡμερ-ίς, oak with edible acorns; βασιλ-ίς, queen

-ιτ- (-ις, f., -ι, n.) primary paroxytone: χάρ-ις, grace; μέλ-ι, honey

-ωτ- (-ως, m.) primary paroxytone; γέλ-ως, laughter

-ω- (-ω, -ως, m. or f.) primary, feminine oxytone, masculine paroxytone: ἠχ-ώ, sound; ἥρ-ως, warrior

-ερ- (-ηρ, m.) primary oxytone: ἀ-ήρ, atmosphere; αἰθ-ήρ, ether

-ορ- (-ορ, -ωρ, n.) primary paroxytone: ἄ-ορ, sword; πέλ-ωρ, prodigy

Agent

-ηυ- (-ευς, m.) primary, verbal or denominative, oxytone: γραφ-εύς, writer

-ευ- (-ευς, m.) secondary denominative oxytone: γραμματ-εύς, scribe

-τερ- (-τηρ, m., -τειρα, f.) primary or secondary verbal oxytone: λυ-τήρ, deliverer; νικ-η τήρ, conqueror

-τορ- (-τωρ, m.,) primary or secondary verbal paroxytone: ῥή-τωρ, orator; νικ-ά-τωρ, conqueror

-τα- (-της, m.) primary oxytone or paroxytone: κρι-τής, judge Secondary (1) verbal, usually oxytone, with short primary vowel or sigma, ναι-έ-της, inhabitant, ἐρα-σ-τής, lover, or with long primary vowel, νικ-η-τής, conqueror, or with long primary vowel and sigma, ὀρχ-η-σ-τής, dancer; (2) denominative, generally paroxytone, οἰκ-έ-της, servant, δεσ-μώ-της, prisoner. From these words, the stem vowel has come to remain attached to the suffix, giving the agent suffixes, -ιτης, -είτης, -ώτης, ιώτης.

Means or Instrument

τρο- (-τρος, m., -τρα, f., -τρον, n.) primary or secondary verbal, feminine and neuter usually paroxytone: δαι-τρός, knife; ῥή-τρα, agreement; βάκ-τρον, staff; ἄρ-ο-τρον, plough

-τλο- (-τλος, m., -τλον, n.) primary, usually paroxytone: ἄν-τλος, bucket; χύ-τλον, liquid
(-τλη, f.) secondary verbal, usually paroxytone: ἐχέ-τλη, handle

-θρο- (-θρον, n.) primary, usually paroxytone: ἄρ-θρον, joint
(-θρα, f.) secondary verbal, usually paroxytone: κοι-μή-θρα, chamber

-θλο- (-θλη, f., -θλον, n.) primary (or secondary?) paroxytone: θύς-θλον, sacred implement; γενέ-θλη, race

-κο- (-κη, f.) primary paroxytone: θή-κη, box

-ευ- (ευς, m.) secondary verbal or denominative oxytone: ἀμελγ-εύς, milkpail

Result

-ματ- (-μα, n.) primary or secondary, verbal (secondary rarely denominative) accent recessive: φράγ-μα, palisade; σῶ-μα, body; δή-λη-μα, bane

-ες- (-ος, n.) primary, mostly paroxytone: βέλ-ος, dart; ἕλ-ος, marsh

Place

-τηριο- (-τήριον, n.) primary or secondary verbal proparoxytone: δικασ-τήριον, court house

-ειο- (-εῖον, n.) primary or secondary denominative paroxytone: ψιλ-εῖον, prairie

-ων- (-ων, m.) primary or secondary denominative oxytone: ἀμπελ-ών, vineyard

-τρα- (-τρα, f.) primary or secondary verbal paroxytone: παλαί-σ-τρα, place for wrestling

Action

-τι- (-τις, f.) primary verbal, usually paroxytone: φά-τις, speech

-σι- (-σις, f.) primary or secondary verbal, usually paroxytone: φύ-σις, nature; ἀφάν-ι-σις, disappearance

-σια- (-σια, f.) usually secondary verbal paroxytone: δοκιμα-σία, testing

Quality

-ια- (-ια, f.) primary or secondary denominative, mostly paroxytone: ἁρμον-ία, harmony

-ος-, -ες- (-ως, m. or f.) primary oxytone: ἠ-ώς, dawn

(-ος, n.) primary recessive: βάρ-ος, weight, ἔρευθ-ος, redness

-τητ- (-της, f.) secondary denominative paroxytone: λεπτ-ό-της, thinness; whence -οτητ- (-ότης, f.) παντ-ότης, universality

-συνα- (-συνη, f.) secondary denominative paroxytone: σωφρο-σύνη, prudence

State or Object

-δον- (-δων, f.) secondary verbal oxytone: ἀλγ-η-δών, suffering

-μονο- (-μονη, f.) primary verbal oxytone: χαρ-μονή, joy

-τυ- (-τυς, f., -τυ, n.) primary: βρω-τύς, meat; ἄσ-τυ, town

-υδ, -υθ- (-υς, f.) primary, often oxytone: χλαμ-ύς, cloak

Diminutives

-ιο- (-ιον, n.) primary or secondary denominative paroxytone: σπορ-ίον, little spore. Various suffixes of stems have become attached to this diminutive, giving the common diminutive suffixes, -αριον, -ιδιον, -υδριον, -υλλιον, -υφιον, all forming neuter proparoxytones.

-ισκο- (-ισκος, m., -ισκη, f.) primary or secondary denominative paroxytone: νεαν-ίσκος, youth; παιδ-ίσκη, little girl. This suffix sometimes combines with -ιον to form a suffix -ισκιον, neuter proparoxytone; ἀσπιδ-ίσκιον, small shield.

Patronymics

-δα- (-δης, m.) secondary denominative paroxytone

-δ- (-ς (δς) f.) secondary denominative oxytone

Stems of the first declension add the suffix directly: Βορεά-δης, son of Boreas; Βορεά-ς, daughter of Boreas.

Stems of the second declension replace ο of the stem with ι: Πριαμ-ίδης, son of Priam; Πριαμ-ίς, daughter of Priam. Those in -ιο, however, change ο to α, giving the suffixes -ιάδης and -ιάς.

Stems of the third declension insert ι before the suffix, ευ dropping the υ before ι: Κεκροπ-ίδης, a son of Cecrops; Κεκροπ-ίς, daughter of Cecrops.

ADJECTIVE SUFFIXES

General

ες (-ης, m., f., -ες, n.) primary, rarely secondary denominative oxytones: ψευδ-ής, false; εὐγεν-ής, well-born; λιπ-αρ-ής, persistent

-ο-, -α- (-ος, m., -η, -α, -ος, f., -ον, n.) primary or secondary (denominative when secondary) always oxytone, except in compounds: ψιλ-ός, ή, όν, bare; ξηρ-ός, ά, όν, dry; βού-νομος, ον, grazed by cattle

-αδ- (-ας, m., f.) primary oxytone: σπορ-άς, scattered; λογ-άς, selected

-ιδ- (-ις, f.) secondary denominative oxytone, feminines of nouns or adjectives, most having become substanives: Δελφ-ίς, Delphian

Ownership or Relation

-ιο- (-ιος, -ια, -ιον) primary denominative proparoxytone: στύγ-ιος, hateful: secondary denominative; (1) the stem vowel may be elided before ι, as θαλάσσ-ιος, marine, from θάλασσα, or (2) it may be retained, as δίκα-ιος, just, ὅμο ιος, similar, whence arise new forms of the same suffix, i.e., -αιος, -οιος, -ειος, ῳος, etc.

-σιο- (-σιος) arose from adding -ιο- to stems in -τι, but is now regularly used as a suffix; θαυμά-σιος, wonderful.

-ιδιο- (-ιδιος) arose from attaching -ιο- to stems in -ιδ-, but has become a regular suffix (especially frequent in the neuter to form diminutives): θαλασσ-ίδιος, marine.

-κο- (-κος, η, ον) secondary denominative oxytone: φυσι-κός, natural: whence has probably come -ικο- (-ικός), πολεμ-ικός, warlike, δερ-ματ-ικός, cutaneous; whence -τικο- (-τικός), especially applied to nouns of agent in -της. The addition of -κός to stems in -ια has given the suffixes -ιακός, and -ακός; to stems in -υ, -υκός.

Material

-ινο- (-νο-) (-ινος, η, ον) primary or secondary denominative proparoxytone: δρύ-ινος, oaken; ξύ-λ-ινος, wooden

The modification of the initial vowel of the suffix has produced the suffix -ηρο- (-ηρος), which is a secondary verbal oxytone, πετ-ε-ηνός, winged.

-ιο- (-εος, -εα, -εον) secondary denominative proparoxytone: the nominative form arose from primary stems in -ε, and the intervocalic ι was then elided, ἀργυρε-ιο-ς – ἀργύρεος, silver; μολύβδ-εος, leaden.

-ινεο- (-ινεος) secondary denominative paroxytone: formed by adding -ιο to -ινο; φηγ-ινέος, oaken.

Quality

-μο- (-μος) primary oxytone: θερ-μός, hot. The addition of this suffix to stems in -τι, δρα-στι-μός, active, has produced a secondary denominative suffix -ιμος, ἐδ-ώδ-ιμος, eatable, and this, by further combination, has given -άλιμος, εἰδ άλιμος, beautiful.

-ρο- (-ρος) primary, nearly always oxytone: λαμπ-ρός, bright; ἐρυθ-ρός, red: secondary, mostly denominative, usually oxytone: φαν-ε-ρός, plain, whence the suffix -ηρος: κυματ-ηρός, billowy.

-λο- (-λος) primary, nearly always oxytone: δει-λός, timid. Secondary denominative oxytone: σιγ-η-λός, silent, whence the suffixes -ηλός, -ωλός, etc.; ἀπατ-ηλός, deceitful; ἁμαρτ-ωλός, used to sin.

Fulness

-εντ- (-εις, -εσσα, -εν) secondary denominative, usually paroxytone, the feminine proparoxytone: χαρί-εις, graceful; πτερό-εις, winged, whence the suffixes, -όεις, -ήεις; σκι-όεις, shady, δενδρ-ήεις, woody.

-νο- (-νος) primary oxytone: σεμ-νός, holy

-ινο- (-ινος) primary and secondary oxytone: πεδ-ινός, quite level; ὀρ-ε-ινός, mountainous, whence -εινός; εὐδι-εινός, quite cheerful.

-ρι- (-ρις) primary paroxytone, ἴδ-ρις, skilful

-αλεο- (-αλεος) secondary paroxytone: ῥωμ-αλέος, strong; ψωρ-αλέος, itchy

-μον- (-μων, -μον) primary, usually paroxytone: ἴδ-μων, skilful

Ability or Fitness

-ικο- (-ικος) secondary verbal oxytone: γραφ-ικός, able to write; ἀρχ-ικός, fit to rule

-τικο- (-τικος) secondary verbal oxytone: πρακ-τικός, practical

-ιμο- (-ιμος) primary or secondary, mostly verbal, proparoxytone: πότ-ιμος, drinkable; θανά-σ-ιμος, deadly, hence -σιμος?

-ιμαιο- (-ιμαῖος), ὑποβολ-ιμαῖος, spurious

Time

-ινο- (-ινος) usually secondary, denominative oxytone: ἡμερ-ινός, of day; ὀπωρ -ινός of late summer; χθεσ-ινός, of yesterday

Likeness

-ωδε- (-ώδης, m., f., -ῶδες, n.) secondary denominative paroxytone, arising from εἶδος, τό, form, in composition as the last term, whence the form -οειδής, and, by contraction, -ώδης; λιμν -ώδης, like a marsh, marshy. This suffix is often used to indicate fulness, also.

Verbal: Capability or Obligation

-το- (-τος) primary or secondary verbal oxytone: σχισ-τός, split; κλυ-τός, renowned; φιλ-η-τός, loved

-τεο- (-τεος) seconc'a·y verbal paroxytone: φιλ-η-τέος, lovable

COMPOSITION

Greek exhibits two types of composition, syntactic and non-syntactic. Syntactic composition is the union under a single accent of two words, one being merely a modifier of the other and in the case demanded by this relation. Such forms arise often from juxtaposition, for reasons of convenience, and are not, properly speaking, compounds, e. g., κυνός-βατος, dog thorn (κύων, κυνός, dog), μυος-ωτίς, mouse ear (μῦσ, μυός, mouse). The subordinate word is usually in the genitive, though, rarely, it may occur in practically any case. In non-syntactic composition, the two terms of the compound are morphologically coordinate, though the one is usually subordinated to the other in meaning. The second word is attached to the stem of the first in the same way that secondary suffixes are added to stems, with the very important exception that the final vowel of the first member has become a universal thematic vowel, or connective, e.g., μακρο-σπορα, κορυνη-φορα. Non-syntactic composition is the only real composition. It is so overwhelmingly prėdominant in Greek that it alone needs to be taken into account. Indeed, syntactic composition must be sedulously avoided by biologists, if confusion is to be prevented, and the few syntactic compounds already in existence in nomenclature should be made to conform to the rules for non-syntactic composition.

Compound words consist of three elements, the first term, the connecting vowel, and the last term. For reasons of convenience, the last term will be considered first, then the connective, and, finally, under the first term, will be given a detailed exposition of composition in the different classes of words.

THE LAST TERM

The last term is always a nominal stem, i. e., noun, adjective, or verbal adjective. The form of the last term is necessarily determined by its character, as follows:

I. If the last term is a noun, it may (1) stand without change, and the resulting compound is properly a substantive, though Greek often employs such words as adjectives, or (2) it may take adjectival endings, according to its declension, and the resulting compound is an adjective. Again, in Greek, practically all compound adjectives may be used as substantives.

II. If the last term is an adjective or verbal adjective, it may stand without change in the resulting compound, but usually it becomes an adjective of two terminations (-**ος**, m., f., -**ον**, n., rarely, -**ης**, m., f., -**ες**, n.). The adjective may take substantive suffixes, in which case the compound will, of course, be a noun.

The following examples will illustrate the form of the last term and the character of the resulting compound.

I.1. The last term is a noun, undergoing no change.[1]

ποδο-σπορα, ἡ (**ποῦς**, **ποδός**, ὁ, foot, **σπορά**, ἡ, seed) foot-spore

αἱματο-κοκκος, ὁ (**αἷμα**, **αἵματος**, **τό**, blood, **κόκκος**, ὁ, berry) blood-berry

ὀφιο-σταφυλη, ἡ (**ὄφις**, **ὄφιος**, **ὄφεως**, ὁ, snake, **σταφυλή**, ἡ, bunch of grapes) briony

[1] For accent of compounds, see Buttmann, 292.

ξιφο-θηκη, ἡ (ξίφος, ξίφεος, τό, sword, θήκη, ἡ, box) scabbard
περι-βλημα, τό (περί, around, βλῆμα, βλήματος, τό, throw) covering
μικρο-καλύβη, ἡ (μικρός, small, καλύβη, ἡ, hut) small hut.

I.2. The last term is a noun, changed to an adjective, usually by a suffix. The various changes of the noun depend upon its declension to a large extent.[1]

a. If the final term is a noun of the first or second declension (stem in -α or -ο, nominative, -ης, -ας, -ος, masculine, -η, -α, -ος, fem., -ον, neut.) the compound adjective will terminate in -ος, masc. and fem., -ον, neut.

εὐ-τοξοτ-ος, -ον (εὖ, good, τοξότης, ὁ, archer) with good archers
καλλι-νεανι-ος (καλλι-, beautiful, νεανίας, ὁ, youth) beautifully youthful
πολυ-λογ-ος (πολύς, much, λόγος, ὁ, word) talkative
λευκο-κομ-ος (λευκός, white, κόμη, ἡ, hair) white-haired
εὐρυ-χωρ-ος (εὐρύς, broad, χώρα, ἡ, space) roomy
τραχυ-οδ-ος (τραχύς, rough, ὁδος, ἡ, road) with rough roads
βαθυ-φυλλ-ος (βαθύς, thick, φύλλον, τό, leaf) thick-leaved, leafy

b. If the final term is a noun of the third declension with the stem in any consonant except ν, ρ, δ, or -ες, the compound adjective ends in -ος, -ον.

μελανο-φλεβ-ος (μέλας, μέλανος, black, φλέψ, φλεβός, ἡ, vein) black-veined
μικρο-μαστιγ-ος (μικρός, short, μάστιξ, μάστιγος, ἡ, whip) short-ciliate
πολυ-ορνιθ-ος (πολύς, many, ὄρνις, ὄρνιθος, ὁ, ἡ, bird) abounding in birds
πυκνο-σαρκ-ος (πυκνός, thick, σάρξ, σαρκός, ἡ, flesh) with firm flesh
ἀ-σωματ-ος (ἀ-, without, σῶμα, σώματος, τό, body) incorporeal
χρυσο-στομ-ος (χρύσεος, golden, στόμα, στόματος, τό, mouth) golden-mouthed

[1] This account has been largely based upon Miller, Scientific Names of Latin and Greek Derivation, 134.

c. If the final term is a noun of the third declension with the stem in ν, ρ, or δ (nom. σ), the compound retains this form, i. e., it is properly a noun used adjectively. Sometimes the noun is inflected in two genders, e. g., -ων, -ον, or -ωρ, -ορ, or, more rarely, it takes the adjective termination, -ος, -ον.

μακρο-χειρ (μακρός, long, χείρ, χειρός, ἡ, hand) long-armed
αὐτο-χθων, -ον (αὐτός, self, χθών, χθονός, ἡ, ground) native
αὐτο-χθον-ος, -ον, country and all
σκληρο-πους (σκληρός, hard, ποῦς, ποδός, m., foot) hard-footed
κακο-πους, -πουν (κακός, bad, ποῦς, ποδός, m., foot) with bad feet

d. If the final term is a noun of the third declension with the stem in -ες (gen. -εος, nom. -ης, m. f. -ος, n.), the compound adjective will terminate in -ης, masc. and fem., -ες, neut.

θεο-γενης, -ες (θεός, ὁ, God, γένος, γένεος, τό, race) born of God
τειχο-μελης (τεῖχος, τείχεος, τό, wall, μέλος, μέλεος, τό, music) walling by music
πολυ-ανθης (πολύς, much, ἄνθος, ἄνθεος, τό, flower) blossoming

e. If the final term is a neuter noun of the third declension with the stem in -ατ, nom. -ας, the compound adjective as a rule ends in -ως (contraction of -αος for -ατος) masc. and fem., -ων neut., or, rarely, in -ος, -ον.

μεγαλο-κερ-ως, -ων (μέγας, μεγάλου, large, κέρας, κέρατος, τό, horn) large horned
πολυ-τερ-ως (πολύς, much, τέρας, τέρατος, τό, wonder) full of wonder
μονο-κερωτ-ος (μόνος, single, κέρας, τό, horn) with one horn
ὀρθο-κερ-ος, -ον (ὀρθός, upright, κέρας, τό, horn) with upright horns
γλυκυ-κρε-ος (γλυκύς, sweet, κρέας, κρέως (κρέατος) τό, meat) sweet-meated

f. If the final term is a noun of the third declension with the stem in the vowel ι, or υ (-ις, -υς, nom. m., f., -ι, -υ, neut.), it retains this form; rarely it terminates in -ος, -ον.

[illegible] (πολύς, many, ἰχθύς, ἰχθύος, ὁ, fish) abounding in [illegible] πολυ-ιχθυ-ος, -ον

πυκνο-δρυς (πυκνός, thick, δρῦς, δρυός, ἡ, oak) beset with oaks
μελαν-δρυ-ος, ον (μέλας, μέλανος, black, δρῦς, oak) dark with oak leaves
χρυσ-οψις (χρύσεος, golden, ὄψις, ὄψεως, ἡ, appearance) looking like gold
φυγο-πολις (φύγος, fleeing, πόλις, πόλεως, ἡ, city) fleeing from a city

II.1. If the last term is an adjective, the compound usually becomes an adjective of two terminations, -ος, -ον; rarely, there is no change. In case it is used substantively, it may appear in any gender at the coiner's pleasure. The compound may, moreover, pass into a noun by the addition of a substantive suffix.

στενο-μακρος, ον (στενός, narrow, μακρός, long) long and narrow
μελανο-φαιος, ον (μέλας, μέλανος, black, φαιός, dusky) dark gray
λευκο-ερυθρος, ον (λευκός, white, ἐρυθρός, red) whitish red
ἑτερο-γλαυκος, ον (ἕτερος, different, γλαυκός, gray) with one eye gray
λευκο-μελας, αινα, αν (λευκός, white, μέλας, black) whitish black
ὀξυ-γλυκυς, εια, υ (ὀξύς, sour, γλυκύς, sweet) sourish sweet
ἑτερο-φων-ια (ἑτερό-φωνος, of different voice) difference of tone
ζηλημο-συνη (ζηλήμων, jealous) jealousy

II.2. If the last term is a verbal adjective (in -ος, -τος, or -τεος), it may retain the active ending, -ος, -ον, or the passive ending, -τος, or -ης, -ες, may be substituted for either.

δια-στροφ-ος, ον (διαστρέφω, to twist about) twisted
περι-τροπ-ος (περιτρόπω, to turn round) turned round
περι-φερ-ης, ες (περιφέρω, to carry round) revolving
δυσ-μαθ-ης (δυσμαθέω, to be slow in learning) hard to learn
δυσ-τακ-τος, ον (τάσσω, to arrange) disordered, irregular
ἀ-λεπιδω-τος (λεπιδόομαι, to be scaly) not covered with scales

THE CONNECTIVE (THEMATIC VOWEL)

The connective in Greek compounds was originally the final vowel of the stem, or, in imparisyllabics, the vowel of the genitive. The connective -ο- was originally, then, characteristic of nouns of the second declension, and of many stems of the third. By analogy, it spread to stems of the first declension, and the remaining stems of the third, and, finally, even to verbal stems. The overwhelming predominance of the connective **-o-** makes it advisable to disregard the use of the thematic vowel of each declension as a connective in making new compounds, and may be considered sufficient warrant for its insertion in compounds already constructed upon the basis of another thematic connective. Such duplicates as Corynephorus and Corynophorus should be avoided for the sake of uniformity in spelling, if for no other reason, but when they are the names of different genera, as in the present case, they are altogether unfortunate. Moreover, alternative connectives of this sort will always furnish occasion for similar blunders on the part of those not thoroughly conversant with the principles of Greek word-formation. The connectives which may be properly used with the different classes of first terms of compound words are shown in the following list of examples. The diversity of classical usage in this matter is a cogent argument for the use of **-o-** as a connective in all cases where the first term is a nominal stem, if not, indeed, everywhere that a connective is required.

1. First declension: stem in **-α**, nom.; **-α**, **-η**, fem.; **-ας**, **-ης**, masc.

   θαλασσ(α)-ο-φυλλον (**θάλασσα**, **ἡ**, sea, **φύλλον**, **τό**, leaf) Thalassophyllum

   κεφαλ(η)-ο-στιγμα (**κεφαλή**, **ἡ**, head, **στίγμα**, **τό**, mark) Cephalostigma

   σκια-φιλη (**σκιά**, **ἡ**, shadow, **φίλος**, loved) Sciaphile

   θηλη-φορα (**θηλή**, **ἡ**, nipple, **φορά**, **ἡ**, carrying) Thelephora

   κλεπτ(η)-ο-φυτον (**κλέπτης**, **ὁ**, thief, **φυτόν**, **τό**, plant) Cleptophytum

2. Second declension: stem in **-ο**, nom., **-ος**, masc. and fem., **-ον**, neut.

   ἀσκο-λεπις (**ἀσκός**, **ὁ**, leathern bag, **λεπίς**, **ἡ**, scale) Ascolepis

   ῥαβδο-κρινον (**ῥάβδος**, **ἡ**, rod, **κρίνον**, **τό**, lily) Rhabdocrinum

   βαλαν(ο)-η-φορος (**βάλανος**, **ἡ**, acorn, **φορός**, carrying) Balanophorus

   ῥοδο-δενδρον (**ῥόδον**, **τό**, rose, **δένδρον**, **τό**, tree) Rhododendrum

3. Third declension:

(*a*) Stem extending in an explosive, i.e., any consonant except σ, μ, ν, λ, ρ, or -ματ.

ῥαδικο-φυλλον (ῥάδιξ, ῥάδικος, ἡ, branch, φύλλον, τό, leaf) Rhadicophyllum

κηλιδ(ο)-ανθος (κηλίς, κηλῖδος, ἡ, spot, ἄνθος, τό, flower) Celidanthus.

κερατο-στομα (κέρας, κέρατος, τό, horn, στόμα, τό, mouth) Ceratostoma

ἀσπιδ(ο)-η-φορος (ἀσπίς, ἀσπίδος, ἡ, round shield, φορός, bearing) shield bearing

κερας-φορος (κέρας, τό, horn, φορός, bearing) bearing horns

μελι-κοκκος (μέλι, μέλιτος, τό, honey, κόκκος, ὁ, berry) Melicoccus

αἰ-πολος (αἴξ, αἰγός, ὁ, ἡ, goat, -πολός (-κολέω, dwell) goat-herd

(*b*) Stem ending in a nasal or a liquid (ν, λ, ρ).

ἀκτινο-στροβος (ἀκτίς, ἀκτῖνος, ἡ, ray, στρόβος, ὁ, whirling) Actinostrobus

δαιμονο-ρωψ (δαίμων, δαίμονος, ὁ, divinity, ῥώψ, ἡ, bush) Daemonorops

θιν(ο)-ανθη (θίς, θινός, ὁ, ἡ, heap, dune, ἄνθη, ἡ, bloom) Thinanthe

ἀκμο-θετον (ἄκμων, ἄκμονος, ὁ, anvil, θετός, placed) anvil-block

ἁλο-σταχυς (ἅλς, ἁλός, ἡ, sea, στάχυς, ὁ, spike) Halostachys

ἁλι-θριδαξ (ἅλς, ἡ, sea, θρίδαξ, ἡ, lettuce) Halithridax

θηρο-φονον (θήρ, θηρός, ὁ, beast, φονός, slaying) Therophonum

γαστρο-χειλος (γαστήρ, γαστρός (-έρος), ἡ, belly, χεῖλος, τό, lip) Gastrochilus

γαστερ(ο)-ανθος (γαστήρ, ἡ, belly, ἄνθος, τό, flower) Gasteranthus

πυρ-φορον (πῦρ, πυρός, τό, fire, φορός, bearing) Pyrphorum

πυρι-φλογος (πῦρ, τό, fire, φλογός, blazing) flaming with fire

(*c*) Stem ending in -ματ-, nom., -μα, neuters

γραμματο-θηκη (γράμμα, γράμματος, τό, line, θήκη, ἡ, box) Grammatothece

δερματο-βλαστος (δέρμα, δέρματος, τό, skin, βλαστός ὁ, sprout) Dermatoblastus

φυματο-στρωμα (φῦμα, φύματος, τό, tumor, στρῶμα, τό, bed) Phymatostroma

στομ(ο)-αρρηνα (στόμα, στόματος, τό, mouth, ἄρρην, ἄρρεν, male) Stomarrhena

στομα-λιμνη (στόμα, τό, mouth, λίμνη, ἡ, lake) salt water lake

(*d*) Stem ending in -ες; nom., -ος, -ης, gen., -εος, mostly neuters.

βελο-στεμμα (βέλος, βέλεος, τό, dart, στέμμα, τό, wreath) Belostemma

ἑλο-φυτον (ἕλος, ἕλεος, τό, marsh, φυτόν, τό, plant) Helophytum

ὀρο-φακη (ὄρος, ὄρεος, τό, mountain, φακῆ, ἡ, lentil) Orophace

ὀρεο-δοξα (ὄρος, τό, mountain, δόξα, ἡ, glory) Oreodoxa

ὀρες-βιος (ὄρος, τό, mountain, βιός, living) living in the mountains

ὀρεσι-τροφος (ὄρος, τό, mountain, τροφός, nurtured) mountain-nurtured

ὀρε-σκιος (ὄρος, τό, mountain, σκιός, shadowed) over-shadowed by mountains

ξιφ-η-φορος (ξίφος, ξίφεος, τό, sword, φορός, bearing) armed with the sword

βελε-η-φορος (βέλος, βέλεος, τό, arrow, φορός, bearing) bearing arrows

(*e*) Stem ending in -ι or -υ; nom., -ις, -υς, masc., fem., -ι, -υ, neut.

πολιο-δενδρον (πόλις, πόλιος, πόλεως, ἡ, city, δένδρον, τό, tree) Poliodendrum

πολια-νομος (πόλις, ἡ, city, νομός, dealing out) civic magistrate

πολι-πορθος (πόλις, ἡ, city, πορθός, destroying) destroyer of cities

τιγρο-ειδης (τίγρος, τίγριος, ἡ, tiger, εἶδος, τό, form) spotted

ὀφιο-σκοροδον (ὄφις, ὄφιος, ὄφεως, ὁ, snake, σκόροδον, τό, garlic, Ophioscorodum

ὀιο-πολος (ὄϊς, ὄϊος, ὁ, ἡ, sheep, -πολός (-πολέω, dwell) shepherd

ἰχθυο-μεθη (ἰχθύς, ἰχθύος, ὁ, fish, μέθη, ἡ, strong drink) Ichthyomethe

νεκυο-στολος (νέκυς, νέκυος, ὁ, dead body, στολός, ferrying) ferrying the dead

νεκυ-η-πολος (νέκυς, ὁ, dead body, -πολός, dwelling among) having to do with the dead

βοτρυ-φορος (βότρυς, βότρυος, ὁ, cluster of grapes, φορός, bearing) bearing grapes

βου-πλευρον (βοῦς, βοός, ὁ, ἡ, ox, πλευρόν, τό, rib) Bupleurum

4. Verbal stems. When the first term is a verbal stem, it enters into composition with a thematic -ε (the form of the second person singular present imperative of -ω verbs); or with a sigmatic stem, -σι, resembling the sigmatic stem of aorists. The influence of analogy has been felt here also, in that both connectives occasionally yield to the -ο- of noun stems, and, more rarely, ε and ι of the verbal stems interchange or assimilate.

φερε-βοτρυς (φέρω, bear, βότρυς, βότρυος, ὁ, bunch of grapes) bearing bunches of grapes
λυσι-θριξ (λύω, loose, θρίξ, τριχός, ἡ, hair) with loose hair
φερεσ-βιος (φέρω, bear, βίος, ὁ, life) bearing life
περσε-φονη (φέρω, bear, φονή, ἡ, death) Persephone, bringer of death
περσε-πολις (πέρθω, destroy, πόλις, ἡ, city) sacker of cities
ἀρχι-θαλασσος (ἄρχω, rule, θάλασσα, ἡ, sea) ruling the sea
λιπο-σκιος (λείπω, leave, σκιά, ἡ, shade) shadowless
ῥιψο-κινδυνος (ῥίπτω, throw, κίνδυνος, ὁ, risk) venturesome

THE FIRST TERM

The first term of a compound may be a nominal stem (noun, pronoun, or adjective), an indeclinable particle (adverb, preposition, or inseparable particle), or a verbal stem. The form of the first term will be that of its stem if this ends in -ο; if the stem ends in -α, -ο- will be substituted as the connective, and if it ends in -ι, -υ, or a consonant, -ο- will be added as a connective. The connective is omitted in the case of an indeclinable particle, and it is regularly elided before an initial vowel of the last term. In the following examples intended to show the form in which first terms of various categories should enter into composition, the effect of analogy is extended over all first terms of compound words which take a connective, with the exception of adjectives in -υς, -εια, -υ, and verbal stems. Its use might well be extended to verbals upon the analogy of λείπω, which regularly enters into composition in the form, λιπο-, but verbal first terms are rare in scientific compounds, and are rather to be discouraged on account of the alternatives to which they are certain to give rise.

From the standpoint of the biologist, the application of the connective -ο- might well have been made universal, but in the case of adjectives in -υς, the use of the thematic -υ as connective is so invariable that the addition of an -ο, as it is found in noun stems of the same sort, was felt to be unwarranted.

I. Nouns.

1. First declension; nominative singular feminine, -α, -η; masculine, -ας, -ης. The stems of this declension are all originally in -α, which is often modified into -η. In feminines, the stem is identical with the nominative singular; in masculines, the stem is obtained by dropping the termination, -σ, of the nominative.[1]

πετρ(α)-ο-φιλη (πέτρα, ἡ, rock, φίλος, loved, loving) Petrophile

ἡμερ(α)-(ο)-ανθος (ἡμέρα, ἡ, day, ἄνθος, τό, flower) hemeranthus, -um 25

μαχαιρ(α)-(ο)-ανθηρα (μάχαιρα, ἡ, dagger, ἀνθηρός, flowery) Machaeranthera 8

ζων(η)-ο-θριξ (ζώνη, ἡ, girdle, θρίξ, ἡ, hair) Zonothrix 2

βορε(α)-ο-φυτον (βορέας, ὁ, north wind, φυτόν, τό, plant) Boreophytum

ἱπποτ(η)-ο-φυλλον (ἱππότης, ὁ, horseman, φύλλον, τό, leaf) Hippotophyllum

γεω-πυξις (γῆ, γέα, ἡ, earth, πυξίς, ἡ, box) Geopyxis 38: γεο-, 1; 9 in γη-

2. Second declension. The stems of this declension terminate in -ο, rarely in -ω, and are obtained for composition by dropping σ of the nom. sing. of masc. and fem. and ν of the neuter.

βιο-φυτον (βίος, ὁ, life, φυτόν, τό, plant) Biophytum 43

ζεφυρ(ο)-ανθος (ζέφυρος, ὁ, west wind, ἄνθος, τό, flower) Zephyranthus

ποταμο-γειτων (ποταμός, ὁ, river, γείτων, ὁ, ἡ, neighbor) Potamogiton 18

δροσο φορος (δρόσος, ἡ, dew, φορός, bearing) Drosophorus 8

[1] The first number after a compound indicates the number of times the proper stem is found in composition in the Greek lexicon as the first term. Other numbers indicate the frequency of alternatives.

ὑαλο-σειρα (ὕαλος, ἡ, glass, σεῖρα, ἡ, band) Hyalosira 7

φηγο-πτερις (φηγός, ἡ, beech, πτερίς, ἡ, fern) Phegopteris 1

ἱματι(ο)-ανθος (ἱμάτιον, τό, outer garment, ἄνθος, τό, flower) Himatianthus 12

ἰο-δραβη (ἴον, τό, violet, δράβη, ἡ, sort of mustard) Iodrabe 4

συκο-μορφη (σῦκον, τό, fig, μορφή, ἡ, form) Sycomorphe 18

ὀστο-θηκη (ὀστέον, ὀστοῦν, τό, bone, θήκη, ἡ, box) place for bones 20; ὀστεο-, 6

λαγω-χειλος (λαγώς, ὁ, hare, χεῖλος, τό, lip) Lagochilus 20; λαγο-, 6

τα(ω)-ουρα (ταῶς, ὁ, peacock, οὐρά, ἡ, tail) Taüra

3. Third declension. These may be either consonant or vowel stems. The stem is derived most readily by dropping the ending, -ος, of the genitive singular.

A. Consonant stems.

(1) Stem ending in an explosive, i.e., any consonant except σ, μ, ν, λ, ρ, and τ in -ματ, nom. -μα; nominative singular ending in a double consonant, ψ or ξ, or in σ.

(*a*) Stem in a labial, π, β, φ; nominative in ψ (labial + σ).

ῥιπ-ο-γονατιον (ῥίψ, ῥιπός, ἡ, rush, γονάτιον, τό, small joint) Rhipogonatium

φλεβ-ο-χιτων (φλέψ, φλεβός, ἡ, vein, χιτών, ὁ, frock) Phlebochiton 13

κατηλιφ-ο-μορφη (κατῆλιψ, κατήλιφος, ἡ, ladder, μορφή, ἡ, form) Cateliphomorphe

(*b*) Stem in a palatal, κ, γ, χ; nominative in ξ (palatal + σ).

ἀλωπεκ-(ο)-ουρος (ἀλώπεξ, ἀλώπεκος, ἡ, fox, οὐρά, ἡ, tail) Alopecurus 1

φλογ-(ο)-ακανθος (φλόξ, φλογός, ἡ, flame, ἄκανθος, ὁ, spiny plant) Phlogacanthus 14

ὀνυχ-ο-πεταλον (ὄνυξ, ὄνυχος, ὁ, claw, πέταλον, τό, leaf) Onychopetalum 2

(*c*) Stem in a dental, τ, δ, θ, or ντ, νθ, κτ; nominative in σ, rarely in ρ

φωτ-ο-φοβος (φώς, φωτός, τό, light, φοβός, fearing) Photophobus 27:1

ἱμαντ-ο-χαιτη (ἱμάς, ἱμάντος, ὁ, thong, χαίτη, ἡ, long hair) Himantochaete 10 ὀδοντ-ο-κυκλος (ὀδούς, ὀδόντος, ὁ, tooth, κύκλος, ὁ, ring) Odontocyclus 18

νυκτ-ο-μυκης (νύξ, νυκτός, ἡ, night, μύκης, ὁ, mushroom) Nyctomyces 28:33 in νυκτι-

κερατ-ο-στυλις (κέρας, κέρατος, τό, horn, στυλίς, ἡ, pillar) Ceratostylis 16:3

στεατ-(ο)-οπια (στέαρ, στέατος, τό, tallow, ὀπός, ὁ, juice) Steatopia 3

ὑδατ-ο-φορα (ὕδωρ, ὕδατος, τό, water, φορά, ἡ, a carrying) Hydatophora 22

κλειδ-ο-νημα (κλείς, κλειδός, ἡ, key, hook, νῆμα, τό, thread) Clidonema 6

χλαμυδ-ο-μονας (χλαμύς, χλαμύδος, ἡ, mantle, μονάς, ἡ, unit) Chlamydomonas 8

κορυθ-(ο)-αιολον (κόρυς, κόρυθος, ἡ, helm, αἰόλος, nimble) Corythaeolum

ἑλμινθ-ο-σταχυς (ἕλμινς, ἕλμινθος, ἡ, worm, στάχυς, ὁ, spike) Helminthostachys 2

(2) Stem ending in a liquid, ν, λ, ρ; nominative in the same consonant, or σ.

(*a*) Stem in ν, nominative in ν, or σ.

χιον-ο-φιλη (χιών, χιόνος, ἡ, snow, φίλος, loving) Chionophile 17

κλων-ο-σταχυς (κλών, κλωνός, ὁ, shoot, στάχυς, ὁ, spike) Clonostachys

χην-ο-ποδιον (χήν, χηνός, ὁ, ἡ, goose, πόδιον, τό, small foot) Chenopodium 13

δελφιν-(ο)-αστρον (δελφίς, δελφῖνος, ὁ, dolphin, ἄστρον, τό, star) Delphinastrum 3

κτεν-(ο)-οδους (κτείς, κτενός, ὁ, comb, ὀδούς, ὁ, tooth) Ctenodus 2

(*b*) Stem in λ, nominative in λ, or σ.

ἁλ-ο-δικτυον (ἅλς, ἁλός, ἡ, sea, δίκτυον, τό, net) Halodictyum 4:79 in ἁλι-

(*c*) Stem in ρ, nominative in ρ.

ἀνδρ-ο-πωγων (ἀνήρ, ἀνδρός, ὁ, man, πώγων, ὁ, beard) Andropogon 119

ἀστερ-(ο)-ομφαλος (ἀστήρ, ἀστέρος, ὁ, star, ὀμφαλός, ὁ, navel) Asteromphalus 19 γαστρ-ο-καρφη (γαστήρ, γαστρός, ἡ, belly, κάρφη, ἡ, dry scale) Gastrocarphe 1:16 in γαστρο-

θηρ-ο-μορφη (φήρ, θηρός, ὁ, beast, μορθή, ἡ, form) Theromorphe 44

πυρ-ο-λειριον (πῦρ, πυρός, τό, fire, λείριον, τό, lily) Pyrolirium 49:64 in -ι-; 11 in ρ-; 4 in -η-

(3) Stem in -ματ, nominative in -μα.

A special study of this class of imparisyllabics has been made for the sake of determining just what warrant existed in Greek for making the stem of the oblique cases the invariable form for composition, These were thought to constitute a very fair criterion on account of their wide extension, and for the further reason that they furnish a large number of alternatives, since the nominative might readily be supposed to represent a first declensional stem in -α. The Greek lexicon contains 1,782 neuters in -μα, largely secondary stems, though there are also many primary ones, these being by far the most frequent in composition. Of these 1,782 neuters, 231 are found in composition or derivation as the first term, occurring altogether in 969 derivatives. Of the 231, 208 occur in derivatives only in the proper stem form in -ματ, being used 555 times. Eleven words, ἄγαλμα, glory (13:1), ἅρμα, chariot (24:1), δέρμα, skin (9:1), θαῦμα, marvel (14:13), θεώρημα, theory (2:1), κέρμα, small change (3:1), κύμα, wave (19:11), κῶμα, coma (2:1), ὄνομα, name (20:12), σῶμα, body (48:5) and φλέγμα, flame (12:4), show the alternative stems, ἀγαλματ- and ἀγαλμ-, though the former is preponderant, occurring in 166 derivatives, while the latter is found in only 51. Six words of this class, αἷμα, blood (32:72), ἔρμα, prop, (2:3), πῶμα, drink (3:4), σπέρμα, seed (16:22), στόμα, mouth (3:24), and χεῖμα, cold (1:4), occur more frequently in the shortened form, αἱμ-, the frequency being 129 to 57. Three only, δήλημα, bane (1), στάλαγμα, drop (1), and φράγμα, fence (3), are found invariably in the shortened form, while three, ἐπίπωμα, παρέγκυμα, and σίγμα occur once in each form. To summarize the foregoing: of 231 neuters in -μα, which furnish stems for compounds or derivatives, 208 always appear in the proper stem form, -ματ, 11 occur more frequently in this form, 6 more frequently in the shortened form, -μ, 3 always in this short form, while 3 occur once in either form. Of 969 words derived from these neuters, 781 show the proper stem in -ματ, while 188 have the shortened stem in -μ. Again, it must be borne in mind that, while these alternative stems are a source of growth rather than a misfortune to the language, in nomenclature they must always lead to confusion, as analogy will sooner or later produce doublets, such as αἱματίσπερμα and αἱμίσπερμα, in the case of every stem which enters into composition. The marked preponderance of the proper stem in compounds of this group has been considered ample warrant for extending this stem to all compounds formed from neuters in -μα. If further warrant were needed, it is found in the fact that every neuter of this class shows the proper stem in the oblique cases, its disappearance in certain compounds being due to the use of the shortened nominative form, a use arising to a large extent out of ignorance.

αἱματ-ο-χαρις (αἷμα, αἵματος, τό, blood, χάρις, ἡ, grace) Haematocharis

δερματ-ο-κυβη (δέρμα, δέρματος, τό, skin, κύβη, ἡ, head) Dermatocybe

πωματ-ο-δερρις (πῶμα, πώματος, τό, drink, δέρρις, ἡ, leather coat) Pomatoderris

θαυματ-ο-πτερις (θαῦμα, θαύματος, τό, wonder, πτερίς, ἡ, fern) Thaumatopteris

σπερματ-ο-χνοος (σπέρμα, σπέρματος, τό, seed, χνόος, ὁ, foam) Spermatochnous

στοματ-ο-θηκιον (στόμα, στόματος, τό, mouth, θήκιον, τό, little box) Stomatothecium

σωματ-(ο)-αγγειον (σῶμα, σώματος, τό, body, ἀγγεῖον, τό, vessel) Somatangium

(4) Stem in -ες, genitive -εος (-εσος), nominative usually in -ος, mostly neuters. The form for composition is obtained by dropping -εος of the genitive, or -ος of the nominative.

κερδ-(ο)-ουρα (κέρδος, κέρδεος, τό, trick, οὐρά, ἡ, tail) Cerdura 4

βελ-ο-περονη (βέλος, βέλεος, τό, dart, περόνη, ἡ, point) Beloperone 7:1 in -εη

ἀγγ-ο-φορα (ἄγγος, ἄγγεος, τό, vase, φορά, a carrying) Angophora 1

χειλ-ο-σκυφος (χεῖλος, χείλεος, τό, lip, σκύφος, τό, cup) Chiloscyphus 2

ἀνθ-ο-φυκος (ἄνθος, ἄνθεος, τό, flower, φῦκος, τό, seaweed) Anthophycus 46:2

φυκ-ο-φυτον (φῦκος, φύκεος, τό, seaweed, φυτόν, τό, plant) Phycophytum 3

B. Vowel stems (in -ι or -ύ).

(1) Stem in -ι, nominative in -ις, masc. and fem., -ι, neut. The stem is obtained by dropping -ος of the genitive, or -ς of masc. or fem. nominative.

ὀψι-(ο)-ανθος (ὄψις, ὄψιος, ὄψεως, ἡ look, ἄνθος, τό, flower) Opsianthus

ὀφι-ο-καρυον (ὄφις, ὄφιος, ὄφεως, ὁ, snake, κάρυον, τό, nut) Ophiocaryum 20

πεπερι-ο-φυλλον (πέπερι, πεπέριος, τό, pepper, φύλλον, τό, leaf) Peperiophyllum 1 in -ο

τροπι-ο-λεπις (τρόπις, τρόπιος, ἡ, keel, λεπίς, ἡ, scale) Tropiolepis

φυσι-ο-γλωχις (φύσις, φύσιος, φύσεως, ἡ, nature, γλωχίς, ἡ, point) Physioglochis 14

(2) Stem in -υ, nominative in -υς, (-αυς, -ευς, -ους) masc. and fem., -υ, neut. The stem is obtained by dropping -ος of the genitive, or -ς of the nominative.

δρυ-ο-πτερις (δρῦς, δρυός, ἡ, oak, πτερίς, ἡ, fern) Dryopteris 11:4 in -υ

μυ-(ο)-ουρα (μῦς, μυός, ὁ, mouse, οὐρά, ἡ, tail) Myura 18:2 in μυσ-

πιτυ-(ο)-οψις (πίτυς, πίτυος, ἡ, pine tree, ὄψις, ἡ, look) Pityopsis

σταχυ-ο-βοτρυς (στάχυς, στάχυος, ὁ, spike, βότρυς, ὁ, bunch of grapes) Stachyobotrys 10:7 in -υη; 1 in -υ

Here are usually placed βοῦς, γραῦς, ναῦς. These originally had the stem in a consonant, digamma, as βοϝ-, but the digamma was lost, leaving third declension stems in -ο and

-α. The proper stem is obtained by dropping -ος of the genitive, though these stems are quite irregular in the matter of composition.

βο-ο-γληνος (βούς, βοός, ὁ, ox, γλήνη, ἡ, eyeball) ox-eyed 17: 96 in βου-; 3 in βο-

γρα-ο-λογια (γραῦς, γραός, ἡ, old woman, λογία, ἡ, speech) gossip 6

Anomalous nouns. A few nouns are included here, in which the stem has been more than usually reduced in the nominative, e. g., γάλα, γόνυ, γυνή, δόρυ, μέλι.

γαλακτ-(ο)-ανθος (γάλα, γάλακτος, τό, milk, ἄνθος, τό, flower) Galactanthus 19

γονατ-ο-ζυγον (γόνυ, γόνατος, τό, knee, ζυγόν, τό, yoke) Gonatozygum 2:8 in γονυ-

γυναικ-ο-τροχας (γυνή, γυναικός, ἡ, wife, τροχάς, ἡ, shoe) Gynaecotrochas 35:1 in γυνο-

δορατ-ο-λωμα (δόρυ, δόρατος, τό, stem, spear, λῶμα, τό, fringe) Doratoloma 6:21 in δορι-; 17 in δορυ-

μελιτ-ο-ξυλον (μέλι, μέλιτος, τό, honey, ξύλον, τό, wood) Melitoxylum 10:46 in μελι-

II. *Adjectives*

1. First and second declension: stems in -α, or -ο, nom. sing., -ος, masc., -ος, -η, -α, fem., -ον, neut. The stem (ending in its proper thematic vowel, -ο, which is also the connective) is readily obtained by dropping -σ of the nominative singular masculine.

μακρο-χλαινι (μακρός, ά, όν, long, large, χλαῖνα, ἡ, cloak) Macrochlaena 99

ὀρθο-μερις (ὀρθός, ή, όν, straight, μερίς, ἡ, part) Orthomeris 92

ἑτερο-ραχις (ἕτερος, α, ον, different, ῥάχις, ἡ, back) Heterorrhachis 125

λευκο-βρυον (λευκός, ή, όν, clear, white, βρύον, τό, alga, moss) Leucobryum 114

στενο-λοφος (στενός, ή, όν, narrow, λόφος, ὁ, crest) Stenolophus 46

χρυσο-νεριον (χρύσεος, η, ον, golden, νήριον, τό, oleander) Chrysonerium μεσ-(ο)-ανθεμον (μέσος, η, ον, middle, ἄνθεμον, τό, flower) Mesanthemum 125

ἁπλο-ταξις (ἁπλόος, η, ον, simple, τάξις, ἡ, array) Haplotaxis 9

2. First and third declension. The feminines of this group follow the first declension, the masculines and neuters the third. According to the termination of the stem of the latter, there are two groups, the one with the stem in a consonant, the other with vowel-stem.

a. Stem in a consonant, usually -ν or -τ, masc. in -ν or -ς, neut. in -ν. The stem is readily derived by dropping -ος of the genitive singular masculine.

μελαν-ο-στικτος (μέλας, αινα, αν, μέλανος, black, στικτός, pricked) Melanostictus 44: μελαν-, 70

παντ-ο-φιλος (πᾶς, πᾶσα, πᾶν, παντός, all, φίλος, loving) Pantophilus 114: παν-, 483

τερεν-ο-χρως (τέρην, εινα, εν, τέρενος, smooth, χρώς, ὁ, skin) with smooth skin 1

b. Stem in the vowel -υ; nom. sing. masc. in -υς, neut. in -υ. The stem is obtained by dropping -ς of the nominative singular masculine. Adjectives in -υ, unlike nouns of this class, do not take the connective -ο-. The use of the stem vowel, -υ, as connective is so nearly absolute that it has seemed unwise to extend the connective -ο- to this class.

ἀμβλυ-νοτος (ἀμβλύς, εῖα, ύ, blunt, νότος, ὁ, south wind) Amblynotus 3

βαθυ-φυτον (βαθύς, deep, low, φυτόν, τό, plant) bathyphytum 86

βαρυ-ξυλον (βαρύς, heavy, ξύλον, τό, wood) Baryxylum 115

βραδυ-πιπτον (βραδύς, slow, πιπτός, fallen) Bradypiptum 25

βραχυ-οδους (βραχύς, short, small, ὀδούς, ὁ, tooth) Brachyodus 52

γλυκυ-ριζα (γλυκύς, sweet, ῥίζα, ἡ, root) Glycyrrhiza 30: 2 in γλυκο-

δριμυ-φυλλον (δριμύς, sharp, φύλλον, τό, leaf) Drimyphyllum 3

εὐρυ-βασις (εὐρύς, wide, broad, βάσις, ἡ, step, base) Eurybasis 66

ἡδυ-οσμος (ἡδύς, sweet, ὀσμή, ἡ, smell) Hedyosmus 58

θηλυ-χιτων (θηλύς, female, tender, χιτών, ὁ, frock) Thelychiton 40

ἰθυ-θριξ (ἰθύς, straight, θρίξ, ἡ, hair) Ithythrix 32

ὀξυ-ακανθα (ὀξύς, sharp, ἄκανθα, ἡ, thorn) Oxyacantha 137

παχυ-ουρα (παχύς, thick, stout, οὐρά, ἡ, tail) Pachyura 30

πλατυ-κωδων (πλατύς, wide, broad, κώδων, ὁ, ἡ, bell) Platycodon 51

πολυ-γαλα (πολύς, much, many, γάλα, τό, milk) Polygala 960

τραχυ-στημων (τραχύς, rough, rugged, στήμων, ἡ, thread) Trachystemon 16:3 in τραχ-

3. Third declension: stem endings various; nominative in -ης, -ες, -ων, -ον, -υς, -υν, -ις, -ι, -ωρ, -ορ; in one ending, -ας, -ης, -ις, -υς, or in the form of all nouns from which adjectives are derived without change (see last term). The form for composition is best obtained by dropping -ος of the genitive singular (-ους of contracts, nom. -ης, -ες).

πληρ-(ο)-ασκος (πλήρης, ες, full, filled, ἀσκός, ὁ, leathern bag) Plerascus 3

ψευδ-ο-λινον (ψευδής, ες, false, λίνον, τό, flax, thread) Pseudolinum 178:2 in -η

πεπον-ο-σικυος (πέπων, ον, πέπονος, ripe, σίκυος, ὁ, gourd) a kind of gourd

τροφι-(ο)-ανθος (τρόφις, τρόφιος, well-fed, ἄνθος, τό, flower) Trophianthus 1

ἀρρεν-(ο)-αχνη (ἄρρην (ἄρσην), εν, ἄρρενος, male, ἄχνη, ἡ, down) Arrhenachne 22

νομαδ-ο-κυστις (νομάς, νομάδος, roaming, κύστις, ἡ, bladder) Nomadocystis 1

συγκλυδ-ο-σπορα (σύγκλυς, σύγκλυδος, brought together, σπορά, ἡ, seed) Synclydospora 1

Irregular Adjectives

μεγαλ-ο-χλοη (μέγας, μεγάλη, μέγα, μεγάλου, great, χλόη, ἡ, grass) 181:16 in μεγα-

III. *Numerals*

1. Cardinals. εἷς, δύο, τρεῖς, τέσσαρες are declined; the numbers from πέντε through ἑκατόν are indeclinable. The first four numerals rarely appear in the normal form in composition:

εἷς is represented by μόνος, η, ον, alone, one, δύο, by δι-, τρεῖς by τρι-, and τέσσαρες by τετρα-. The last three very rarely elide, and only before ι or α. The numerals from πέντε to δώδεκα should terminate in -α when compounded.

μονο-θριξ (μόνος, η, ον, one, θρίξ, ἡ, hair) Monothrix

δι-ῳον (δι-, two- ᾠόν, τό, egg) Dioum

τρι-κερατιον (τρι-, three-, κεράτιον, τό, little horn) Triceratium

τετρ(α)-ακτις (τετρα-, four-, ἀκτίς, ἡ, ray) Tetractis

πεντα-πελτη (πέντε, five, πέλτη, ἡ, small shield) Pentapelte

ὀκτα-βλεφαρις (ὀκτώ, eight, βλεφαρίς, ἡ, eyelash) Octablepharis

δεκα-ραφη (δέκα, ten, ῥάφη, ἡ, seam) Decarraphe

2. Ordinals, and the higher cardinals in -οι, enter into composition in the same manner as adjectives in -ος.

πρωτο-κοκκος (πρῶτος, η, ον, first, κόκκος, ὁ, berry) Protococcus

μυριο-φυσα (μυρίος, α, ον, numberless, φῦσα, ἡ, bubble) Myriophysa

χιλιο-φυλλον (χίλιοι, αι, α, thousand, φύλλον, τό, leaf) Chiliophyllum

IV. *Indeclinables*

1. Adverbs. These are rare in nominal compounds. They are attached immediately to the second member.

ἀγχι-σπορος (ἄγχι, near, σπορά, ἡ, seed) near of kin

ἀει-χρυσουν (ἀεί, ever, χρύσεος, golden) Aichrysum

ἁπαξ-ανθος (ἅπαξ, once, once only, ἄνθος, flowering) hapaxanthus

ἀρτι θαλης (ἄρτι, just, exactly, θαλλός (*θαλός) ὁ, shoot) just blooming

εὐ-ραμνος (εὖ, right, true, ῥάμνος, ἡ, thorn) Eurhamnus

χαμαι-νηριον (χαμαί, on the ground, νήριον, τό, oleander) Chamaenerium

2. Prepositions. These are attached directly to the second member. The final vowel is elided before an initial vowel (except in περί, and πρό), and if the latter be aspirated, a preceding smooth explosive is roughened. Εκ becomes ἐξ before a vowel.

ἀμφι-δοναξ (ἀμφί, on both sides of, δόναξ, ὁ, reed) Amphidonax

ἀνα-χαρις (ἀνά, on, upward, χάρις, ἡ, grace) Anacharis

δια-φανη (διά, through, φανός, bright) Diaphane

καθ-εδρα (κατά, down, down from, ἕδρα, ἡ, seat) Cathedra
περι-ανθος (περί, around, ἄνθος, τό, flower) Perianthus
ὑπερ-ανθηρα (ὑπερ, over, ἀνθηρός, blooming) Hyperanthera
ὑπο-πιτυς (ὑπο, from under, πίτυς, ἡ, pine) Hypopitys

3. Inseparable particles. These are attached directly to the second member.

ἀ-φυλλον (ἀν- (ἀ-, before a consonant) without, φύλλον, τό, leaf) Aphyllum
ἀν-ειλημα (ἀν-, without, εἴλημα, τό, veil) Anilema
ἀρι-ανθηρα (ἀρι-, very, quite, ἀνθηρός, flowering) Arianthera
δα-σκιος (δα-, intensitive, σκιά, ἡ, shade) thickly shaded, bushy
δυσ-μορφια (δυς-, hard, bad, μορφία, ἡ, form) badness of form
ἐρι-τριχιον (ἐρι, very, much, τρίχιον, τό, little hair) Eritrichium
ζα-καλλης (ζα-, very, κάλλος, τό, beauty) very beautiful
ἡμι-γραφις (ἡμι-, half, γραφίς, ἡ, style, needle) Hemigraphis
νη-πενθης (νη-, without, πένθος, τό, sorrow) Nepenthes

V. *Verbal Stems*

λειψι-θριξ (λείπω (λειψ-) lose, θρίξ, ἡ, hair) having lost the hair
λιπο-θριξ (λείπω (λιπ-), lose, θρίξ, ἡ, hair) wanting hair
ἀρχεσι-μολπος (ἄρχω (αρχεσ-) begin, μολπή, ἡ, song and dance) beginning the strain
ἀρχε-κακος (ἄρχω (ἄρχ-), begin, κακός, bad) beginning mischief

WORD FORMATION IN LATIN

DERIVATION

Latin has developed derivation enormously, while composition has had a relatively feeble development. This is explained by the fact that derivation must have maintained the lead which it doubtless acquired during the formative period of the language, and, also, by the fact that the need for compound words during the classic and post-classic periods was supplied by repeated and extensive borrowing from Greek. Derivation has, in consequence, a much greater importance for the nomenclator than composition, a condition quite contrary to that which prevails in Greek. As in the latter language, nominal derivatives are formed by adding noun or adjective suffixes to roots or stems.

NOUN SUFFIXES

Agent

-TOR (-SOR), m., -TRIX, f., primary or secondary verbal; rarely denominative: *da-tor,* giver; *ora-tor,* speaker; *ton-sor,* barber; *via-tor,* traveler; *impera-tor,* commander; *pisca-tor,* fisherman

-ES (stem -T, -IT, -ET), m. or f., primary: *ped-es,* walker; *com-es,* companion; *tram-es,* pathway

-O (stem -ON) m., primary: *combib-o,* pot-companion; *ger-o,* porter

-ARIUS, m., primary or secondary, verbal or denominative: *ferr-arius,* smith; *pisc-arius,* fishmonger

Means or Instrument

-BULUM, n., primary or secondary, usually verbal: *pa-bulum,* fodder; *vesti-bulum,* court

-CULUM (-CLUM), -CRUM, n., primary or secondary, usually verbal: *sar-culum,* hoe; *vin-clum,* fetter; *ful-crum,* support

-BRUM, n., -BRA, f., primary or secondary, verbal: *cri-brum,* sieve; *fla-brum,* blast; *tere-bra,* borer

-TRUM, n., -TRA, f., -TER, m., primary or secondary: *plaus-trum,* wagon; *mulc-tra,* milk-pail; *cul-ter,* knife

-MEN, -MENTUM, n., primary or secondary, usually verbal: *teg-men,* cover; *funda-mentum,* foundation

Action or Result

-OR, -IS, m., -ES, f., -US, n. (stem -OR, -ER), primary: *cal-or,* heat; *lab-or,* toil; *cin-is,* ashes; *pulv-is,* dust; *caed-es,* a setting-down; *nub-es,* cloud; *fun-us,* burial; *on-us,* burden

-TUS (-SUS), mostly masculine, primary: *fruc-tus,* fruit; *pas-tus,* food; *vic-tus,* food; *sen-sus,* perception

-IO (stem -ION), f., secondary verbal, rarely primary: *obsid-io,* siege; *reg-io,* district

-TIO (-SIO), (stem -TION), primary or secondary verbal: *auc-tio,* increase; *huma-tio,* burial; *conver-sio,* alteration

-TURA (-SURA), primary or secondary verbal: *tex-tura,* web; *commis-sura,* joint

-MEN, -MENTUM, n., primary or secondary verbal: *frag-men,* piece; *arma-mentum,* equipment

-MONIUM, n., -MONIA, f., primary or secondary, verbal or denominative: *testi-monium,* testimony; *ali-monia,* support; *sancti-monia,* sanctity

Quality

-IA, -TIA, f., secondary denominative: *audac-ia,* boldness; *patient-ia,* patience; *nigri-tia,* blackness

-IES, -TIES, f., secondary denominative: *pernic-ies,* ruin; *segnities,* laziness

-DO, f., secondary denominative or verbal: *arun-do,* reed; *hirudo,* leech; *dulce-do,* sweetness

-GO, f., secondary denominative or verbal: *albu-go,* whiteness; *lanu-go,* down; *rubi-go,* redness, rust

-TAS, -TUS, f., secondary denominative: *cavi-tas,* hollow; *celeritas,* swiftness; *plenu-tas,* fulness; *senec-tus,* old age

-TUDO, f., secondary denominative: *ampli-tudo,* width; *crassitudo,* thickness

-IUM, -TIUM, secondary denominative or verbal: *auspic-ium,* omen; *hospit-ium,* inn; *servi-tium,* slavery

-ASTER, m., secondary denominative: *ole-aster,* wild olive

Place

-ARIUM, n., primary or secondary denominative: *avi-arium,* poultry-yard; *herb-arium,* place for plants; *virid-arium,* garden

-ETUM, -TUM, n., primary or secondary denominative: *caric-etum,* field of sedges; *fruti-cetum,* thicket; *ros-etum,* rose-garden; *arbus-tum,* grove; *salic-tum,* willow grove

-ILE, n., primary denominative: *bov-ile,* cattle yard; *ov-ile,* sheepfold

-TORIUM, n., secondary denominative: *audi-torium,* lecture room; *ora-torium,* oratory

Diminutives

-ULUS (-OLUS after a vowel), primary or secondary denominative: *glob-ulus,* little globe; *herb-ula,* little herb; *capit-ulum,* little head; *atri-olum,* little hall; *osti-olum,* little mouth

-CULUS, -UNCULUS, primary or secondary denominative; *mus-culus*, little mouse; *nube-cula*, little cloud; *oper-culum*, little lid; *cent-unculus*, cloth of many colors; *orati-un-cula*, little speech

-ELLUS, -ILLUS, primary or secondary denominative: *mis-ellus*, wretch; *lam-ella*, small leaf; *pat-ella*, small dish; *penic-illus*, hair pencil; *osc-illum*, little face

-UNCIO, secondary denominative: *hom-uncio*, manikin

The gender of diminutives is regularly that of the stem to which they are attached.

Patronymics. These are formed by the regular Greek suffixes, which have given rise in Greek to adjectives that have become nouns in Latin.

ADJECTIVE SUFFIXES

Ownership or Relation

-ANUS, -ENUS, -INUS, primary or secondary denominative: *pag-anus*, rustic; *ser-enus*, calm; *mar-inus*, of the sea

-ACUS, -ICUS, primary or secondary denominative: *pausi-acus*, olive-colored; *hepat-icus*, liver-colored

-ALIS, -ELIS, -ILIS, -ULUS, primary or secondary denominative: *litor-alis*, of the shore; *hum-ilis*, lowly; *ed-ulis*, edible

-ARIS, -ARIUS, -TORIUS, primary or secondary denominative: *milit-aris*, martial: *lamin-arius*, blade-like; *desul-torius*, of a vaulter

-ATUS, -ITUS, -UTUS, primary or secondary denominative: *ped-atus*, having a foot; *turr-itus*, turreted; *hirs-utus*, rough

-EUS, -EIUS, -ICIUS, primary or secondary denominative: *frond-eus*, leafy; *pleb-eius*, of the commons; *advent-icius*, foreign

Material

-ACEUS, -ICIUS, primary or secondary denominative: *ochr-aceus*, of ochre; *viol-aceus*, violet-colored; *later-icius*, brick red

-EUS, -IUS, -EIUS, primary or secondary denominative: *lign-eus*, of wood; *ros-eus*, rosy; *aur-eus*, golden; *limon-ius*, lemon yellow; *chalyb-eius*, of steel

-INUS, -INEUS, -GNUS, primary or secondary denominative: *lilac-inus,* lilac-colored; *querc-inus,* oaken; *frax-ineus,* ashen; *fulig-ineus,* soot-black; *abie-gnus,* of fir-wood; *sali-gnus,* of willow

Quality or Fitness

-AX, primary: *ten-ax,* tenacious; *rap-ax,* furious; *vor-ax,* consuming

-IDUS, -ULUS, primary: *flor-idus,* blooming; *morb-idus,* diseased; *cred-ulus,* trustful; *pend-ulus,* hanging

-VUS (-UUS), -IVUS, -TIVUS, primary or secondary verbal or denominative: *decid-uus,* apt to fall; *aest-ivus,* of summer; *fugi-tivus,* fleeing

-IUS, primary or secondary denominative or verbal: *patr-ius,* paternal; *exim-ius,* choice

-ILIS, -BILIS, -TILIS (-SILIS), secondary verbal: *flex-ilis,* flexible; *fac-ilis,* easy; *nota-bilis,* noteworthy; *plica-tilis,* twisted; *fos-silis,* dug up

Fulness

-OSUS, secondary denominative: *form-osus,* beautiful; *silv-osus,* forested; *lim-osus,* muddy

-(O)LENS, -(O)LENTUS, primary or secondary denominative: *grave-olens,* of heavy odor; *succu-lentus,* fresh, full of juice; *lutu-lentus,* muddy

-BUNDUS, -CUNDUS, primary or secondary verbal, rarely denominative: *erra-bundus,* vagrant; *fe-cundus,* fruitful; *rubi-cundus,* ruddy

Place or Origin

-ANUS, -ANEUS, -ENUS, primary or secondary denominative: *mont-anus,* of the mountains; *subterr-aneus,* underground; *terr-enus,* earthy

-ENSIS, primary or secondary denominative: *for-ensis,* of the market place; *padov-ensis,* of Padua

-ESTER, (-ESTRIS), -TER (-TRIS), primary or secondary denominative: *camp-ester,* of the fields; *silv-estris,* forest; *lacus-ter,* of a lake; *palus-tris,* marshy

-TIMUS, secondary denominative: *mari-timus,* of the sea; *fini-timus,* bordering

Time

-ANUS, primary or secondary denominative: *anteluc-anus,* before daybreak; *meridi-anus,* of midday; *cotidi-anus,* daily

-ERNUS, (-TERNUS), -URNUS (-TURNUS), primary or secondary denominative: *hib-ernus,* wintry; *sempi-ternus,* everlasting; *di-urnus,* by day; *noct-urnus,* nocturnal

-NUS, secondary denominative: *autum-nus,* of autumn; *ver-nus,* vernal

Diminutives

-ULUS (-OLUS), secondary denominative: *frigid-ulus,* chilly; *lute-olus,* yellowish

-CULUS, secondary denominative, especially common with the neuter of the comparative: *minus-culus,* somewhat smaller; *crassius-culus,* somewhat thick

COMPOSITION

Latin, like Greek, exhibits two methods of composition, syntactic and non-syntactic. The former, found in such compounds as *aquaeductus, nomenclator, respublica,* and *paterfamilias,* is rare and archaic, and needs to be noticed only to call attention to its use in specific names, such as *urticaefolia, menthaeflora,* etc., which should be treated as non-syntactic and written *urticifolia, menthiflora,* etc. Non-syntactic composition has been developed to a certain extent in Latin, but the language is far inferior to Greek in this regard. Latin has largely obviated the need of composition by a wide extension of derivation, with the result that composition always seems awkward and foreign to the language. Notwithstanding this, Latin has a large number of compounds, mostly adjectives, which have been used by biologists, and new compounds will doubtless be made upon the model afforded by these. It should be borne in mind that derivation by suffixes is the easy and natural method of word formation in Latin, as composition is the natural way in Greek, and a desirable working rule might be formulated to the effect that derivatives be taken from Latin and compounds from Greek. Linne[1] has written as follows upon this point: "Nomina generica ex duobus vocabulis latinis integris & conjunctis composita, vix toleranda sunt. Ejusmodi vocabula, graeca lingua pulcherrima sunt; at Latina non facile eadem admittit. Admissimus nonnulla vocabula latina, sed non ideo in posterum imitanda sunt."

THE LAST TERM

The last term is a noun, adjective, or verbal stem. According to the nature of the last term, its form is as follows:

I. If the last term is a noun, the compound (1) will be a noun:

angi-portus (**angus,* strait, *portus,* harbor) a narrow street
ante-cursor (*ante,* before, *cursor,* runner) forerunner, vanguard
tri-dens (*tri,* three, *dens,* tooth) trident
nemori-cultrix (*nemus, nemoris,* forest, *cultrix,* cultivator) forest-lover
mani-pretium (*manus,* hand, *pretium,* price) workman's pay
albo-galerus (*albus,* white, *galerus,* hat) white hat of a flamen

(2) or the compound will be an adjective, appearing in one of three forms: (1) *us, a, um;* (2) *is, e;* (3) the form of the noun.

(*a*) If the last term belong to the first, second, or fourth declension (stem in *-a*, *-o*, and *-u*, respectively), the compound adjective will regularly take the terminations of the first and second declensions (*us*, m., *a*, f., *um*, n.), or it may take the endings of the third declension (*is*, m., f., *e*, n.)

in-formis (*in*, not, *forma*, form) formless
igni-comus (*ignis,* fire, *coma*, hair) fiery-haired
magni-sonus (*magnus*, great, *sonus*, sound) loud-sounding
multi-vius (*multus,* many, *via*, way) having many ways
albi-cerus, albi-ceris (*albus,* white, *cera,* wax) wax-white

[1] Linné. Philosophia Botanica, 160. 1751.

uni-cornis, uni-cornus (*unus,* one, *cornu,* horn) with one horn

angui-manus (*anguis,* snake, *manus,* hand) with serpent hand

magn-animus , *magn-animis* (*magnus,* great, *animus,* soul) great-souled

long-aevus (*longus,* long, *aevum,* age) of great age

multi-jugus, -jugis (*multus,* many, *jugum,* yoke) yoked many together

multi-meter (*multus,* many, *metrum,* measure) of many feet

(*b*) If the last term belong to the third declension (stem in a consonant or in *-i*) the compound adjective will have the form and inflection of a noun, or its stem may take the endings *us, a, um,* or, more rarely, *is, e.*

aequi-lanx (*aequus,* equal, *lanx,* scale) with equal scale

in-frons (*in,* without, *frons,* leaf) without foliage

nigri-color (*niger,* black, *color,* color) of a black color

multi-pes (*multus,* many, *pes,* foot) many-footed

multi-radix (*multus,* many, *radix,* root) many-rooted

uni-finis (*unus,* one, *finis,* limit) possessing the same termination

semi-bos (*semi,* half, *bos,* ox) half-ox

bi-dens (*bi,* two, *dens,* tooth) two-toothed

aequi-pes, aequi-pedis (*aequus,* equal, *pes,* foot) isosceles

in-orus (*in,* not, *os, oris,* mouth) without a mouth

multi-colorus (*multus,* many, *color, coloris,* color) many-colored

multi-nominis (*multus,* many, *nomen, nominis,* name) many-named

multi-genus, -generus, -generis, (*multus, genus, generis,* kind) of many kinds

multi-laudus (*multus, laus, laudis,* praise) much-praised

aequi-latus, aequi-laterus (*aequus,* equal, *latus, lateris,* side) equilateral

(*c*) If the last term belong to the fifth declension, the compound adjectives will appear in *us, a, um,* or, rarely, in the form of a noun.

levi-fidus (*levis,* light, *fides,* faith) of little faith
per-dius (*per,* through, *dies,* day) throughout the day
ex-spes (*ex,* out of, *spes,* hope) hopeless

II. If the last term is an adjective, it will not be changed, though noun suffixes may be added to it, thus making a substantive.

igni-potens (*ignis,* fire, *potens,* powerful) potent in fire
aequi-par (*aequus,* even, *par,* equal) perfectly equal
semi-sepultus (*semi,* half, *sepultus,* buried) half buried
anim-aequus (*animus,* mind, *aequus,* even) not easily moved
albo-gilvus (*albus,* white, *gilvus,* pale-yellow) whitish yellow
longi-vivax (*longus,* long, *vivax,* tenacious of life) long-lived

III. If the last term is a verbal stem, the compound may be a noun of the first or third declension, or an adjective of three terminations or one termination.

limi-cola (*limus,* mud, *colo,* dwell) a mud-dweller
lapi-cida (*lapis, lapidis,* stone, *caedo,* cut) a stone-cutter
tubi-cen (*tuba,* trumpet, *cano,* sing) a trumpeter
man-ceps (*manus,* hand, *capio,* take) purchaser
frugi-legus (*frux, frugis,* fruit, *lego,* collect) fruit-gathering
herbi-gradus (*herba,* grass, *gradior,* go) going in the grass
multi-fidus (*multus,* many, *findo,* cleave) many-cleft
gemmi-fer (*gemma,* bud, *fero,* bear) bearing buds
spini-ger (*spina,* thorn, *gero,* bear) thorn-bearing

THE CONNECTIVE

The connecting vowel, *-i-*, has been so extended in Latin that the language practically knows no other connective. An *-o-* has found its way into some words after the analogy of Greek compounds, but these, as well as those in which the connecting vowel is *-u-*, are so rare that all connectives other than *-i-* may be entirely disregarded.

THE FIRST TERM

The first term of the compound in Latin may be a nominal stem (noun or adjective), an indeclinable (adverb, preposition, or inseparable particle), or more rarely a verbal stem. Nouns are extremely rare as the first term of Latin compounds, except where the last term is a verbal stem. Adjectives likewise, with the exception of numerals and a few common words such as *aequus, longus, multus,* etc., are rarely found in composition. In consequence, the first terms of Latin compounds are very largely made up of numerals and indeclinables, the latter taking, of course, no connective. In Latin, inflected stems appear almost invariably in the proper stem form. The infrequency of such stems in composition doubtless accounts in a large measure for this uniformity.

NOUNS

lani-pes (*lana,* wool, *pes,* foot) with wool on the feet
limi-genus (*limus,* mud, *gigno,* bring forth) mud-born
grani-fer (*granum,* grain, *fero,* bear) grain-bearing
funi-repus (*funis,* rope, *repo,* creep) rope-dancer
frugi-ferens (*frux, frugis,* fruit, *ferens,* bearing) fruitful
corpori-cida (*corpus, corporis,* body, *caedo,* cut) butcher
ori-putidus (*os, oris,* mouth, *putidus,* fetid) with vile mouth
corni-frons (*cornu,* horn, *frons,* forehead) with horned forehead
lacu-turris (*lacus,* lake, *turris,* tower) a kind of cabbage
fidei-commissum (*fides,* trust, *committo,* commit) a bequest in trust

ADJECTIVES

laeti-ficus (*laetus,* glad, *facio,* make) gladdening
soli-vagus (*solus,* alone, *vagor,* wander) wandering alone
tardi-gradus (*tardus,* slow, *gradior,* walk) slow-paced
atri-capillus (*ater,* black, *capillus,* hair of the head) black-haired
grandi-scapius (*grandis,* great, *scapus,* stem) having a large stem
levi-caulis (*levis,* smooth, *caulis,* stem) smooth-stemmed
pleuri-laterus (*plus, pluris,* more, *latus, lateris,* side) with several sides
serpenti-pes (*serpens,* creeping, *pes,* foot) serpent-footed

NUMERALS. *Unus, centum,* the higher cardinals and all ordinals appear in the stem form with *-i-* as the connective, the remaining cardinals are unchanged: ***duo,*** two, is replaced by ***bi-, tres,*** three, by ***tri-,*** and ***quattuor,*** four, by *quadri-.*

uni-folius (*unus,* one, *folium,* leaf) one-leaved
bi-fidus (*bi,* two, *findo,* cleave) cleft into two parts
tri-furcus (*tri,* three, *furca,* prong) three-pronged
quadri-jugis (*quadri,* four, *jugum,* yoke) yoked in fours
quinque-partitus (*quinque,* five, *partio,* divide) five-parted
septem-nervus (*septem,* seven, *nervus,* nerve) seven-nerved
decem-remis (*decem,* ten, *remus,* oar) ten-oared
centi-ceps (*centum,* hundred, *caput,* head) hundred-headed
primi-formis (*primus,* first, *forma,* form) original
quinti-ceps (*quintus,* fifth, *caput,* head) having five peaks

INDECLINABLES

Adverbs and adverbial prefixes:

semper-florium (*semper,* always, *flos,* flower) evergreen
paen-ultimus (*paene,* nearly, *ultimus,* last) last but one
per-magnus (*per,* very, *magnus,* large) very large
prae-longus (*prae,* very, *longus,* long) very long
in-divisus (*in,* not, *divido,* divide) undivided
sub-globosus (*sub,* rather, somewhat, *globosus,* spherical) nearly spherical

Prepositions:

ab-normis (*ab,* away from, *norma,* rule) irregular
ac-clivus (*ad,* to, *clivus,* slope) steep
ante-pes (*ante,* before, *pes,* foot) forefoot
circum-scissus (*circum,* around, *scindo,* split) split around
com-mutabilis (*cum,* together with, *mutabilis,* changeful) subject to change
de-jugis (*de,* from, *jugum,* yoke) sloping
ex-aridus (*ex,* out, *aridus,* dry) dried out
in-fuscus (*in,* in, *fuscus,* dark) dark-brown
inter-nodium (*inter,* between, *nodus,* knot) internode
ob-stipus (*ob,* towards, *stipes,* stalk) bent to one side
super-nans (*super,* above, *no,* swim) swimming at the top

Inseparables:

ambi-formis (*ambi*, around, *forma*, form) of doubtful form
dis-calceatis (*dis*, without, *calceo*, put on shoes) barefooted
re-formatus (*re*, back, again, *firmo*, fix) re-establish
se-jugis (*se*, apart, *jugum*, yoke) disjoined
ve-grandis (*ve*, out, not, *grandis*, large) not very large, small

ALTERNATIVES

Duplicate names or terms may arise in nomenclature from alternative words, stems, connectives, or terminations, or from the alternative use of nouns, adjectives, and verbal stems from the same root to form the last term of a compound. In Latin, alternatives cause little trouble because of the slight development of composition, and for the reason that Latin derivatives are largely specific. In Greek, the confusion arising from alternatives is great, and it is imperative that composition be made to conform to certain definite rules. An observance of the following rules in making compounds will aid greatly in preventing the occurrence of real duplicates, as well as the occurrence of extremely similar, though perfectly distinct compounds, which are a source of vexation to many biologists.

(1) When the language shows two or more alternative words, such as χάσμα and χάσμη, γράμμα and γραμμή, γραφίς and γραφή, the more primitive word should be chosen. As this involves a considerable knowledge of Greek, the only safe plan is for the coiner to make sure that his proposed compound does not appear in an alternative form. This may be readily done at the same time that he assures himself that the word does not already exist in his science in the form in which he proposes it.

(2) Duplicates arising from alternative stems and connectives are readily avoided by observing the rule already suggested, viz., that the proper connective is always -*o*- in Greek and that words always enter into composition in the full stem form.

(3) Since a root may often appear in the first term of the compound as a noun, adjective, or verbal stem, it is advisable that the coiner of a name should avoid using a root already found in either of its other forms in composition with the same last term.

(4) The last term of a generic name should be a noun. Although such compounds usually become adjectives in Greek, the confusion which thus arises from alternative endings or gender terminations can only be avoided by restricting such compounds to the form and gender of the noun of the last term, i. e., the last term of a compound should always remain unchanged.

(5) Verbals should be invariably avoided in compounding nouns and adjectives, i. e., in all the composition found in nomenclature. A compound of identical or similar meaning can always be secured by employing a noun or adjective, and the use of verbal stems, in many ways peculiar, should be left to the philologist.

(6) The repeated use of different suffixes in connection with the same generic compound, or indeed with the same last term, should be carefully avoided.

(7) The alternative termination of Latin compound adjectives in *-us* and *-is*, though hardly productive of any real difficulty, might well be avoided for the sake of the biologist who does not understand that these are merely alternative endings. Saint-Lager[1] has suggested that terminations in *-us* be assimilated to *-is*, and this suggestion might well be carried out, although the *-us* termination has the slight advantage of indicating gender somewhat more definitely.

ACCENT

The accent of all Greek and Latin derivatives in science is determined by the accentuation of Latin, since all Greek words after transliteration are governed, of course, by the usual rules of accent for Latin words. These are as follows:

(1) The ultimate is never accented.

(2) In words of two or more syllables the accent is on the penult when this is long; when the penult is short, the antepenult is accented.

GENDER

The gender of a name is the gender of the last term in its proper language, whether the termination conform or not. The

[1] Saint-Lager. Chapitre de Grammaire à l'usage des botanistes. 1892.

gender of the primitive may be changed at any time, of course, by the addition of a proper suffix of a different gender. The most frequent mistakes in the matter of gender occur in connection with Greek neuters in -α, -ας, and -ος; thus, γάλα, milk, and κρέας, meat, are usually regarded as feminines, when found as the last term of a compound, and ἄνθος, flower, as masculine.

CORRECTION LIST

In this list are included a number of the more common generic names which show improper formation. No attempt has been made to make the list exhaustive, as this would be an idle expenditure of time until there is a wider appreciation of the necessity of placing nomenclature upon a classical basis. The names given here serve simply as examples of the malformations which abound throughout biological nomenclature. The duplicate names which arise from malformations or from alternatives are discussed under section IV.

I. Compound with improper stem, often also with faulty connective.

Acianthus = Acidanthus (ἀκίς, ἀκίδος, ἡ, point)
Acilepis = Acidolepis
Acispermum = Acidosperma (σπέρμα, σπέρματος, τό, seed)
Acleisanthus = Acleanthus (ἀκλεής, inglorious)
Acrosanthus = Acranthus (ἄκρος, α, ον, at the point, highest)
Agrostistachys = Agrostidostachys (ἄγροστις, ιδος, ἡ, grass)
Amblirion = Amblylirium (ἀμβλύς, blunt, dull)
Amianthium = Amiantanthium (ἀμίαντος, ον, pure)
Chimophila = Chimatophila (χεῖμα, χείματος, τό, cold, frost)
Chiococca = Chionococcus (χιών, χιόνος, ἡ, snow)
Chiogenes = Chionogenes
Chroococcus = Chrotococcus (χρώς, χρωτός, ὁ, skin, color)
Coleosanthus = Coleanthus (κολεός, ὁ, sheath)
Cybianthus = Cybanthus (κύβη, ἡ, head)
Cynosurus = Cynura (κύων, κυνός, ὁ, dog)
Dasanthera = Dasyanthera (δασύς, shaggy, cfr. δάσος, shagginess)

Dermosporium,= Dermatosporium (δέρμα, δέρματος, τό, skin)
Dermocybe = Dermatocybe
Eilemanthus = Ilematanthus (εἴλημα, εἰλήματος, τό, veil)
Epigynanthus = Epigynaecanthus (γυνή, γυναικός, ἡ, woman)
Galanthus = Galactanthus (γάλα, γάλακτος, τό, milk)
Galarhoeus – Galactorrheus
Gasteranthus – Gastranthus (γαστήρ, γαστρός, ἡ, belly)
Geropogon – Gerontopogon (γέρων, γέροντος, ὁ, old man)
Gigandra – Gigantandra (γίγας, γίγαντος, ὁ, giant)
Gynopogon – Gynaecopogon (γυνή, γυναικός, ἡ, woman)
Haemocarpus – Haematocarpus (αἷμα, αἵματος, τό, blood)
Hedysarum = Hedyarum (ἡδύς, sweet, but ἡδύσαρον, Diosc.!)
Helixanthera – Helicanthera (ἕλιξ, ικος, twisted)
Homalobus – Homalolobus (ὁμαλός, even, equal), hardly Homolobus (ὁμός, one and the same)
Ilysanthus – Ilyanthus (ἰλύς, ἡ, mud)
Iondraba – Iodrabe (ἴον, τό, violet)
Ionactis – Iactis
Isonanthus – Isanthus (ἴσος, equal)
Kalosanthus – Calanthus (καλός, beautiful; better, Callianthus, καλλι-)
Korycarpus – Corythocarpus (κόρυς, κόρυθος, ἡ, helmet)
Lacistema – Lacidostema (λακίς, λακίδος, ἡ, rent)
Leontostomium – Leontostomatium (στόμα, στόματος, τό, mouth)
Lepargyraea = Lepidargyraea (λεπίς, λεπίδος, ἡ, scale, also λέπος, τό, scale)
Lepicystis = Lepidocystis
Lepisanthus = Lepidanthus
Manisuris = Manura (μανός, rare, porous)
Megacephalum = Megalocephalum (μέγας, μεγάλου, great)
Megapterium = Megalopterium
Megasanthus = Megalanthus
Melasanthus = Melananthus (μέλας, μέλανος, black)
Melianthus = Melitanthus (μέλι, μέλιτος, τό, honey)
Melilotus = Melitolotus
Myosurus = Myura (μῦς, μυός, ὁ, mouse)

Namaspora = Namatospora (*νάμα, νάματος, τό*, stream)
Nemacladus = Nematocladus (*νῆμα, νήματος, τό*, thread)
Nemastylis = Nematostylis
Onygena = Onychogenes (*ὄνυξ, ὄνυχος, ὁ*, nail, claw)
Oonopsis = Oopsis (*ᾠόν, τό*, egg)
Ophispermum = Ophiosperma (*ὄφις, ὄφιος, ὁ*, snake)
Pachysandra = Pachyandra (*παχύς*, thick)
Pachysanthus = Pachyanthus
Peliosanthus = Pelianthus (*πελιός*, livid)
Pholistoma = Pholidostoma (*φολίς, φολίδος, ἡ*, horny scale, spot
Pyxipoma = Pyxidopoma (*πυξίς, πυξίδος, ἡ*, a box of boxwood)
Raphiolepis = Rhaphidolepis (*ῥαφίς, ῥαφίδος, ἡ*, needle)
Regmandra = Rhegmatandra (*ῥῆγμα, ῥήγματος, τό*, fracture)
Rhexantha = Rhexianthe (*ῥῆξις, ῥήξεως, ἡ*, rending, rent)
Salpianthus = Salpinganthus (*σάλπιγξ, σάλπιγγος, ἡ*, war-trumpet)
Salpixanthus – Salpinganthus
Schismoceras – Schismatoceras (*σχίσμα, σχίσματος, τό*, cleft; cfr. *σχισμή, ἡ*)
Scolosanthus – Scolanthus (*σκῶλος, ὁ*, stake; better, Scolopanthus from *σκόλοψ, ὁ*)
Spermacoce – Spermatococe (*σπέρμα, σπέρματος, τό*, seed)
Spermolepis – Spermatolepis
Stachysanthes = Stachyanthus (*στάχυς, στάχυος, ὁ*, spike)
Stemastrum = Stematastrum (*στῆμα, στήματος, τό*, stamen)
Stigmanthus = Stigmatanthus (*στίγμα, στίγματος, τό*, point; cfr. *στιγμή, ἡ*)
Thisantha = Thinanthe (*θίς, θινός, ὁ, ἡ*, sand)
Thrixspermum = Trichosperma (*θρίξ, τριχός, ἡ*, hair)
Toxylon = Toxoxylum (*τόξον, τό*, bow)
Tremanthus = Trematanthus (*τρῆμα, τρήματος, τό*, hole)
Trichosanthes = Trichanthus (*θρίξ, τριχός, ἡ*, hair)

II. Compounds with improper connective.
Acanthopphippium = Acanthephippium (*ἄκανθα, ἡ*, thorn, *ἐφίππιον, τό*, saddle)
Actegiton = Actogiton (*ἀκτή, ἡ*, headland, seacoast)
Actephila = Actophila

Aloexylon = Alo(o)xylum (ἀλόη, ἡ, aloe)
Amaracarpus = Amarocarpus (ἀμάρα, ἡ, trench)
Amb yocarpum = Amblycarpum (ἀμβλύς, blunt)
Amecarpus = Amocarpus (ἄμη, ἡ, mattock)
Ammodenia = Ammadenia (ἄμμος, ἡ, sand, ἀδήν, ὁ, gland)
Ampelygonum = Ampelogony (ἄμπελος, ἡ, vine)
Amphiorhox = Amphirrhox (ἀμφί, around, on both sides)
Andripetalum = Andropetalum (ἀνήρ, ἀνδρός, ὁ, man)
Andriopetalum = Andropetalum (cfr. ἀνδρίον, τό, manikin)
Anthephora = Anthophora (ἄνθος, τό, flower, cfr. ἄνθη, ἡ)
Artanema = Artonema (ἄρτος, ὁ, loaf of bread)
Batheogyne = Bathygyne (βαθύς, deep, high)
Beloanthera = Belanthera (βέλος, τό, dart)
Blephariglottis = Blepharidoglottis (βλεφαρίς, ίδος, ἡ, eyelash, cfr. βλέφαρον, τό, eyelid)
Blepharispermum = Blepharidosperma
Botryceras = Botryoceras (βότρυς, βότρυος, ὁ, cluster of grapes)
Brachyolobus = Brachylobus (βραχύς, short)
Chorioactis = Choriactis (χόριον, τό, membrane)
Coreopsis = Coriopsis (κόρις, κόριος, ὁ, bug)
Corispermum = Coriosperma
Cypripedium = Cypridopedium (Κύπρις, ιδος, ἡ, Aphrodite)
Dacrymyces = Dacryomyces (δάκρυον, τό, tear, cfr. δάκρυ, τό)
Dacrycarpos = Dacryocarpus
Dasiphora = Dasyphora
Dasyochloa = Dasychloe (δασύς, shaggy)
Dictyderma = Dictyoderma (δίκτυον, τό, fishing net)
Endespermum = Endosperma (ἔνδον, within)
Gaiadendron = Gaiodendrum (γαῖα, ἡ, earth; better, Geodendrum from γῆ)
Glyphyllaea = Glyphidophyllaea (γλυφίς, γλυφίδος, ἡ, notch)
Graphephorum = Graphophorum (γραφή, ἡ, drawing)
Halicoccus = Halococcus (ἅλς, ἁλός, ἡ, sea)
Halyseris = Haloseris
Harpaecarpus = Harpocarpus (ἅρπη, ἡ, bird of prey, sickle)
Harpechloa = Harpochloe
Hebeanthe = Hebanthus (ἥβη, ἡ, youth)

Hebeloma = Heboloma
Heleocharis – Helocharis (ἕλος, ἕλεος, τό, marsh)
Heleochloa = Helochloe
Helichrysum – Heliochrysum (ἥλιος, ὁ, sun)
Hyoscyamus – Hyocyamus (ὗς, ὑός, ὁ, ἡ, swine)
Ixonanthes – Ixanthes (ἰξός, ὁ, mistletoe)
Lachnastoma – Lachnostoma (λάχνη, ἡ, down)
Menispermum = Menosperma (μήνη, ἡ, moon)
Napeanthus – Napanthus (νάπη, ἡ, woody dell)
Nomaphila – Nomophila (νομός, ὁ, pasture)
Opegrapha – Opographe (ὀπή, ἡ, opening)
Oreocarya = Orocarya (ὄρος, ὄρεος, τό, mountain)
Oreodoxa = Orodoxa
Pachyospora = Pachyspora (παχύς, thick)
Pellacalyx = Pellocalyx (πέλλα, ἡ, leather)
Pentstemon = Pentastemon (πέντε, πεντα-, five)
Phleboanthe = Phlebanthe (φλέψ, φλεβός, ἡ, vein)
Phoradendron = Phorodendrum (φώρ, φωρός, ὁ, thief)
Retiniphyllum = Rhetinophyllum (ῥητίνη, ἡ, resin of the pine)
Retinispora = Rhetinospora
Rhamphicarpa = Rhamphidocarpus (ῥαμφίς, ῥαμφίδος, ἡ, hook)
Scaphespermum = Scaphosperma (σκάφος, τό, hollow)
Sciaphila = Sciophile (σκιά, ἡ, shade)
Stictyosiphon = Stictosiphon (στικτός, ή, όν, pricked)
Stylipus = Stylopus (στῦλος, ὁ, pillar, cfr. στυλίς, ἡ)
Telipogon – Telopogon (τέλος, τό, end)
Thallisphaera – Thallosphaera (θαλλός, ὁ, young shoot)
Thelebolus = Thelobolus (θηλή, ἡ, nipple)
Thelephora = Thelophora
Thelesperma – Thelosperma
Theleophytum – Thelophytum
Xyloaloe = Xylaloe (ξύλον, τό, wood)
Zygnema – Zygonema (ζυγόν, ὁ, yoke)

III. Compounds with improper ending.

For reasons already given, the following compounds are corrected to conform to the primitive form of the last terms. Since it may not be clearly understood that certain adjective endings are incorrect, while others are correct but confusing, a summary is given of the correct endings for compound adjectives based upon nouns of the different declensions:

First and second declension: -ος, -ον

Third declension:

Stems in ν, ρ, δ: no change, rarely -ων, or -ωρ, or -ος, -ον

Stems in any other consonant, or in -ες: -ος, -ον

Stems in -ατ, nom. -ας: -ως, -ων, or -ος, ον

Stems in -ι or -υ: no change, or -ος, -ον

Acanthobotrya = Acanthobotrys (βότρυς, ὁ, a cluster of grapes)

Acanthocarpa = Acanthocarpus (καρπός, ὁ, fruit)

Acrodryon = Acrodrys (δρύς, ἡ, oak. Acrodryus, -um is permissible but rare)

Acrospermum = Acrosperma (σπέρμα, τό, seed)

Acrotriche = Acrothrix (θρίξ, τριχός, ἡ, hair)

Amphilophis = Amphilophus (λόφος, ὁ, crest)

Botrytis = Botryites (βοτρυίτης, ὁ, like grapes)

Callistachya – Callistachys (στάχυς, ὁ, spike)

Calycera – Calliceras (κέρας, τό, horn)

Ceratocaulis – Ceratocaulus (καυλός, ὁ, stalk)

Centrophyta – Centrophytum (φυτόν, τό, plant)

Chamaemeles – Chamaemelum (μῆλον, τό, apple)

Cheilococca – Chilococcus (κόκκος, ὁ, berry)

Cyanotis – Cyanus or Cyanotus (οὖς, ὠτός, τό, ear)

Cyclopeltis – Cyclopelte (πέλτη, ἡ, small shield)

Cyrtorrhyncha – Cyrtorrhynchus (ῥύγχος, τό, snout)

Dasytricha = Dasythrix (θρίξ, τριχός, ἡ, hair)

Desmophlebis = Desmophleps (φλέψ, ἡ, vein)

Didiplis = Didiplus (διπλός, διπλόος, twofold)

Distichlis = Distichus (στίχος, ὁ, row)

Epistemum = Epistemon (στήμων, ὁ, warp) or Epistema (στῆμα, τό, stamen)

Euchaetis = Euchaete (χαίτη, ἡ, long hair)

Gigandra = Gigantaner or Gigantandrus (ἀνήρ, ἀνδρός, ὁ, man)

Glossocomia – Glossocome (κόμη ἡ, hair)

Glycyosmis = Glycyosme (ὀσμή, ἡ, smell)

Grammitis = Grammatites (γράμμα, ατος, τό, line, -ιτης)
Gyrocerus = Gyroceras (κέρας, τό, horn) or Gyroceros (κερώς, ών, horned)
Haplolepidea = Haplolepis (λεπίς, ίδος, ἡ, scale) or Haplolepidus
Haplophlebia = Haplophleps (φλέψ, ἡ, vein)
Hedycrea = Hedycreas (κρέας, τό, meat)
Hippuris = Hippura (οὐρά, ἡ, tail, but ἵππυρις, horse-tailed)
Hydrangea = Hydrangium (ἀγγεῖον, τό, vessel)
Lagopoda = Lagopus (πούς, ποδός, ὁ, foot)
Lagotis = Lagus or Lagotus (οὖς, ὠτός, τό, ear)
Leptis = Leptus (λεπτός, fine, thin)
Leptostachya = Leptostachys (στάχυς, ὁ, spike)
Lycianthes = Lycianthus (ἄνθος, τό, flower)
Macroscepis = Macroscepe (σκέπη, ἡ, cover)
Melancranis = Melanocranus (κράνος, τό, helmet)
Monochila = Monochilus (χεῖλος, τό, lip)
Myagrum = Myagra (μυάγρα, ἡ, mouse-trap)
Nemacaulis = Nematocaulus (καυλός, ὁ, stalk)
Neurada = Neuras (νευράς, άδος, ἡ, a plant)
Odontoptera = Odontopterum (πτερόν, τό, feather)
Oncogastra = Oncogaster (γαστήρ, γαστρός, ἡ, belly)
Pachnolepia = Pachnolepis (λεπίς, ίδος, ἡ, scale)
Pentachaeta = Pentachaete (χαίτη, ἡ, long hair)
Phaenopoda = Phaenopus (πούς, ποδός, ὁ, foot)
Rhizobotrya = Rhizobotrys (βότρυς, ὁ, bunch of grapes)
Rhyncholopha = Rhyncholophus (λόφος, ὁ, crest)
Sciuris = Sciurus (σκίουρος, ὁ, shadow-tail, squirrel)
Sclerostomum = Sclerostoma (στόμα, τό, mouth)
Scotophylla = Scotophyllum (φύλλον, τό, leaf)
Therofon = Therophonum (θηροφόνος, ον, killing wild beasts)
Xylostyla = Xylostylus (στῦλος, ὁ, pillar)
Zygopeltis = Zygopelte (πέλτη, ἡ, shield)

A consideration of duplicates arising out of the above cases, or from alternative words, will be found in section IV.

III

Terms are invalid unless properly transliterated; retroactively, all improper transliterations are to be corrected.

"Nomina generica Graeca Latinis literis pingenda sunt." Critica Botanica 127.

"The strict Latin orthography can not be too rigorously insisted upon; consistency will in no other way be attainable." Miller, Scientific Names 127.

The following table shows the proper transliteration of Greek vowels, diphthongs, and consonants into Latin. For the sake of uniformity, alternative transliterations (such as a for η final, e for ει), are avoided.

α = a	αι = ae	β = b	ρ = r
ε = e	αυ = au	γ = g	τ = t
η = e	ει = i	= n before κ, [γ, χ	φ = ph
ι = i	ευ = eu	δ = d	χ = ch
ο = o	ηυ = eu	ζ = z	ψ = ps
= u in final -ος, [ον	οι = oe	θ = th	γκ = nc
υ = y	ου = u	κ = c	γχ = nch
ω = o	υι = yi	λ = l	γγ = ng
	ᾳ = a	μ = m	ῥ = rh, ῤῥ = rrh
	ῃ = e	ν = n	ʽ = h
	ῳ = o	= m in final -ον	
		ξ = x	
		π = p	
		σ, ς = s	

Medial ʽ (*h*) arising from word-formation is to be transliterated, thus preventing elision of a preceding vowel, unless its presence is already shown by aspirating the preceding consonant, as in ἐφήμερα. Latin usage is variable in this particular, since words already compounded in Greek, in which the aspirate was not visible, were transliterated into Latin as they stood, while in other words in which the presence of the aspirate was felt or known, the latter was transliterated. In scientific words it is important that the rough breathing be rendered by *h*, not only in order that the terms of a compound may be readily recognized, but also to avoid the possible confusion of two compounds otherwise exactly alike.[1]

The following list will serve to illustrate the more frequent errors in transliteration, and their correction.

Adenocaulon – Adenocaulum (or, much better, Adenocaulus, the last term being καυλός, ὁ, stalk); Lachnocaulum
Agropyron – Agropyrum (πυρός, ὁ, wheat)
Aerophyton – Aerophytum (φυτόν, τό, plant); Petrophytum
Ampelodesmos = Ampelodesmus (δεσμός, ὁ, band)
Amphicarpon = Amphicarpum (better, Amphicarpus from καρπός, ὁ, fruit)
Acrospeira = Acrospira (σπεῖρα, ἡ, knot, coil)
Amorpha = Amorphe (μορφή, ἡ, form)
Arachnion = Arachnium (ἀράχνιον, τό, spider's web)
Apios = Apius (ἄπιος, ἡ, pear, pear-tree)
Aplopappus = Haplopappus (ἁπλόος, simple)
Arctostaphylos = Arctostaphylus (better, Arctostaphyle from σταφυλή, ἡ, bunch of grapes)
Astrebla = Astreble (στρέβλη, ἡ, roller, στρεβλός, ή, όν, twisted)
Batodendron = Batodendrum (δένδρον, τό, tree); Linodendrum, Phorodendrum, Rhododendrum, Toxicodendrum
Blepharoneuron = Blepharoneurum (νεῦρον, τό, fibre, nerve)
Brachychaeta – Brachychaete (χαίτη, ἡ, hair)
Callirhoe = Callirrhoe; Glycyrrhiza, Coralliorrhiza, etc.
Chaetochloa = Chaetochloe (χλόη, ἡ, grass); Echinochloe, Eriochloe, Helochloe, Leptochloe, Scolochloe
Chamaecladon = Chamaecladus (κλάδος, ὁ, shoot)
Chamaenerion = Chamaenerium (νήριον, τό, oleander)
Chamaerhodos = Chamaerhodus (better Chamaerhodum, ῥόδον, τό, rose)
Chionyphe = Chionohyphe (ὑφή, ἡ, web)
Cheiranthes = Chiranthus (χείρ, ἡ, hand); Chiromyces
Coilomyces = Coelomyces (κοῖλος, hollow)

[1]Linné. Critica Botanica, 129. 1737.
Dall, W. H. Nomenclature in Zoology and Botany, 55. 1877.
Kuntze, Otto. Revisio Generum Plantarum, 3:354. 1893.
Miller, Walter. Scientific Names of Latin and Greek Derivation. *Proc. Cal. Acad. Sci.*, 1:127. 1897.

Cyperus = Cypirus (*κύπειρος*, ὁ, marsh plant)
Corypha = Coryphe (*κορυφή*, ἡ, head, top)
Dasylirion = Dasylirium (*λείριον*, τό, lily)
Diospyros = Diopyrus (*πυρός*, ὁ, wheat)
Dolichos = Dolichus (*δολιχός*, long)
Eleocharis = Helocharis (*ἕλος*, τό, marsh)
Elodea = Helodes (*ἑλώδης*, marshy)
Gyrotheca = Gyrothece (*θήκη*, ἡ, box); Heterothece, Tetragonothece
Haplymenium – Haplohymenium (*ὑμένιον*, τό, little membrane)
Helicoon = Helicoum (*ᾠόν*, τό, egg)
Hemicarpha – Hemicarphe (*κάρφη*, ἡ, scale, better Hemicarphus, *κάρφος*, τό, scale)
Hydrocleis – Hydroclis (*κλείς*, ἡ, hook, key)
Hydrodictyon – Hydrodictyum (*δίκτυον*, τό, net)
Korycarpus = Corythocarpus (*κόρυς*, *κόρυθος*, ἡ, helmet)
Lecanidion = Lecanidium (*λεκανίδιον*, τό, dish, pan)
Lycopersicon = Lycopersicum (*περσικόν*, τό, peach)
Metroxylon = Metroxylum (*ξύλον*, τό, wood); Stereoxylum
Microthyrion = Microthyrium (*θύριον*, τό, little door)
Opegrapha = Opographe (*γραφή*, ἡ, drawing)
Orophaca = Orophace (*φακῆ*, ἡ, lentil)
Potamageton = Potamogiton (*γείτων*, ὁ, neighbor)
Protalos = Protohalus (*ἅλς*, *ἁλός*, ἡ, sea)
Prinos = Prinus (*πρίνος*, ἡ, evergreen oak)
Rhodospatha = Rhodospathe (*σπάθη*, ἡ, broad blade)
Sicyos – Sicyus (*σίκυος*, ὁ, common cucumber)
Spirodela = Spirodele (*δῆλος*, visible)
Steirochaete = Stirochaete (*στεῖρα*, ἡ, beam of a keel); Stironema
Stenospermation = Stenospermatium (*σπερμάτιον*, τό, little seed)
Symphoricarpos – Symphoricarpus (*καρπός*, ὁ, fruit)
Symplocos – Symplocus (*σύμπλοκος*, entwined)
Syndesmion = Syndesmium (better, Syndesmus, *σύνδεσμος*, ὁ, band)

IV

Of two or more similar terms, the earliest alone is valid, unless they show an essential difference in root, suffix, or prefix; differences of spelling, gender, or alternative termination are insufficient. Retroactively, the earliest name, if not already in the proper form, is to be corrected, while all others fall.

"Nomina generica, simili sono exeuntia, ansam praebent confusionis." Critica Botanica 43.

"Nomina generica ex aliis nominibus genericis, cum syllaba quadam in fine addita, conflata, non placent." Ibid. 38.

Similar generic names have long constituted a grave source of confusion in biology. Nearly every writer upon botanical nomenclature has appreciated this fact, and has suggested some method of obviating the difficulty. Linné[1] pointed out clearly the way by which all such duplicates and apparent duplicates might be avoided, but in the subsequent rapid development of taxonomy his precepts were lost sight of or ignored. The Paris Code, though silent on this matter, unintentionally aggravated the situation by the unfortunate reservations of Article 66. In passing, it should be noted how signally the purpose of this scholarly article has been defeated by the presence of an unimportant exception. The provision that "every botanist is authorized to rectify the faulty names or terminations, unless it be a question of a very ancient name current under its incorrect form," obviously made exception only for names given by Aristotle, Theophrastus, Dioscorides, Pliny, and other Greek and Roman writers upon plants. But this exception has since been persistently misunderstood, or purposely extended to cover any incorrect name of any degree of currency whatsoever, and has finally found expression in the absurd dictum that "the original form of a name is to be retained no matter how incorrect it may be." This feeling seems to have had some influence upon the treatment of similar generic names in the Berlin Rules and in the Rochester Code. Though the statement of the rule is different, the treatment is practically identical in both. According

[1] Critica Botanica, 39, 43.

to the former[1], "similar names are to be conserved, if they differ ever so little in the last syllable; if they only differ in the mode of spelling, the newer one must fall." Also, "there are to be conserved *Adenia* as well as *Adenium, Apios* as well as *Apium, Chloris* as well as *Chlorea* and *Chlora, Danae* as well as *Danais, Hydrothrix* as well as *Hydrotriche, Silvaea* as well as *Silvia*, etc.; we doubt that there is any scholar who will confound them. On the contrary, *Tetraclis* and *Tetracleis, Oxythece* and *Oxytheca, Epidendrum* and *Epidendron, Oxycoccus* and *Oxycoccos, Asterocarpus* and *Astrocarpus, Peltostema* and *Peltistema* are only different modes of spelling the same word, and the newer one is to be rejected if they name different genera." The Rochester Rules[2] provide that "Similar generic names are not to be rejected on account of slight differences, except in the spelling of the same word; for example, *Apios* and *Apium* are to be retained, but of *Epidendrum* and *Epidendron, Asterocarpus* and *Astrocarpus*, the latter is to be rejected." In both codes, it will be noticed that similar names are to be rejected only when the difference is merely one of transliteration of the ending, or, very rarely, of connective. A difference of gender termination or of alternative ending is considered sufficient to warrant retention, even though this difference results from incorrect formation, as in Hydrotriche.

Both rules are equally far from any classical warrant, and, in consequence, neither code can furnish a logical or accurate basis for the treatment of similar terms. In formulating a rule for these, however, it is impossible to give serious consideration to the views of mere logophiles, who would make wholesale rejections on the basis of slight or fancied similarities. Thus, it has been suggested that Micranthus and Micranthemum are so similar as to warrant the rejection of one, while of Macranthe and Megalanthe, Glycyphila and Glycyphylla, one should be rejected because the first two are practically identical in meaning, and the last two in pronunciation! Between the two extremes there

[1] Vorschläge zur Ergänzung der "Lois de la Nomenclature Botanique." Berlin, 1892.

[2] *Bull. Torr. Bot. Club*, 19:290. 1892.

is but one logical position, namely, similar terms are identical in nomenclature when as Greek and Latin words they exhibit no essential differences. Thus, Cerastium, Ceratium, and Ceratia are merely different forms of a Greek word κεράτιον, and are homonyms, while Lecane, Lecanarium, and Lecanidium are different words, the last two being formed upon the first by the use of suffixes. Frequent affixation of the same stem should be carefully avoided, however, regardless of the validity of the resulting derivatives.

Such Greek words as ἄνθος, γράφις, κεφαλή, and their relatives, which are extremely frequent in nomenclature, will serve very well to show the difference between homonyms and similar yet valid terms. Besides many compounds, the lexicon shows twenty derivatives of the root ἀνθ-: of these, the following seventeen are sufficiently distinct to justify their use: ἄνθος, τό, flower; ἀνθήλη, ἡ, panicle; ἄνθεμον, τό, flower; ἀνθέμιον, τό, floweret; ἀνθίον, τό, floweret; ἀνθέρικος, ὁ, flower of asphodel; ἀνθεών (ἀνθών), ὁ, flower bed; ἀνθεμώδης, flower-like; ἀνθεμόεις, flowery; ἀνθεμωτός, adorned with flowers; ἀνθηδών, ἡ, bee; ἀνθηρός, flowery; ἀνθηρίτης, ἡ, bloom; ἄνθησις, ἡ, full bloom; ἀνθητικός, blossoming; ἄνθινος, blooming; ἀνθοσύνη, ἡ, bloom. Ἄνθη, ἡ, full bloom, flower, should be avoided in composition, since it is identical with ἄνθος when used as a first term, and is confusing as a last term; ἄνθειον is identical with ἀνθίον, and ἀνθεμίς too near ἄνθεμον to be fortunate. The root γραφ- shows two series of derivatives, one based upon the root, and the other upon the stem γραμματ-. Of the latter, γράμμα, τό, letter, picture, γραμματεῖον, τό, document, γραμματείδιον, τό, small tablets, γραμματεύς, ὁ, scribe, and γραμματική, ἡ, written character, are different, while γραμμή, ἡ, stroke, line, is to be regarded as a mere alternative of γράμμα. In the first series, γραφή, ἡ, drawing, γραφεῖον, τό, pencil, γραφείδιον, τό, pencil, and γραφικός, graphic, are distinct, but γραφίς, ἡ, stylus, and γράφος, τό, letter, are alternatives. Of the derivatives of κεφαλή, ἡ, head, κεφαλίς, ἡ, little head, should be avoided, but the following are distinct; κεφάλιον, τό, little head; κεφαλίδιον, τό, little head; κεφαλίνη, ἡ, head of the tongue; κεφαλικός, of the head; κεφάλαιος, of the head; κεφαλαιώδης, chief. Nomenclature would, however, become very much involved for anyone but the philologist, if all the proper derivatives of such roots as the above were to find a place in it. Such a condition can be readily avoided if proposers of terms will take the trouble to acquire a Greek vocabulary.

I. Homonyms.

These arise from alternative forms of the same root or stem, from mere differences of spelling, transliteration, gender or alternative ending, or from differences produced by erroneous connectives or terminations.

Aceras Pers. 1807
  Acerates Elliott 1817
Aceratium DC. 1824
  Aceratia F. Müll. 1854
Acetabulum Tourn. 1700
  Acetabula Fries 1822
Acetabularia Lamx. 1816
  Acetabularium Endl. 1836
Achlys DC. 1821
  Achlya Nees 1823
Adenia Forsk. 1775
  Adenium Roem. & Schult. 1819
Adenogyne Klotzsch 1841
  Adenogynum Rchb. & Zoll. 1856
Adenophorus Desvaux 1808
  Adenophora Fisch. 1823
Apios Boerh. 1720
  Apium Hoffm. 1814
Calanthe R. Br. 1821 (mel. Calanthus)
  Kalosanthes Haworth 1821
  Calanthea DC. 1824
  Calosanthes Blume 1826
Callitriche L. 1751 (cor. Callithrix)
  Calythrix R. Br. 1819
  Calothrix Ag. 1824
Cerastium L. 1737 (cor. Ceratium)
  Ceratia Adans. 1763
  Ceratium Alb. & Schwein. 1805
Chamaedrys Tourn. 1700
  Chamaedryon Seringe 1825
Chamaemelum Tourn. 1700
  Chamaemeles Lindl. 1822
  Chamaemela DC. 1837
Chlora Adans. 1763
  Chloraea Lindl. 1826
  Chlorea Nyland. 1854
Coleosanthus Cassini 1817 (cor. Coleanthus)
  Coleanthus Seidl 1817
Dasanthera Raf. 1819 (cor. Dasyanthera)
  Dasianthera Presl 1831
Dermatocarpon Eschw. 1824 (cor. Dermatocarpum)
  Dermatocarpus Miers 1852
  Dermocarpa Crouan 1856
Desmanthus Willd. 1805
  Desmosanthes Blume 1825
Dicera Forst. 1776 (cor. Diceras)

Diceras Endl. 1840
Dictyanthes Rchb. 1837 (mel. Dictyanthus)
Dictyanthus Decaisne 1844
Drimys Forst. 1776
Drimia Jacq. 1786
Epiphegus Spreng. 1820
Epiphagus Rylands 1843
Eremanthis Cassini 1827 (cor. Eremanthus)
Eremanthus Lessing 1829
Eremanthe Spach 1836
Erythranthus Hanstein 1853
Erythranthe Baillon 1858
Eurotia Adans. 1763
Eurotium Link 1809
Gamochilum Walpers 1839
Gamochilus Lestib. 1841
Glyciphylla Raf. 1819 (cor. Glycyphylla)
Glycyphylla Steven 1834
Glyphia Cassini 1818
Glyphaea Hook. f. 1846
Gonatobotrys Corda 1839
Gonatobotryum Sacc. 1879
Gonyanthes Blume 1823 (cor. Gonatanthus)
Gonatanthus Klotzsch 1840
Grammocarpus Ser. 1825 (cor. Grammatocarpus)
Grammatocarpus Presl 1831
Heterocladia Decaisne 1841
Heterocladium Schimp. 1852
Hippobroma G. Don 1834
Hippobromus Eck. & Zeyh. 1836
Holophyllum Lessing 1830
Holophylla G. Don 1837
Isomerium R. Br. 1830
Isomeria Presl 1837
Lecanium Presl 1843
Lecania Massalongo 1853
Lepidocarpus Adans. 1763
Lepidocarpa Blume 1855
Lepidotis Palis. 1805 (cor. Lepidotus)
Lepidota Sterb. 1820
Lepidotus Fries 1836
Lepidotia Rchb. 1841
Lepidotum Dunal 1852
Lepisanthus Blume 1825 (cor. Lepidanthus)
Lepidanthus Nees 1830
Macranthus Poir. 1813
Macranthea Boiss. 1840
Macrantha Bunge 1843
Macropodium R. Br. 1812
Macropodia Fuckel 1869
Marainophyllum Pohl. 1825 (cor. Marantophyllum)
Marantophyllum Miquel 1855
Megalanthe Gaudin 1828 (mel. Megalanthus)
Megasanthus G. Don 1834
Microglossa DC. 1836
Microglossum Sacc. 1884
Microtea Swartz 1788 (mel. Microtes)
Microtis R. Br. 1810
Monochila G. Don 1834 (cor. Monochilus)
Monochilus Fisch. & Meyer 1835

Oliganthes Cassini 1817 (cor. Oliganthus)
Oliganthos Barneoud 1845
Oligotrichum DC. 1805
Oligothrix DC. 1837
Pachypleurum Rchb. 1832
Pachypleuria Presl 1836
Pachypleura Jamb. & Spach 1842
Petrophile Knight & Salisb. 1809
Petrophila R. Br. 1800
Rytiphlaea Ag. 1817 (cor. Rhytidophloeus)
Rhytidofloyos Corda 1845
Salpianthus Humb. & Bonp. 1808 (cor. Salpinganthus)
Salpixantha Hook. 1845
Schismus Palis. 1812 (mel. Schisma)
Schisma DuMort. 1822
Schizanthus Ruiz & Pav. 1794
Schisanthus Haworth 1819
Schistanthe Kunze 1841
Sphaerophorus Pers. 1794
Sphaerophora Blume 1850
Sphaeroplea Ag. 1824 (cor. Sphaeropleum)
Sphaeropleum Link 1826
Stilbe Berg. 1767
Stilbum Tode 1790
Tapinanthus Blume 1824
Tapeinanthus Herbert 1837
Tet andra A. DC. 1845 (cor. rTetraner)
Tessarandra Lindl. 1847
Thrixspermum Lour. 1790 (cor. Trichosperma)
Trichospermum Blume 1825
Trachysperma Raf. 1809
Trachyspermum Link 1821
Trichopteris Necker 1790
Trichipteris Presl 1822
Trichosanthus L. 1737 (cor. Trichanthus)
Trichantha Hook. 1844
Trichanthus Philippi 1857
Xanthoglossa DC. 1837
Xanthoglossum Lindl. 1852

II. Terms classically different, but so similar in form as to be unfortunate. There is not sufficient warrant for the rejection of these, but their formation is to be avoided, if not, indeed, invalidated, for the future.

Acarphaea Harvey & Gr. 1849
Acarpha Griseb. 1856
Chlora Adans. 1763
Chloris Swartz 1788
Danae Medic. 1787
Danais Vent. 1799
Galax L. 1753
Galactia P. Br. 1756
Galaxia Thunb. 1782
Gliocladium Corda 1840
Gloeocladia, J. Ag. 1842 (cor. Gloeocladium)
Glyphis Achar. 1814
Glyphia Cassini 1818
Hydrophila Ehrhart 1780 (cor. Hydrophile)

Philydrum Gaertn. 1788 (mel. Philohydrum)
Isomerium R. Br. 1830
Isomeris Torr. & Gr. 1838
Ixianthes Benth. 1836 (mel. Ixianthus)
Ixanthus Griseb. 1839
Lepanthes Swartz 1799 (mel. Lepanthus)
Lepisanthus Blume 1825 (cor. Lepidanthus)

Rhaphidospora Nees 1832
Rhaphiospora Körb. 1855
Syncephalum DC. 1837
Syncephalis Van Tieghem 1875
Theriophonum Blume 1835
Therofon Raf. 1836 (cor. Therophonum)
Xanthiopsis DC. 1836
Xanthopsis DC. 1837

III. Similar terms distinct classically and nomenclaturally

Actinostemon Klotzsch 1841
Actinostemma Lindl. 1847
Alectra Thunb. 1784
Alectryon Gaertn. 1788
Brachylobos DC. 1821 (cor. Brachylobus)
Brachylobium C. A. Meyer 1841
Calopogon R. Br. 1813
Calopogonium Desvaux 1826
Ceramianthemum Donati 1750
Ceramianthe Rchb. 1831 (mel. Ceramianthus)
Cladodes Lour, 1790
Cladodium Bridel 1826
Diceratium Lagasca 1815

Diceras Endl. 1840
Eritrichium Gaudin 1828
Eriothrix Rchb. 1828
Glechoma L. 1737 (cor. Glechonoma)
Glechon Spreng. 1827
Haplocarpha Lessing 1831
Haplocarpaea Endl. 1838
Micranthus Wendland 1798
Micranthemum Michx. 1803
Stylidium Swartz 1807
Stylis Poir. 1817
Trachypodium Leman 1828
Trachypus Reinw. & Hornsch. 1829

V

Terms are invalid unless properly spelled; retroactively, improper spellings are to be corrected.

Apart from its application to improper formations, this thesis is of secondary importance. It is given place here merely to emphasize again the fact that nomenclature in all its aspects must rest upon a classical basis, a repetition rendered imperative for the reason that many biologists and more than one code still regard the Latin of Linné as the model. In Greek, a large number of incorrect spellings have arisen from the careless practice of dropping one or more letters at the end of a word, or from the arbitrary change of the termination. The names of Theophrastus and Dioscorides, especially, have suffered mutilation, and should be restored to the original form, while the correction of later misspellings should be made upon the basis of the classical form of the terms of the compound. In the rare cases in which the spelling of a Greek word has been changed in Latin, the Greek form should prevail.

VI

Terms are invalid if they exceed six syllables in length; retroactively, the correction of sesquipedalian words must never take place by contraction or mutilation.

"Nomina Generica Sesquipedalia, enunciatu difficilia, vel nauseosa, fugienda sunt." Critica Botanica 133.

The practice of biologists with respect to the formation of extremely long terms has been so exemplary that the present rule scarcely requires postulation. Its justification may be found in the fact that inconveniently long words, more or less frequent a century ago, still appear occasionally, and that such words, if there were no definite sentiment or legislation against them, might again become frequent as the supply of primitives and short compounds becomes exhausted. It is more or less unsatisfactory to limit the length of a word by the number of syllables, since these vary greatly in length in different stems, but this is undoubtedly better than limitation by the number of letters. It is a question whether nomenclature would not gain more than it would lose, if the maximum length of words were placed at five syllables, though the number of changes necessitated would probably render such a rule inacceptable. Naturally, the present rule should not be made operative in the case of names of groups above the genus.

VII

Hybrid terms are invalid: retroactively, Greek-Latin hybrids are to be corrected upon the basis of the Greek element, but all vernacular and personal hybrids fall.

"Nomina generica ex vocabulo graeco & latino, similibusque, Hybrida, non agnoscenda sunt." Critica Botanica 28.

"Everyone is bound to reject a name in the following cases: . . . (4) When it is formed by the combination of two languages." Paris Code, Article 60. 1867.

"The possibilities of the field he has opened up for us are indeed great, witness: Smithia, Smithago, Johnsmithotoma, Igsmithia, Smithalga, Smithodendron. I dwell on this because it seems to me that botanical Latin is impure enough already without such gratuitous monstrosities." Pound. *American Naturalist,* 26:147. 1892.

"An unhappy feature of Dr. Kuntze's work, and one in vindication of which I can say nothing, is his method of constructing new names for genera. Perhaps in some distant century, when self-repeating history may have brought the return of times when scientists were mostly men of clear ethics, solid learning, and refined tastes, some such reform in plant nomenclature as that which M. Saint-Lager in these times vainly advocates will be carried into effect. If, before the advent of that good time, Dr. Kuntze's *Radlkofertoma* and *Schweinfurthafra* shall have become current for certain genera, they will be the first to be rejected." Greene. Pittonia, 2:277. 1892.

The indifference of many biologists to a classical standard for nomenclature reaches its logical culmination in the formation of hybrid words. Botanists especially are practically unanimous in condemning hybrids, but, in spite of this fact, carelessness and ignorance are steadily increasing the number of illegitimate words. It is unnecessary to prove that hybrids are as unfortunate in nomenclature as in philology, but it is necessary that particular attention be given to them in order that they may be avoided, or at least corrected. No biologist of any real attainment can afford to stand sponsor for a hybrid name, when a trifling expenditure of time will yield a word of pure birth.

For the sake of clearness, hybrids may be divided into two classes: (1) Greek-Latin hybrids, in which one element is Greek and the other Latin; (2) vernacular hybrids, in which one element is from a modern tongue, while the other is classical, usually Latin. Each class shows hybrids in which both terms are independent words, and those in which one term is an affix. There is no essential difference between these as hybrids, but the distinction is an important one, because words of the second group are rarely recognized as hybrids on account of the slight familiarity of biologists with classical methods of derivation. The matter presents indeed some difficulty for the philologist, because of the similarity of cognate affixes in Greek and Latin, and because of Greek affixes borrowed by Latin. On account of the difficulty of detecting them, hybrids of this sort are becoming more and more common. The raising of hybrid sectional names in ευ-, ψευδο-, -ωδης, *-ella*, *-astrum*, etc., to the rank of generic names is contributing very largely to this result, as also the endeavor to honor a biologist by attaching all the Latin suffixes in turn to his name.

There has been considerable discussion regarding the treatment of such hybrids as pseudorepens and Eucarex. The contention is made that these words are not hybrids, since these affixes were regularly used by Latin writers, but, as a matter of fact, they are not found in classical Latin outside of borrowed Greek words in which they are a proper affix. It has further been urged that such words are scarcely hybrids, for the reason that pseudorepens does not mean "false creeping," but merely refers to a species of Agropyrum, which is not A. repens. Such argument is mere sophistry, since every compound or derivative which contains a Greek and Latin element, whether independent word or affix, is a hybrid. The only possible exception is found in those rare Greek words which have become so completely domiciled in Latin that their origin is no longer felt.

The correction of hybrids[1] is possible only when the word

[1] Since the above was written, three instances of a similar correction of

arises from the combination of Greek and Latin forms, in which case the cognate or corresponding Greek form is used to replace the Latin element. In the case of vernacular hybrids, such substitution is so rarely possible that it may be entirely disregarded, and all vernacular hybrids, the majority of which are personals, are to be summarily rejected. Such names fall not only because they are hybrids, but also on account of the operation of rules I and VIII. Greek-Latin hybrids, which are current in nomenclature, are to be corrected and followed by double citation of author and reviser, but hybrids proposed in the future, being invalid under the present rule, may be corrected or ignored at the will of the reviser, who alone is to be cited for the new name in either event.

The following list will illustrate the various kinds of hybrids, as well as the method of correction, when this is possible.

I. Greek-Latin hybrids in which both terms are independent words.

Actiniceps – Actinocybe (ἀκτίς, ἀκτῖνος, ἡ, ray, κύβη, ἡ, head)
Aureobasidium – Chrysobasidium (χρύσεος, golden, βασίδιον, τό, pedicel)
Baculospora – Bactrospora (βάκτρον, τό, staff, σπορά, ἡ, seed)
Botrypes – Botryopus (βότρυς, ὁ, cluster of grapes, πούς, ποδός, ὁ, foot)
Callosisperma – Sclematosperma (σκλῆμα, σκλήματος, τό, hardness, σπέρμα, τό, seed)
Claudopus – Loxopus (λοξός, slanting, crooked, πούς, ὁ, foot)
Clavogaster – Rhopalogaster (ῥόπαλον, τό, club, γαστήρ, ἡ, belly)
Clypeosphaeria – Peltosphaeria (πέλτη, ἡ, small shield, σφαῖρα, ἡ, ball)
Fagopyrum = Phegopyrus (φηγός, ἡ, oak, πυρός, ὁ, wheat)
Fimbristylis = Lomatostylis (λῶμα, λώματος, τό, fringe, στυλίς, ἡ, pillar)
Fusicolla = Chytocolla, (χυτός, poured out, κόλλα, ἡ, glue)
Geminispora = Dissospora (δίσσος, double, σπορά, ἡ, seed)

hybrids have been found in Pfeiffer's Nomenclator Botanicus 1:624, 1050, 1640. Catasetum Kunth is corrected to Catachaetum, Diastemella Oersted to Diastemation, and Loroglossum Rich. to Himantoglossum.

Gorgoniceps = Gorgocybe (Γοργώ, όος, ἡ, the Gorgon, κύβη, ἡ, head)
Hemicarex = Hemidonax (ἡμι-, half, δόναξ, ὁ, reed)
Massospora = Mazospora (μᾶζα, ἡ, barley cake, σπορά, ἡ, seed)
Muciporus = Myxoporus (μύξα, ἡ, mucus, πόρος, ὁ, pore)
Nemacola = Nemocolus (νέμος, τό, wooded pasture, -κολός, dwelling)
Nitophyllum = Phaedrophyllum (φαιδρός, bright, φύλλον, τό, leaf)
Nothofagus = Nothophegus (νόθος, spurious, φηγός, ἡ, oak)
Nucleophagus = Caryophagus (κάρυον, τό, nut, φάγος, eating)
Onychosepalum = Onychocalyx (ὄνυξ, ὄνυχος, ὁ, claw, κάλυξ, ἡ, cup of a flower)
Pachyfissidens = Pachyschizodon (παχύς, thick, σχιζοδών, ὁ, split tooth)
Peltigera = Peltophora (πέλτη, ἡ, shield, φορά, ἡ, a carrying)
Phaioclavulina = Phaeocoryne (φαιός, dusky, κορύνη, ἡ, club)
Pseudopeziza = Pseudopezis (ψευδής, false, πέζις, ἡ, stalkless fungus)
Radulotypus = Psectrotypus (ψῆκτρα, ἡ, scraper, τύπος, ὁ, form)
Retiporus = Dictyoporus (δίκτυον, τό, net, πόρος, ὁ, pore)
Scirpodendron = Donacodendrum (δόναξ, δόνακος, ὁ, reed, δένδρον, τό, tree)
Septosporium = Schizosporium (σχίζα, ἡ, cleft, σπόριον, τό, spore)
Verticicladium = Helicocladium (ἕλιξ, ἕλικος, ἡ, whirl, κλαδίον, τό, branch)

II. Greek-Latin hybrids in which one term is an affix.
Anthostomella = Anthostomatium
Bisporella = Disporyllium
Brizula = Brizyllium
Chlorosa = Chlorotes
Coryneliella = Corynisce
Cyphella = Cypharium
Dolicholus = Dolichidium
Eucaprifolium = Euaegophyllum (αἴξ, αἰγός, ὁ, ἡ, goat, φύλλον, τό, leaf)

Eucarduus = Euacantha (ἄκανθα, ἡ, thistle)
Fusidium = Atractidium (ἄτρακτος, ὁ, spindle)
Gaurella = Gauryllium
Glossula = Glossidium
Graphiola = Micrographium (cfr. Graphis, Graphium, Graphidium, Graphyllium)
Hormiactella – Hormiactinium (ὁρμιά, ἡ, fishline, ἀκτίς, ἀκτῖνος, ἡ, ray)
Hypocrella – Hypocreatium
Ilicioides – Dryodes (δρῦς, δρυός, ἡ, oak)
Juncodes – Thryodes (θρύον, τό, rush)
Labridium = Chilidium (χεῖλος, τό, lip)
Lachnella = Lachnium
Lachnellula – Microlachnium
Lithophragmella = Lithophragmatium
Lophiola = Microlophium
Myriactula = Myriactinium
Nasturtioides = Napyodes (νᾶπυ, τό, mustard)
Phaeodiscula = Phaeodiscium
Pholiotella = Pholidotium (φολιδωτός, clad with horny scales)
Polystomella – Polystomatium
Pterula = Pteridium
Rhodiola = Rhodarium
Sphaerosporula = Sphaerosporyllium
Stigmatella – Stigmatium
Struthiola = Struthidium
Tiarella = Tiaryllium
Trichopeltulum = Trichopeltium
Typhula = Typhidium
Zomicarpella = Zomatocarpium (ζῶμα, ζώματος, τό, girded doublet)

III. Vernacular-classic hybrids in which one term is a personal name.

Hybrids of this class lack even the excuse of ignorance. Nomenclature can show but one greater monstrosity, namely, the mutilated vernacular compound. Such personals can not be corrected and must fall irrevocably. Kuntze has been censured unjustly as the originator of the personal hybrid, since the latter was already found in numerous examples, as he himself has shown. But he deserves to be severely censured for greatly extending its use. Such atrocities as Pseudolcskia, Microschwenkia, Gerrardanthus, Pringleophytum, etc., were in existence before the Revisio, but they are lost sight of in the deluge of such foundlings proposed in the latter. The magnitude of Kuntze's offense against a classical nomenclature may be seen from the fact that out of 109 generic names proposed by him, 67 are personal hybrids, and the remainder are almost entirely mutilations, such as Watsonamra, Clarkeinda, Schweinfurthafra, Itoasia, etc. In the first volume of the Revisio,[1] the author gives a variety of methods by which the same botanist may be "honored" *ad nauseam* without increasing homonymy. The whole treatment manifests not only an entire absence of linguistic taste, but also an abiding ignorance of classical philology. Kuntze elsewhere[2] says apologetically, "Ich bin im Griechischen wenig erfahren." It is to be regretted that this feeling did not restrain him from such monstrous treatment of classical stems.

The following lists, though by no means complete, will serve to illustrate the various kinds of vernacular hybrids, all of which are to be rejected.

1. Vernacular-Greek hybrids—personals.

Bakeropteris	Beccariodendron
Balfourodendrum	Beckeropsis
Barleriacanthus	Benthamidium
Barlerianthus	Blumeodendrum
Barleriopsis	Buforrestia
Barleriosiphon	Caloknightia
Barleriotes	Chamaesaracha
Beccarianthus	Chamissomneia

[1] Kuntze, O. Reviso Generum Plantarum, 1:51. 1891.
[2] Ibid., 3:214. 1893.

Christiastrum
Cordierites
Cyphokentia
Diserneston
Doellochloa
Doniophytum
Dyerophytum
Ellisiophyllum
Englerophoenix
Epizeimeria
Epibrissonia
Eugeniastrum
Eugenioides
Fritschiantha
Gayophytum
Gerrardanthus
Glaziostelma
Grayemma
Hackelochloa
Halterophora
Harveyastrum
Henningsocarpum
Henningsomyces
Huegeliroea
Kentiopsis
Kuhniastera
Kuntzeomyces
Leioclusia
Lenzites
Ludwigiantha
Lyonothamnus
Macounastrum
Macowanites
Mannoglottis
Marilaunidium
Melioschinzia
Microkentia
Microschwenkia
Microweissia
Montagnites
Neobrunia
Neocontarinia
Neograyia
Neohuttonia
Neopeckia
Neoskofitzia
Neowashingtonia
Nesogordonia
Oliverodoxa
Orchidofunkia
Osbeckiastrum
Palaeogrewia
Parabesleria
Parabouchetia
Parapottsia
Phaenohoffmannia
Pleomassaria
Porteranthus
Preussiaster
Pringleophytum
Prockiopsis
Protohopea
Protoventuria
Pseudehretia
Pseudobarleria
Pseudogunnera
Pucciniopsis
Pycnoseynesia
Radlkofertoma
Rhabdoweissia
Roeperocharis
Sarcolippia
Schmitzomia
Schroeteriaster

Sibbaldiopsis
Silvianthus
Siphoneugenia
Smithantha
Stahlianthus
Stanhopeastrum
Sternhopeastrum
Thalianthus
Thileodoxa
Thouanidium
Tulasnodea
Urbanodendrum
Uroskinnera
Weinmannodora
Wittmackanthus
Zieridium

2. Vernacular-Latin hybrids—personals.

Absolmsia
Agardhina
Algogrunowia
Algorichtera
Arcangelina
Balfourina
Bartramidula
Baumanniella
Beccariella
Benthamistella
Berkeleyna
Bisboeckelera
Brocchinia
Caruelina
Cohnidonum
Cookeina
Crepinula
Delpinoina
Detonina
Dillwynella
Drudeola
Eremicella
Errerana
Fabreola
Flueckigera
Forsteronia
Freynella
Friesula
Fuckelina
Gerrardina
Gibberinula
Greeneina
Grisebachiella
Harziella
Hemsleyna
Hendersonula
Hodgsoniola
Hofmeisterella
Hookerina
Hostana
Jacksonago
Julella
Karstenula
Kickxella
Koehneago
Knyaria
Kuetzingina
Latzinaea
Magnusina
Massariella
Massarina
Montagnula
Mohlana
Mortierella
Munkiella
Neilrichina

Nicholsoniella
Nylanderaria
Nymanina
Octavianina
Oliveriana
Oudemansiella
Patouillardiella
Peckiella
Peckifungus
Penzigina
Peyritschiella
Pfeifferago
Phillipimalva
Pringsheimina
Richterago
Saccardinula
Saccardoella
Scopulina
Stephanina
Thozetella
Triumfettaria
Urbanisol
Velloziella
Vernonella
Voglinoana
Warscewiczella
Weddellina
Wettsteiniella
Wildpretina
Wingina
Winterella
Winterina
Zukalina

IV. Vernacular-classic hybrids—impersonal.

Calamovilfa
Camphoromyrtus
Galedragon
Gelatinosporium
Iguanura
Liquidambar
Obaejacoides
Sphaeropezia
Tacsonia
Talinastrum
Talinellum
Tamarindus
Toluifera
Vauanthes

VIII.

Vernacular names are invalid; this rule is retroactive.

"Nomina generica primitiva nemo sanus introducit." Critica Botanica 22.

"Nomina generica, quae ex Graeca vel Latina lingua radicem non habent, rejicienda sunt." Ibid. 48.

"Not to draw names from barbarous tongues, unless those names be frequently quoted in books of travel, and have an agreeable form that adapts itself readily to the Latin tongue, and to the tongues of civilized countries." Paris Code, Article 28.

The vernacular name has long been the refuge of the unlettered or indifferent systematist, and will doubtless continue to be while there are biologists of this kind. The arguments against the use of vernacular terms are so obvious and cogent that they would not be dwelt upon were it not for the contradictory provisions of the Paris Code. As in so many other questions of nomenclature, Linné's pronouncement should have been regarded as final by the framer of the Code. But, as in more than one place, the Code admits a fatal exception. It is absurd to base biological nomenclature in any degree upon books of travel, and it is futile to think that an author who speaks any vernacular tongue whatever, no matter how crude and uncouth, would find it either harsh or disagreeable. Some biologists have endeavored to improve vernacular names by shortening them or by adding a Latin suffix, but such a remedy is worse than the original trouble. Correction by translation, as Chenanthus for Gansblum, is occasionally possible, and in such cases might be more fortunate than the rejection of a name. The fundamental fact still remains, however, that nomenclature is already essentially classical, and should in the future be made completely so. Vernacular names have no place in it. This condition can be made to prevail only by rejecting all such names whether past or future.

Anagrams, if they be considered words at all, are vernacular, since they are neither Greek nor Latin. They are the ultimate product of puerility or illiteracy in nomenclature. Such a series as Filago, Gifola, Ifloga, Logfia, and Oglifa throws a clear light upon the good sense and linguistic taste of the authors concerned. One might better make names after the fashion of Carroll, or take names from the "hog-Latin" of childhood. All other mutilations, like anagrams, are unpardonable offenses against nomenclature, and are to be summarily rejected.

I. Anagrams.

Alibum (Liabum)	Behuria (Hubera)
Amida (Madia)	Beriesia (Siebera)
Anogra (Onagra)	Blitrydium (Tryblidium)
Baziasa (Sabazia)	Galpinsia (Salpingia)

Gandriloa (Oligandra)
Gelfuga (Fluggea)
Gifola (Filago)
Gosela (Selago)
Ifloga (Filago)
Lagatea (Galatea)
Lebidiera (Briedelia)
Narthecium (Anthericum)
Neoceis (Senecio)
Nepera (Spennera)
Norysca (Ascyron)
Obaejaca (Jacobaea)
Oglifa (Filago)
Parosela (Psoralea)
Phledinium (Delphinium)
Ranugia (Anguria)
Trelotra (Rottlera)
Trilisa (Liatris)

II. Vernacular mutilations.

Andreoskia
Beccarinda
Berkleasmium
Bolusafra
Brittonamra
Cavanilla
Clarkeinda
Cosimibuena
Dickneckera
Durandeeldea
Elidurandia
Fregirardia
Gomortega
Gonzalagunia
Hallomuellera
Hasskarlinda
Isidrogalvia
Itoasia
Kinginda
Kurzamra
Kurzinda
Lippomuellera
Maximowasia
Meyerafra
Muelleramra
Razumovia
Ridleyinda
Schinzafra
Sebschauera
Schweinfurthafra
Watsonamra

IX

A name is not valid unless its etymology and application are clearly indicated: this rule is not retroactive.

"Nomina generica, quae Characterem essentialem, vel faciem plantae exhibent, optimae sunt." Critica Botanica 97.

"Botanists who have generic names to publish show judgment and taste by attending to the following recommendations: . . . . (2) To give the etymology of each name." Paris Code, Article 28.

The desirability of being able to know the etymology and application of each generic and specific name is obvious, but the rule given above will work advantageously in other matters also. An author who cites accurately the derivation of a proposed name will be much less apt to err in its construction, while the necessity for indicating its application will bring about greater accuracy in the choice of characters. Desirable as it might be, it is futile to demand that names show a proper degree of relevancy, or to reject them because they are more or less inapplicable. In matters of taste, it is both possible and highly desirable to have a standard, but it is idle to expect that it will be either appreciated or followed by the majority. Since names are to be rejected if improperly constructed, it is imperative that the exact etymology be given in each case, in order that their validity may be readily ascertained. A name then would stand or fall by its given etymology. It is extremely unsatisfactory to say of a name, for example, "from the Greek for flower;" the exact form of the Greek or Latin stem employed should be given.

X

The termination of family, ordinal, class, and branch names shall be uniform within each group: tribes shall terminate in *-inae*, families in *-aceae*, orders in *-ales*, classes in *-eae*, and branches in *-phyta*.

"The names of divisions and subdivisions, of classes and subclasses, are drawn from their principal characters. They are expressed by words of Greek and Latin origin, some similarity of form and termination being given to those that designate groups of the same nature." Paris Code, Article 18.

The designation of all groups of the same rank by means of a common suffix is at present merely a convenience, but with the increasing minuteness of systematic work and the growing tendency toward segregation, it will soon become a necessity. Subdivisions and superdivisions will need to be set off from tribes, families, orders, and classes, and the terminations for the latter must be definitely fixed in order to secure a basis for distinguishing the next group above and below. The number of possible divisions above the genus is fifteen, which makes it impossible that each should receive a distinct suffix. The most satisfactory method, then, will be to fix the designations for the five main groups, and to indicate sub and super divisions by prefixes, or by slight variations of the proper suffix. A further reason for this is found in the fact that cognate suffixes can alone be used, since generic names are either Greek or Latin, and that proper cognate suffixes are few. In fact, they are practically exhausted by the five principal groups, *-alis, -ales,* being, indeed, very hard to justify as a termination for Greek stems. It should be noted that *-phyta* is merely the neuter plural of the Greek word, *φυτόν, τό,* plant, and can be attached only to Greek stems.

The following examples will illustrate the operation of the above rule.

Protophyta: Schizophyceae: Nematogenales: Nostocaceae: Aulosirinae

Phycophyta: Chlorophyceae: Conjugatales: Zygnemataceae: Mesocarpinae

Carpophyta: Ascomyceteae: Discomycetales: Pezizaceae: Sarcoscyphinae

Bryophyta: Hepaticeae: Jungermanniales: Jungermanniaceae: Aploziinae

Pteridophyta: Filiceae: Filicales: Polypodiaceae: Onocleinae

Spermatophyta: Angiospermateae: Glumales: Graminaceae: Festucinae

XI

In proposing generic names, the following rules are to be observed:

(1) The name shall be a Greek substantive, i. e., not a simple adjective.

(2) A single generic name may be founded upon the name of a botanist. Such names are only to be formed by adding *-ia* to cognomina ending in a consonant and *-a* to cognomina in a vowel or *-r*, except in the case of names already Latinised, in which case the termination is first dropped.

(3) Personal generic names shall be bestowed only in recognition of eminent services in botany.

(4) Anagrams and geographical names are invalid.

(5) Double generic names are invalid.

Generic names should in the future be formed exclusively from Greek, as simple Latin nouns suitable for plant names have been practically exhausted, and the formation of compound terms in Latin is awkward. Greek nominal stems of all sorts, simple or compound, with the exception of simple adjectives, such as μακρός, μέγας, etc., are readily available. The proposal of generic names in honor of rulers, patrons, collectors, friends, and relatives should be severely discountenanced. Furthermore, duplicates of the same personal, as Saccardaea, Saccardia, Pasaccardoa, Saccardoella, Saccardinula, and Beccaria, Beccariella, Beccarianthus, Beccardinda, and Beccariodendron must be regarded as invalid, because their terminations are no longer significant endings, but mere variations, and also because they are hybrids. Anagrams, as has been pointed out before, fall because they are vernacular, or mutilated, or both. Geographical names are almost invariably vernacular also. Double generic names, such as Dens-canis and Bursa-pastoris are compounded syntactically and are hence invalid, while others, such as Genisto-Spartium and Lilio-Narcissus are mere hybrids.

XII

pecific names, the following rules are to be observed:
[illegible] be a Greek or Latin adjective, referring to a character or
[illegible], or to its habitat.
ive specific names are to be avoided.
ves, superlatives, and geographical adjectives are invalid; not

djectives and genitives are invalid; not retroactive.
ic name is invalid if the same as the generic name; retroactive.

e above rules are of primary importance, but their ll materially improve the nomenclature of species. t the best usage at the present time, but need to be order that they may be more generally followed. me should not only mean something, but should rect and evident application to some characteristic r habitat. In this connection, the necessity for the is, though there will doubtless be dissent from the geographical and personal names. In support of ken on geographical names, it is sufficient to cite nadensis," "carolinianus," "pennsylvanicus," "vir-, of Linnaeus, Gronovius, Elliott, and others, for the country over, and the names "coloradensis," issouriensis," etc., of more recent writers for spe-mpletely ignore the political limits of their native pias syriaca L. is a classical example of the value al names for species. The logical outcome of geo-ies is seen in such absurdities as Crataegus nd Panicum auburne, and, when combined with a of illiteracy, in such nomenclatural atrocities as orado and C. shallotte. The genus Crataegus fur-cing proof that nomenclatural and taxonomic in-hand in hand.

e of naming species after persons has absolutely mmend it. As a rule, personal specific names are a mistaken desire to honor some one, or of mere day is long past in which a biologist can be hon-hing his name to a species, and the honoring of is not the province of nomenclature. It can not be gainsaid that the use of personal names for species does obviate the necessity of knowing the species of a genus sufficiently well to avoid homonyms, but it is clear that such knowledge might at least make for more thorough systematic work. With respect to doublets, it is greatly to be regretted that the original rule of the Rochester Code was not permitted to stand. The Madison amendment has not only resulted in numerous absurd one-word binomials, but has actually weakened the cause of priority by making the latter override all considerations of accuracy and taste.

BIBLIOGRAPHY

Allen and Greenough. Latin Grammar. 1893.

Andrews, Lewis, and Short. A new Latin Dictionary. 1882.

Barnhart, J. H. Family Nomenclature. *Bull. Torr. Bot Club*, *22*:1. 1895.

Buttmann, Philip. A Greek Grammar. 1851.

Clements, F. E. A System of Nomenclature for Phytogeography. Engler's Jahrb., 31:b. 70. 1902.

Curtius, George. Principles of Greek Etymology. 1875.

Dall, A. H. Nomenclature in Zoology and Botany.

DeCandolle, Alphonse. Laws of Botanical Nomenclature. 1868.

Dieterich, Karl. Untersuchungen zur Geschichte der Griechischen Sprache. 1898.

Gray, Asa. Some Points in Botanical Nomenclature. *Am. Journ. Sci.*, 26:417. 1883.

Goodwin. A Greek Grammar.

Harms, H. Die Nomenclaturbewegung der letzten Jahre. *Eng. Bot. Jahrb.*, 23:b. 56. 1897.

Henry, Victor. A Short Comparative Grammar of Greek and Latin. 1890.

Kuntze, O. Revisio Generum Plantarum, I, II. 1891.

Ibid., III, 1. 1893.

Ibid., III, 2. 1898.

Liddel and Scott. A Greek-English Lexicon.

Linné, Carl von. Critica Botanica. 1737.

Philosophia Botanica. 1751.

List of Pteridophyta and Spermatophyta. 1893-4.

Miller, Walter. Scientific Names of Greek and Latin Derivation. *Proc. Cal. Acad. Sci.*, 1:115. 1897.

Pfeiffer, Ludwig. Nomenclator Botanicus. 1873.

Proceedings of the Madison Botanical Congress. 1894

Smith and Hall. A Copious and Critical English-Latin Dictionary. 1899.

Yonge, C. D. An English-Greek Lexicon. 1899.

ERRATUM

At foot of pages, numbers 321 to 384 inclusive, should be 1 to 64.

CONTENTS OF THE UNIVERSITY STUDIES, Vol. II

No. 1

1. *Additional Notes on the New Fossil, Daemonelix*
By ERWIN HINCKLEY BARBOUR

2. *On the Decrease of Predication and of Sentence Weight in English Prose*
By G. W. GERWIG

3. *Mirabeau, an Opponent of Absolutism*
By F. M. FLING

No. 2

1. *History of the Discovery and Report of Progress in the Study of Daemonelix*
By ERWIN HINCKLEY BARBOUR

2. *Notes on the Chemical Composition of the Silicious Tubes of the Devil's Corkscrew, Daemonelix*
By THOMAS HERBERT MARSLAND

3. *On the Continuity of Chance*
By ELLERY W. DAVIS

4. *The Bacon-Shakespeare Controversy—A Contribution*
By CARSON HILDRETH

5. *Generalization and Economic Standards*
By W. G. LANGWORTHY TAYLOR

No. 3

1. *Topical Digest of the Rig-Veda*

2. *Spanish Verbs with Vowel Gradation in the Present System*
By A. H. EDGREN

3. *The Oath of the Tennis Court*
By F. M. FLING

No. 4

1. *Influence of the Breton Deputation and the Breton Club in the Revolution (April–October, 1789)*
By CHARLES KUHLMANN

2. *The Mercantile Conditions of the Crisis of 1893*
By FRANK S. PHILBRICK

Single numbers of the STUDIES may be bought for $1.00 each. Address all correspondence regarding purchase or exchange to

UNIVERSITY OF NEBRASKA LIBRARY

JACOB NORTH & CO., PRINTERS, LINCOLN.

Vol. III. April, 1903 No. 2

University Studies

Published by the University of Nebraska

CONTENTS

I. The Degree of Accuracy of Statistical Data
Carl C. Engberg. 87

II. The Anomalous Dispersion and Selective Absorption of Fuchsin
W. B. Cartmel. 101

III. Mallophaga from Birds of Costa Rica, Central America
M. A. Carriker, Jr. 123

LINCOLN, NEBRASKA

Entered at the post-office in Lincoln, Nebraska, as second-class matter, as University Bulletin, Series 8, No. 9

UNIVERSITY STUDIES

VOL. III APRIL, 1903 No. 2

I.—*The Degree of Accuracy of Statistical Data*

BY CARL C. ENGBERG

I

This paper is written as a protest against the unnecessary refinement of statistical computations as carried out by the biometricians of to-day. These practices are well illustrated by the case of the college freshman, who in his zeal and desire for absolute accuracy used π to fifteen decimal figures in the determination of the size of the micrometer divisions in his microscope, although the change in focus necessary even for the other eye of the same observer will, in many cases, alter the size of the object observed by as much as half a division. While this is an extreme example, it is not much worse than the performances of all inexperienced computers, or even than those of many distinguished mathematicians who are experienced computers, but who have had their practical sense killed by impractical theories.

In order to give greater weight to my remarks, I shall discuss cases taken from the works of prominent biometricians, especially those of Professor Karl Pearson, the originator and developer of the science.

II

Distribution of 8,689 cases of enteric fever received into the Metropolitan Asylums Board Hospitals, 1871–93. Karl Pearson, *Mathematical Theory of Evolution.* Phil. Trans. A., vol. 186, pp. 390–91.

AGE	NUMBER OF CASES	AGE	NUMBER OF CASES
Under 5	266	40–45	163
5–10	1143	45–50	98
10–15	2019	50–55	40
15–20	1955	55–60	14
20–25	1319	60–65	8
25–30	857	65–70	4
30–35	503	70–75	1
35–40	299		

The distribution here given for the 13 cases above 60 is what Professor Pearson "considers" the most probable distribution for those years. It seems that the hospital authorities "lumped" their old patients, a practice fatal to accurate statistics.

For convenience, we shall take five years as the unit. We then get for the first four moments about the vertical through —2.5 years:[1]

$$\nu_1=4.294,\ \nu_2=22.341,\ \nu_3=137.051,\ \nu_4=967.682.$$

Transforming to moments about the centroid vertical by means of the relations

$$\mu_1=0$$

$$\mu_2=\nu_2-\nu_1^2+{}^1/_6$$

$$\mu_3=\nu_3-3\nu_1\nu_2+2\nu_1^3$$

$$\mu_4=\nu_4-4\nu_1\nu_3+6\nu_1^2\nu_2-3\nu_1^4+\nu_2-\nu_1^2+\frac{1}{15},$$

we get:

$$\mu_2=4.071,\ \mu_3=7.599,\ \mu_4=69.314.$$

[1] In practical computations, a vertical near the mean is used, but as I had worked out these moments for another purpose long before this paper was thought of, I preferred to use them here rather than to do the whole work over again.

Putting $\beta_1=\frac{\mu_3^2}{\mu_2^3}$ and $\beta_2=\frac{\mu_4}{\mu_2^2}$ we find:

$\beta_1=.856$ and $\beta_2=4.182$;

whence the critical function

$$3\beta_1-2\beta_2+6=.203>0,$$

and the equation of the theoretical frequency curve takes the form

$$y=y_0\left(1+\frac{x}{a_1}\right)^{m_1}\left(1-\frac{x}{a_2}\right)^{m_2}.$$

As the formulae used in determining the constants in the above equation are the best argument for excessive accuracy, I shall give them.

$$r=\frac{6(\beta_2-\beta_1-1)}{3\beta_1-2\beta_2+6}$$

$$\epsilon=\frac{r^2}{4+\frac{1}{4}\beta_1(r+2)^2/(r+1)}$$

$$b=\sqrt{\mu_2}\,\frac{\{\beta_1(r+2)^2+16(r+1)\}^{\frac{1}{2}}}{2}$$

$$m_1=m_1'-1 \quad \text{and} \quad m_2=m_2'-1$$

where m_1' and m_2' are the roots of $m'^2-rm'+\epsilon=0$.

$$a_1+a_2=b \quad \text{and} \quad \frac{a_1}{a_2}=\frac{m_1}{m_2}$$

$$y_0=\frac{\alpha}{b}\,\frac{(m_1+m_2+1)\sqrt{m_1+m_2}}{\sqrt{2\pi m_1 m_2}}\,e^{\frac{1}{2}\left(\frac{1}{m_1+m_2}-\frac{1}{m_1}-\frac{1}{m_2}\right)}$$

α=the number of cases.

$$\text{Skewness}=\tfrac{1}{2}\sqrt{\beta_1}\,\frac{r+2}{r-2}\equiv A.$$

$$d=\sqrt{\mu_2}\,A.$$

We then get for the values of the constants:

$r= 68.78$	$d= .965$
$e=24'.3$	$A= .479$
$b= 74.14$	
$m_1= 2.757$	$m_2=64.02$
$a_1= 3.056$	$a_2=71.08$
$y_o=1883$	

The centroid vertical is at 18.97 years.

Carrying the work out to six decimal figures, Professor Pearson finds for these constants:

$$\mu_2=4.070554, \ \mu_3=7.598196, \ \mu_4=69.379605$$

$$3\beta_1-2\beta_2+6=.1935$$

$r= 72.28642$	$d= .98643$
$e= 259.78912$	$A= .488922$
$b= 77.28312$	
$m_1= 2.79291$	$m_2=67.49351$
$a_1= 3.07801$	$a_2=74.20511$
$y_o=1890.83$	

The centroid vertical is at 18.9691 years; i. e., .29382 unit from 15-20. To compute the run of a disease to a twenty-thousandth part of a year is rather fine work, especially when five years is the unit.

A comparison of the two sets of results shows a considerable difference in the values of the constants. Theoretically the latter is the more correct set of values; practically the former is the better. The equations given above apparently necessitate a high degree of approximation, for high powers of the moments and other constants are involved, and the successive powers of an approximate decimal fraction are correct to fewer and fewer decimal places; so that, if we want the values of the constants correct to two or three decimals, we must start with at least six decimal figures in the values of the moments. This is true if we are dealing with six-place data, but suppose we have before us only three-place data? We can then at best make a guess at the fourth figure, but can tell absolutely nothing about the following figures. Here, then, we are compelled to follow a rule which ought to be more widely observed; namely, use the degree of approximation warranted by the data, and let the answer take care of itself. Against this we have the contention that our data are not merely three- or four-place data, but are of any desired degree of approximation, for

$$\nu_s = \frac{\Sigma(y_r . r_s)}{\Sigma y_r},$$

and the division here indicated may be carried out as far as we please. Further, the best theoretical curve is necessarily defined as the one which fits the given observations best. Under these pleas have been committed many outrageous crimes against common sense laws in computation, and by the greatest of masters. It will be seen later, however, that a very slight change in the data or even the slightest change in the unit of groupings, in most cases affects the value of the moments in the third or fourth place. Under these circumstances, we ought to get sensibly as good a fit with three or four figures as with five or six, but with only a fraction of the work. Furthermore, neither curve can coincide with the polygon of observations, and as they differ somewhat in shape, in this place the one, in that the other may be the better fit. If the degree of approximation warranted by the data has been used, the chances are that, on the whole, the one curve will be as good a fit as the other.

To determine the degree of accuracy of the above data, I let the number of cases from 15–20 years be 2018 instead of 2019, a very insignificant change. Computing the ν's we get:

$$\nu_1 = 4.294, \ \nu_2 = 22.343, \ \nu_3 = 137.072, \ \nu_4 = 967.866.$$

These give for the moments about the centroid vertical

$$\mu_2 = 4.072, \ \mu_3 = 7.595, \ \mu_4 = 69.347.$$ [1]

This change is well within the probable error of the number of cases for the given period, and hence this set of values of the

[1] Had the actual distribution of the 13 cases above 60 been in any way different from the one assumed by Professor Pearson, a much greater change in the moments would have occurred.

ν's and μ's is as probable as the previous one. Thus we can, by very slight manipulations of our data, obtain a new set of constants which differ considerably from the old set, but which must give rise to a curve equally as good as the first one. To carry out the computations to six or more places, when an exceedingly slight accidental change in the data will make a proportionately large change in the constants, is a pronounced case of "saving at the spigot and spending at the bung."

In fig. 1 are drawn the curves obtained, using six and three decimal figures respectively. A comparison of the two curves with the frequency polygon shows the one to be as good a fit as the other.

III

The distribution of dorsal teeth on the rostrum of 915 specimens of Palaemonetes Varians from Saltram Park, Plymouth. Professor Karl Pearson, *The Mathematical Theory of Evolution.* Phil. Trans., A., vol. 186, pp. 403–4.

TEETH	CASES
1	2
2	18
3	123
4	372
5	349
6	50
7	1

The centroid vertical lies .314 of a tooth beyond 4. i. e., at 4.314 teeth. The moments about the vertical through 4 are:

$$\nu_1 = .3137,\ \nu_2 = .8426,\ \nu_3 = .4973,\ \nu_4 = 1.9705.$$

These give to the moments about the centroid vertical the following values:

$$\mu_2 = .9109,\ \mu_3 = -.234,\ \mu_4 = 2.6259;$$

whence

$$\beta_1 = .0724,\ \beta_2 = 3.1647,$$

and

$$2\beta_2-3\beta_1-6=.1122>0,$$

or we have a curve of type IV.

Proceeding to determine the other constants, we find:

$$r=111.9.$$

$\nu=110.4$ (ν is positive since μ_3 is negative. Professor Pearson gets these interchanged),

$$a=7.155 \qquad m=56.99.$$

Distance of origin from centroid vertical=7.057.

$$\log y_0=\overline{18}.0822.$$

Thus the equation of the curve is

$$y=y_0 \cos^{113.9}\theta e^{-110.4\theta}.$$
$$x=7.155 \tan \theta.$$

Professor Pearson's constants, obtained by the aid of seven-place logarithm tables are

$\mu_2=.910906$	$\beta_1=.072222$
$\mu_3=.233908$	$\beta_2=3.164684$
$\mu_4=2.625896$	$2\beta_2-3\beta_1-6=.112702$
$r=111.398$	$a=7.16613$
$\nu=109.047$	$m=56.699$

Distance of origin from centroid vertical—7.0149,

$$\log y_0=\overline{18}.4431056,$$

whence the equation of the curve is

$$y=y_0 \cos^{113.398}\theta e^{109.047\theta}.$$
$$x=7.16613 \tan \theta.$$

The two curves are drawn in fig. 2. A comparison of them with the frequency polygon shows that not enough is gained in accuracy by carrying more than four decimal figures to warrant the extra expenditure of time and labor.

Thus this problem also verifies the conclusions reached above: that, in the fitting of a theoretical curve to the observations, it is the height of folly to waste time and energy on needless and meaningless refinement of the computations; and that, as a rule, in spite of the high powers of the constants, three or four decimal places will be sufficient to give results as reliable as the data warrant.

IV

The equations

$$y = y_0 \cos^{2m} \theta \, e^{-v\theta}$$
$$x = a \tan \theta$$

have a rather forbidding appearance, and the mere sight of them is often sufficient to deter a man from going farther with the work. If handled properly, however, they are not so bad as they look. Taking logarithms, we have

$$\log \tan \theta = \log x - \log a$$
$$\log y = \log y_0 + 2m \log \cos \theta - vM\theta \log e,$$

where M is the factor which converts from circular to radian measure.

We now take a card and write on it the values of $\log a$, $\log y_0$, $\log (vM \log e)$, and $\log 2m$, thus:

$\log a =$	, $\log (vM \log e) =$
$\log y_0 =$	, $\log 2m =$

The proper use of this card will enable us to plot the curve in one-fifth of the time required did we use no labor-saving devices. Similar methods may be employed in computing other forms of probability curves.

V

Another matter that should be spoken of in this connection is the method of computing the ν's, i. e., the moments about an arbitrary vertical. Professor Pearson, in one of his articles, publishes a table of the first six powers of the integers from 1 up to 30, and refers to this table several times. Professor Davenport, in his *Statistical Methods*, copies this table, and says: "This table is useful in calculating moments." This is not true, except in very rare instances, and then only to a very insignificant extent; but the statement coming from such authorities is likely to unduly influence inexperienced computers and cause them to waste much valuable time. Instead of multiplying the ordinates by the successive powers of the corresponding abscissas to obtain the moments, it will be found much better to proceed as in the following example:

Abscissa	Frequency	$N_1=\Sigma r y_r$	$N_2=\Sigma r.y_r r$	$N_3=\Sigma r.y_r r^2$	$N_4=\Sigma r.y_r.r^3$
−3	2	−3× 2=− 6	−3×− 6= 18	−3× 18=− 54	−3×− 54=162
−2	18	−2× 18=− 36	−2×− 36= 72	−2× 72=−144	−2×−144=288
−1	123	−1×123=−123	−1×−123=123	−1×123=−123	−1×−123=123
0	372	0×372= 0	0× 0= 0	0× 0= 0	0× 0= 0
1	349	1×349= 349	1× 349=349	1×349= 349	1× 349=349
2	50	2× 50= 100	2× 100=200	2×200= 400	2× 400=800
3	1	3× 1= 3	3× 3= 9	3×9 = 27	3× 27= 81

No doubt this is the method generally followed by expert biometricians but as the only reference books on the subject in existence recommend the long and roundabout way, I feel justified in calling attention to the matter.

VI

In vol. I, part IV, of *Biometrika,* Professor Pearson generalizes what he has called "Francis Galton's Difference Problem;" that is, he has obtained a general law giving the difference between individuals competing for a set of prizes. The work is very brilliant, and shows well Professor Pearson's mathematical genius. Nevertheless, in his application to a special case, he destroys the usefulness of his results by using seven-place logarithms and computing the constants to six decimal places. Four-place tables are as accurate as any statistics we have, or ever can obtain on the subject, and, using these, the work of computing the constants may be done in hours where Professor Pearson's method requires days, and the results would be as reliable as the data warrant.

VII

In vol. I, part III, of *Biometrika,* Professor Pearson shows how to fit Makeham's curve to mortality statistics, a work of great advantage to actuaries, as it gives a general rule for fitting. Makeham's formula is

$$l_x = ks^x (g)^{c^x},$$

where l_x denotes the number who attain the age of x, and k, s, g, and c are constants to be determined from known data.

Here, even if we take the origin at the middle of the range, x will still receive values as large as 30 or 35, and hence great accuracy is necessary. Professor Pearson, in fact, carries out his approximations to twelve decimal figures, using a large Brunsviga, as logarithm tables are necessarily out of the question. A comparison of the values of some of the constants derived by different processes gives us results which would be ludicrous were they not pathetic.

For c, Professor Pearson finds:

$$c = 1.098,096,393,273,$$

while King and Hardy, by a method of averages, get:

$$c = 1.095,612,204.$$

Now putting

$$\beta = \frac{4\left(\frac{20\mu_3}{l^3} - \frac{3\mu_1}{l}\right)}{\frac{12\mu_2}{l^2} - 1},$$

where l is the range of the mortality table, here taken to be the 60 years, from 25 to 85, he finds:

$$\beta = .718,529,308,595.$$

"By a rougher quadrature process for the whole range from 20 to 90," he finds:

$$\beta = .801,086,783.$$

The value of β, as computed from King and Hardy's c for the range from 17 to 88 years, is:

$$\beta = .804,162,5$$

The difference in the values of both β and c is due partly to difference of range, partly to method of computation, and hence it is not to be expected that they should agree absolutely. Still, there is no excuse for carrying out the computations to twelve decimals, when different methods give results which differ in the first figure. Again, the constants can not be more accurate than the data upon which they are based, and no mortality statistics at our disposal are correct even to half the number of places used by Professor Pearson. From trials I have made, I am sure that an eight-place logarithm table would give, on the whole, fully as good results as those obtained by Professor Pearson, especially as, in spite of the difference in values of the constants, the improvement of his results over those obtained by the method of averages "is not very sensible." It is to be deplored that Professor Pearson should mar the effectiveness of his work by his desire for pseudo-accuracy.

VIII

In vol. I, part II, of *Biometrika,* Mr. W. Palin Elderton, an actuary, gives tables for testing curve fitting. These tables, which are six-place tables, have been computed by the aid of eight-, ten-, and eleven-place logarithm and integral tables, involving an enormous expenditure of time and energy. Now it so happens in most cases where a large number of observations is involved that these tables are of no value. For instance, in the first case discussed in this paper, Professor Pearson finds the average percentage error to be 5.75, which he considers by no means bad. If, however, we apply Mr. Elderton's test, we find the probability, that in a random sampling a deviation system as great or greater shall exist, to be about .000000, that is, not once in a million samplings would he get a greater deviation.

This probability P is given by the equation

$$P=\sqrt{\frac{2}{\pi}}\int_{\chi}^{\infty} e^{-\frac{1}{2}\chi^2}d\chi+\sqrt{\frac{2}{\pi}}e^{\frac{1}{2}\chi^2}\left(\frac{\chi}{1}+\frac{\chi^3}{1.3}+\frac{\chi^5}{1.3.5}+\ldots\ldots\ldots\ldots+\frac{\chi^{n-3}}{1.3.5\ldots(n-3)}\right)$$

for n even, and by

$$P=e^{-\frac{1}{2}\chi^2}\left(1+\frac{\chi^2}{2}+\frac{\chi^4}{2.4}+\frac{\chi^6}{2.4.6}+\ldots\ldots+\frac{\chi^{n-3}}{2.4.6\ldots(n-3)}\right)$$

for n odd, where n is the number of groups, and

$$\chi^2=S\frac{(m_r-m_r')^2}{m_r}=\text{sum}\left(\frac{\left\{\begin{array}{c}\text{squares of differences of theo-}\\ \text{retical and observed frequency}\end{array}\right\}}{\text{theoretical frequency}}\right)$$

The trouble with this formula for χ^2 is that it assumes that, if, say, 1,000 observations distributed in n groups give a mean error of 4 per cent, 16,000 observations distributed in the same way shall give a mean error of 1 per cent. Now some objects are more variable than others, so that it is easily conceived that 1,000 observations on one plant or animal give as accurate a result as 16,000 observations on another more variable plant or animal, but whatever the value of P in the first case, it is very large as compared with the P of the second case. One example is sufficient to show this.

CASE I

Group	Observed frequency	Calculated frequency	$m_r - m_r'$	$(m_r - m_r')^2/m_r$
1	1	1	0	0
2	8	7	—1	.1
3	50	48	—2	.1
4	120	123	3	.1
5	195	200	5	.1
6	211	200	—11	.6
7	150	155	5	.2
8	100	104	4	.2
9	40	38	—2	.1
10	2	1	—1	1.0
Totals....	877	877	0	$\chi^2=2.5$

Mr. Elderton's table for $\chi^2=2.5$ and $n=10$ gives $P=.98$ about.

CASE II

Group	Observed frequency	Calculated frequency	$m_r - m_r'$	$(m_r - m_r')^2 m_r$
1	3	3	0	0
2	25	24	—1	0
3	400	385	—15	.6
4	1000	1025	25	.6
5	1997	1970	—27	.4
6	2095	2015	—80	3.2
7	1431	1475	44	1.3
8	955	990	35	1.2
9	375	395	20	1.0
10	20	19	—1	.1
Totals....	8301	8301	0	$\chi'^2=8.4$

Mr. Elderton's table gives $P'=.49$ about.

Thus, although the second set of observations has a less percentage error in frequency than the former, its probability is only half as large. This difficulty is partially obviated by dividing χ'^2 by the square root of the ratio of the two frequencies and entering the table with this new value of χ'^2. This, however, gives no absolute measure of the probability. When, therefore, the rapid growth of the numerator, as compared with the denominator of the fraction

$$\frac{(m_r - m_r')^2}{m_r},$$

for large values of m_r is sufficient even to destroy the value of the table, it is sheer folly to fool with ten-place logarithms. Mr. Elderton would have conferred a much greater benefit on biometricians had he put less time on the "small differences in the eleventh place," and more on the securing of an absolute measure of the probability.[1]

This paper has not been written in a fault-finding spirit by a detractor of the new science of biometry, but by a teacher of the science, in the hope of doing something to help reverse the popular process of swallowing the camel, but straining at the gnat.

[1]A measure of this probability, which takes no account of the variability of an object, is of no value whatsoever.

II.—*The Anomalous Dispersion and Selective Absorption of Fuchsin*[1]

BY W. B. CARTMEL

Although fuchsin is the substance in which anomalous dispersion was first observed, and although it shows this phenomenon much more decidedly than any other substance upon which it has been found possible thus far to make anything like reliable measurements, its optical constants have not yet been fully determined except by indirect methods. Very complete absorption curves have indeed been given for solutions of different concentration, but not for solid fuchsin.

It was therefore thought that it might be of interest to determine the absorption and dispersion directly, both of these upon the same identical fuchsin; and as a very good determination of the dispersion curve has already been given by Pflüger,[2] who measured the deviation produced by a thin wedge of solid fuchsin, it was decided to redetermine this, using interferential means. This would present the advantage of a redetermination by a different method, and, furthermore, the methods of absorption and dispersion could both be made upon the same fluid.

Films were therefore prepared in the usual way by dipping glass plates into an alchoholic solution of fuchsin and allowing the alchohol to evaporate. The fuchsin upon which the first experiments were made was some that had been purchased for general laboratory purposes, and it was found that the dispersion and absorption had values very much lower than those given by Pflüger. Some fuchsin of the same kind as that which Pflüger had used was therefore imported from Kahlbaum in Berlin, and

[1]Read before the Washington meeting of the American Association for the Advancement of Science.

[2]*Wied, Ann.*, vol. 65, p. 203. 1898.

upon this were carried out the experiments which form the basis of the following paper.

As already pointed out by Wood[1], the great difficulty in determining the dispersion of strongly absorbing substances by interferential methods is that the ray which passes through the substance is so reduced in intensity that it is not capable of causing interference when it meets the undiminished light of the other ray. This may be easily conceived to be the case with fuchsin, when we consider that a film of fuchsin a wave-length thick transmits only five parts in one hundred million of the incident light within the absorption band. Experiments were, therefore, made with the object of reducing, if possible, the intensity of the light in one of the paths of the interferometer, without producing any change in its optical length, in order that the fuchsin might be placed in the more intense beam. However, none of the various plans tried were adopted. It was then decided to use a form of interferometer in which the light does not return upon itself, and for two reasons: First, because in traversing the film twice the diminution in intensity is squared while the retardation is only doubled; and, second, because the enormous reflection from the surface of the film obscures the fringes in the ordinary form of interferometer, but in the type used the reflected light does not reach the observer's eye at all.

After a number of trials to determine the best adjustment, it was found possible, by using this type of interferometer and making the films sufficiently thin, to obtain distinct fringes throughout the spectrum. An unsymmetrical arrangement was first tried, in which the partly silvered plates were very lightly silvered, or not silvered at all, and the fringes observed at a direction at right angles to that at which the light entered the instrument. In this way fringes were obtained from beams of unequal intensity, though the method was finally abandoned for the following, which is more satisfactory.

The partly silvered mirrors of the interferometer are silvered so as to reflect and transmit equally, in order that the light in the two paths may be of equal intensity, and then the fuchsin film is

[1] *Phil. Mag.*, vol. 1, p. 43. 1901.

introduced in one of the paths and an absorbing screen in the other. Good fringes may now be seen because the intensity of the light in the two paths is reduced. The retardation may be measured by the shift of the fringes on the removal of the fuchsin film. But the presence of the absorbing screen causes the fringes to be indistinct when the fuchsin film is removed, so the absorbing screen should have only half the absorption of the fuchsin film, in order that the fringes may be seen with equal distinctness whether the fuchsin film is in or out of the interferometer, and the absorbing screen in both upper and lower halves of the other path, so that with the interferometer adjusted for vertical fringes two sets were seen, one above the other, but one set displaced with respect to the other.

Sunlight from a slit *S*, fig. 1, was brought to a focus by means of the lens *L*, so that an image of the slit fell upon the glass plate upon which the fuchsin had been deposited, and thus light of very great intensity was concentrated upon a strip of the film only a millimeter wide, a portion narrow enough so that the thickness could be determined definitely. With a wider film only an average thickness could have been obtained.

The light, after leaving the interferometer, was brought to a focus upon the slit of a small spectroscope by means of the lens, *L'*. By observing through the eye-piece of the spectroscope, spectral bands could be seen.

At this point it may be as well to mention that it takes very careful adjustment of this type of interferometer to be able to observe the bands either with a spectroscope or a telescope. Even with the naked eye it was found that the bands might, under certain circumstances, be seen with the eye in one position, but with the eye nearer or farther from the interferometer they were invisible. Or again, by moving the eye nearer or farther away from the interferometer, a position might be found in which two sets of bands could be seen, crossing one another at right angles, though this only occurred when the instrument was very carelessly adjusted. The following method of adjustment enabled very satisfactory fringes to be produced. A telescope having a fairly well corrected objective was focussed as carefully as pos-

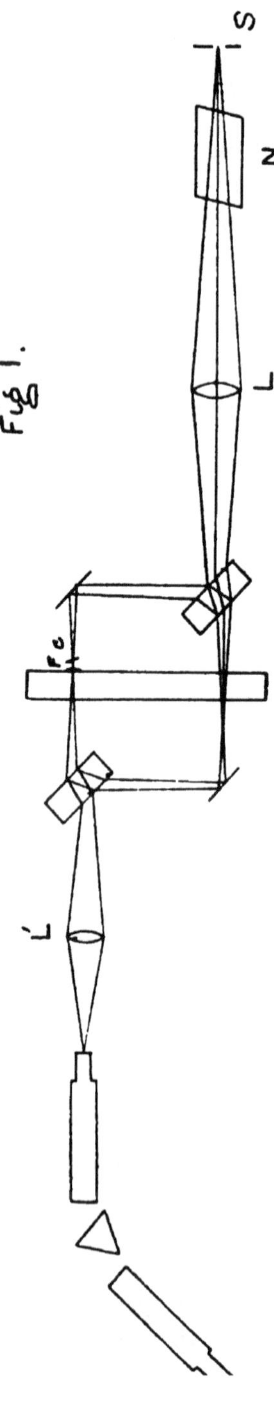

Fig 1.

sible for parallel rays, and then an object at least two or three hundred meters away was observed by means of this telescope in such a way that light from the distant object reached the telescope directly and at the same time by reflection from two of the mirrors of the interferometer. If the mirrors were not quite parallel, two images of the object could be seen, so that all that was necessary to make the mirrors parallel was to adjust them till the two images seen in the telescope coincided. In this way, by comparing one mirror with another, all the mirrors of the instrument could be brought into almost perfect parallelism. The final adjustment was made by observing with the naked eye the reflection of a pointed object held near the instrument in such a way that reflections of the object reached the eye by way of the two paths of the instrument, and the images thus seen brought into coincidence by moving one of the mirrors parallel to itself by a screw motion. When the images are thus brought into coincidence, the two paths of the instrument are equal and the colored fringes of white light may be seen. A simple adjustment of one of the mirrors by trial will widen or narrow down the bands at will, or rotate them through any azimuth, though they can not be so widened or narrowed or rotated through any azimuth if the mirrors are not nearly parallel.

After having arranged the interferometer so that the fringes were all that could be desired, no difficulty whatever was experienced in seeing them with a telescope, but there was still trouble in seeing spectral bands when the apparatus was set up, as shown in the figure. The difficulty was found to be in the lenses, L and L', which, though achromatic and well corrected, were of very short focus, but on changing these for lenses of 25 cm. focal length, the spectral bands were so bright and clear that they could evidently be very much dimmed by the introduction of the fuchsin and still be visible.

In order to determine what kind of spectroscope would give the best results when viewing very faint bands, a trial was made using various spectroscopes. The conclusion reached was that a spectroscope of very low dispersive power, having a low power eye-piece, was the most satisfactory, and therefore that kind was used in this work.

The bands as viewed with the system just described presented the appearance represented diagrammatically in fig. 2, and showed the anomalous dispersion of fuchsin in a general way at a glance.

Fig 2.

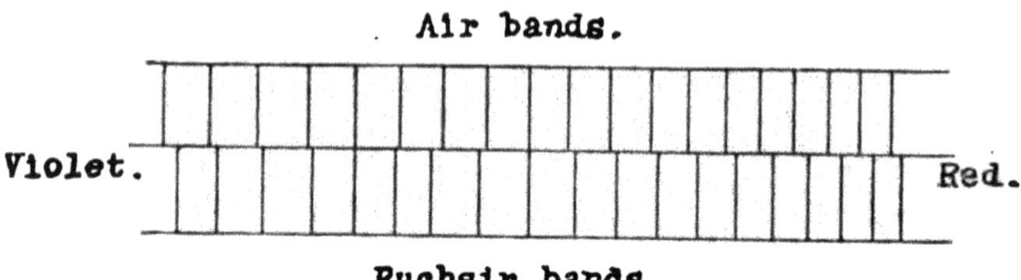

In discussing these bands we will, for the sake of clearness, call those bands affected by the fuchsin the fuchsin bands, and the others the air bands. The bands as shown in fig. 2 correspond to a film too thin to produce a whole band retardation, so, at the two points where two sets of bands coincide, the index of refraction is unity. To determine the index of refraction in any other part of the spectrum, one of the mirrors is moved parallel to itself by means of a screw till a fuchsin band appears in the required portion of the spectrum, unless indeed one is already there. Now, by means of a compensator, an air band is brought into coincidence with the fuchsin band, and the amount of retardation introduced by the compensator is equivalent to the retardation produced by the fuchsin. In the case of a film thick enough to produce more than one band displacement, the fraction of a band is measured by means of the compensator, and the whole number of bands added to this.

It was found that when an attempt was made to produce an arbitrary shift of a definite fraction of a band, either by moving one of the mirrors or rotating a compensator, the bands were disturbed by the mere touching of the mirror or compensator. A very thin mica compensator, thin enough so that it had to be rotated through about 20° to produce a shift of one band, was therefore used, instead of the usual thick glass ones, and with this no such disturbing effect was observed, since an infinitesimal movement of this would not visibly affect the bands. To get around the difficulty of mica having three different indices of refraction, the piece of mica was cut so that one of its axes of elasticity was the axis of rotation. By using plane polarized light, and making the plane of polarization parallel to the axis of rotation of the mica, the retardation introduced by the mica when at different angles was always proportional to just the one index of refraction. It was found incidentally that the introduction of the Nicol prism, *N*, fig. 2, made the bands much more distinct. On rotating the Nicol prism through a right angle, the bands became more confused than with the Nicol out.

The compensation *C*, fig. 1, was made by cutting a thin piece of mica in two and attaching one of the pieces to a device by means of which it could be rotated through an angle and the angle read off. These two pieces of mica were placed in the same path as the fuchsin, the fixed piece and the fuchsin in the lower half of the path, and the movable piece in the upper half, the line of separation of the mica precisely on a level with the edge of the fuchsin; in both upper and lower halves of the other half was a very thin film of fuchsin that served as an absorbing screen. When both pieces of mica were at right angles to the beam of light, their effect was small, because they affected the air bands alike, being of equal thickness. It could be insured that they were equally thick by removing the fuchsin and observing whether the upper and lower spectral bands coincided when both pieces of mica were in the same plane.

It is usual to calibrate a compensator by turning it through different angles, and noting the angles corresponding to different numbers of bands displacement. However, for this work, as it was always less than a single band displacement that had to be measured, it did not help to know the angle corresponding to one band displacement or two bands displacement, because the greatest change of curvature of the calibration curve comes between zero and one.

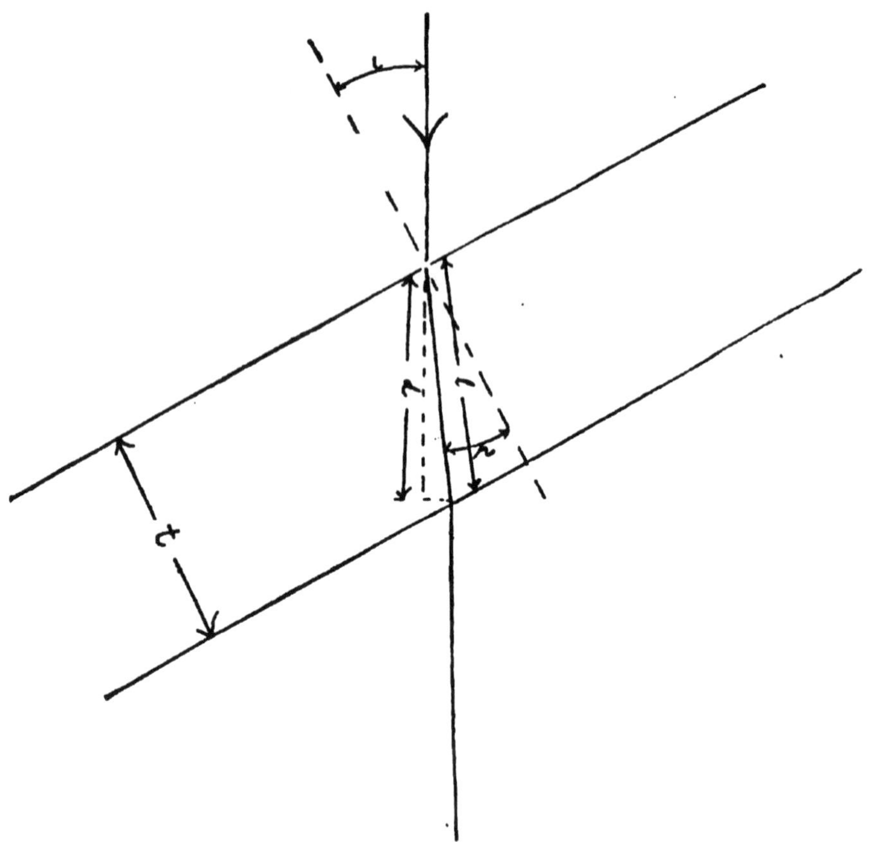

A formula was therefore used which may be deduced as follows: If a piece of mica of thickness t is traversed by a ray which meets it at an angle of incidence i as shown in the sketch, the length of the path of the ray will be increased by $(\mu l - d)$, if the index of refraction of the mica is represented by μ.

Also

$$l = t \sec r = \frac{t}{\sqrt{1 - \sin^2 r}}$$

$$d = l \cos (i - r)$$

$$\mu l - d = l[\mu - \cos (i - r)]$$

which reduces to

$$t\,(\sqrt{\mu^2-\sin^2 i}-\cos i)$$

The introduction of the mica at right angles to the ray will shorten its path by $(\mu-1)t$. The number of wave-lengths retardation produced by a piece of mica in the path of the ray when the mica is moved from a position at right angles to the ray, through an angle i, is the difference of these two quantities, and we have at once

$$n\lambda=t(\sqrt{\mu^2-\sin^2 i}-\cos i)-(\mu-1)t$$

In using a compensator with this type of interferometer, there is a shift of the bands due to the lateral displacement of the beam of light, so it is usual to use a double compensator, one piece in each path, thus overcoming the difficulty by moving both pieces together and giving an equal lateral displacement to each beam. The compensator that was used could be turned through 40°, producing a retardation of three or four bands, before any effect due to this lateral displacement could be observed, and as the greatest shift that had to be measured was less than a band, it was decided to use a single compensator for the sake of convenience.

Using the above formula, the values of i that were observed for one, two, and three bands displacement were found to be consistent. Assuming μ to be constant and equal to 1.58 the value of t was found from the formula, and then by substituting back this value the various values of $n\lambda$, corresponding to different values of i, were found and a curve of $n\lambda$ and i plotted. From this the fraction of a band displacement corresponding to any value of i for any part of the spectrum could be determined. There would be a slight error due to the variation of the index of refraction of the mica for light of different wave-lengths, but as this was less than 1 per cent from each end of the spectrum to the other, the compensator could be calibrated for one part of the spectrum, and a linear correction made in any other part. As a matter of fact, it was calibrated in the red end of the spectrum, and as the index of refraction of fuchsin is nearly unity at the violet end of the spectrum, and as $(\mu-1)$ is the quantity that is measured, it does not matter very much whether the correction is made or not, considering that there are much graver errors which are unavoidable.

Knowing $n\lambda$, the retardation produced by the fuchsin in light of any wave-length, we may determine μ, the index of refraction of fuchsin for light of that wave length, by the formula

$$n\lambda=(\mu-1)t$$

provided we know t, the thickness of the fuchsin film. This was determined by a method which will be described later. The measurements were made upon a film whose thickness was 580 $\mu\mu$, with the exception of those values falling between $\lambda=460$ $\mu\mu$ and $\lambda=590$ $\mu\mu$, which were made upon a film 192 $\mu\mu$ thick.

The results are plotted in plate I, and in the following table they are compared with the values obtained by Pflüger[1] by the prism method, and by Walter[2] by the method of total reflection:

λ	PFLÜGER	WALTER	INTERFERENCE METHOD
A	....	2.019	2.055
a	....	2.086	2.12
706	2.31	....	2.15
B	....	2.161	2.22
Li	2.34	....	2.27
C	....	2.310	2.35
634	....	2.412	2.48
D	2.64	2.684	2.70
535	1.95	....	1.98
E	....	1.912	1.85
F	1.05	1.074	1.05
Sr	0.83	....	0.82
455	....	0.847	0.83
G	1.04	0.95	1.00
425	....	1.00	1.11
H	....	1.32	....

The curve is uncertain within the absorption band and perhaps also to some extent in the violet, but in the red end of the spec-

[1] *Loc. cit.*

[2] *Wied. Ann.*, vol. 57, p. 396. 1896.

trum it is established beyond a doubt that the values given by Pflüger do not apply to the fuchsin upon which my measurements were made. The fact that Pflüger obtained the same values using prisms of different angles seems to show either that the two methods lead to different results, or that the composition of the two samples of fuchsin was different. There are a number of red triphenyal menthane dyes of distinctly different compositions, all of which go under the name of fuchsin. Wernicke[1] obtained the dispersion curve of a fuchsin by the prism method and found the indices to range from 1.31 to 1.90. The one on which the writer's first work was done had indices ranging from less than unity to about two. Kundt experimented on a fuchsin having two absorption bands in the visible spectrum. However, there is no doubt but what the fuchsin in question was made by the same process as that which Pflüger used, though, having been purchased seven years later, it was probably made in a different batch, and hence there is a possibility of the composition being slightly different.

The agreement with Walter in the red is extremely satisfactory. The two curves run uniformly parallel to one another, this writer's being a little higher than Walter's, which may be accounted for by an error in the thickness of a systematic error of some kind. A slight change in the thickness would make the two curves almost identical.

The thickness of the films was determined as follows: A portion of the film is washed away with alcohol, leaving a clean, sharp edge. In general, when a film is thus washed away, there is a concentration of fuchsin at the edge, due to a flowing of the solvent. This is of course undesirable and was eliminated by using a piece of blotting paper with which to clean away the fuchsin. This piece of blotting paper was itself cut to a clean sharp edge and slightly dampened with alcohol. The glass near the film was wiped with it, a new edge of the blotting paper was prepared, dampened, and the operation was repeated until a satisfactory edge of fuchsin was obtained.

Now a plate of glass was placed upon the glass upon which

[1]*Pogg. Ann.*, 155. 1875.

the fuchsin had been deposited. There were thus two layers of air of different thickness, one between the fuchsin and the glass, and the other between the two pieces of glass. The difference in the thickness of the two films of air gives the thickness of the fuchsin. The thickness of the air films may be obtained very accurately by the interference of their films, by a method first used by Wernicke[1] and modified by Wiener.[2] White light was allowed to fall on the two air films, and the reflected light was examined with a spectroscope, by means of which interference bands could be seen in the spectrum.

The arrangement of the apparatus is shown by fig. 3. The glass plates enclosing the air film are shown at *G*. *P* is a totally reflecting prism, one edge of which forms an edge of the slit of the collimator. A Rowland grating, *R*, sent a spectrum into the observing telescope, and the distance from band to band was measured by means of the micrometer eye-piece.

The two glass plates, *G*, were so arranged before the slit that the film of air between the fuchsin and the glass were before the lower part of the slit, and the film of air between the two bare plates before the upper part. By a suitable mechanical contrivance, the plates were adjusted so that the bands seen at the eye-piece were vertical.

On examining the two sets of bands, it was noticed that while those corresponding to the film between the two glass surfaces were regular and became uniformly narrower toward the violet, those corresponding to the air film between the glass and the fuchsin showed an irregularity.

For instance, when there was a film of fuchsin whose thickness was about a fourth of a wave-length of red light, the bands of one set were displaced with respect to the other set by half a band in the red, and this displacement regularly increased, till a point in the blue-green was reached where it was a whole band. At this point there was a sudden change of half a band, and then a regular change took place, going from short to shorter wave-

[1] *Pogg. Ann.* Ergbd. 8, p. 65. 1878.
[2] *Wied. Ann.*, vol. 31, p. 629. 1887.

lengths, till in the violet there was again a whole band displacement.

This brings up the question of phase change. There was evidently about half a band phase change in the blue-green, but in the violet end the red fuchsin reflects like a transparent body,

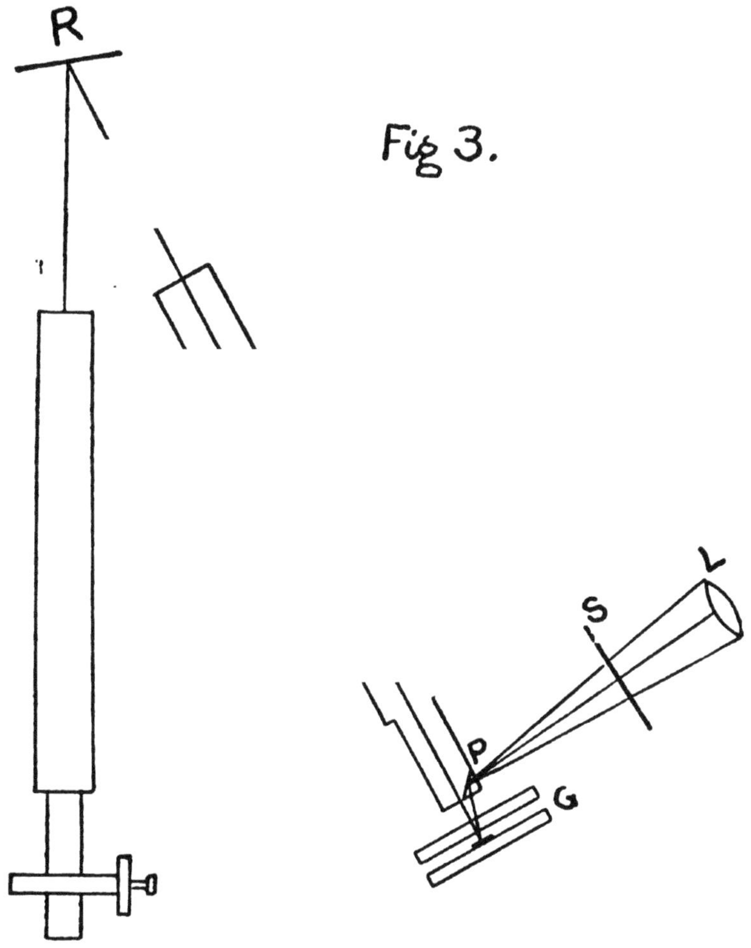

and it is likely that the difference of phase change between it and the glass is zero. Pflüger has gone into this point very thoroughly, and says that from the experiments of Wernicke it is safe to assume that the phase change by reflection from fuchsin is the same as that from glass for wave-lengths longer than 640 $\mu\mu$. The measurements were, therefore, made with the light comprised between the B and C lines. Sunlight was used and a number of gelatin screens interposed at S, fig. 3, so as to keep out light of those wave-lengths not needed, because the films bleach easily.

The thickness of the fuchsin may be determined from the two sets of interference bands in several ways, but the method involved in the following formulas, due to Wiener, was the one used:

We have

$$2t = m\lambda_m = (m+1)\lambda_{m+1}$$

when t is the thickness of the film between the two pieces of glass, λ_m the wave-length corresponding to the center of one band, and λ_{m+1} that of the next band towards the violet. Taking the one of the bands due to the film of air between the glass and the fuchsin, which falls between the mth and $(m+1)^{st}$ of the other bands, and calling it λ'_m we have:

$$2t' = m\lambda'_m$$

where t' is the thickness of the film of air corresponding to this case. From these, two equations follow at once:

$$t' - t = \frac{\lambda_m - \lambda'_m}{\lambda_m - \lambda_{m+1}} \cdot \frac{\lambda_{m+1}}{2}$$

which gives the thickness of the fuchsin $t - t'$ directly.

The thickness of the air films was regulated so that about five or six interference bands fell between the B and C lines. Thus there were several sets of bands upon which independent measurements could be made.

Some idea of the accuracy of the measurement can be obtained from the following set of measurements made upon one of the films, in which is included every measurement that was made upon that film:

WAVE-LENGTH	HALF THE MEASURED SHIFT	THICK-NESS
661	36	625
661	46	615
667	57	610
667	60	606
672	70	602
672	67	604
677	92	585
677	89	588
660	37	623
660	48	612
665	67	608
665	60	605
671	70	601
671	72	604
676	89	587
676	82	594

This makes the probable error of a single observation 8 $\mu\mu$ and of the mean 2 $\mu\mu$, which is about the order of accuracy that interference methods usually give.

There is some little uncertainty sometimes as to whether the shift refers to a fraction of a band or to one or two whole bands plus a fraction of a band. In the above instance, the shift is about eight-tenths of a band, and the total displacement one and eight-tenths, giving a thickness of about nine-tenths. If the displacement had been assumed to be two and eight-tenths bands, the resulting thickness would have been one and four-tenths wave-lengths. To decide this point definitely a number of films were made of different thicknesses. The first of this series was made from a saturated solution, and the rest of the films were made, each succeeding one thinner than the one which preceded it, by continually diluting the solution from which they were deposited. Starting from the thinnest (which was 320 $\mu\mu$ thick), the thickness of each succeeding one was measured till the thickest was reached (617 $\mu\mu$). The continuity of the series furnished data from which the thickness was determined definitely, and observation upon the films with a spectro-photometer confirmed the conclusions.

The same trouble was experienced in the measurement of the retardations with the interferometer, but when the thickness was definitely known a comparison with a film sufficiently thin to show bands all through the spectrum cleared up this point, for with the very thin film the retardation was never so much as a whole wave-length, for reasons already mentioned. The matter might have been cleared up by viewing the fringes of white light in the usual way and observing the shift of the central fringe produced by the fuchsin, but this was found impracticable on account of the selective absorption of the fuchsin film, which made the central fringe look just like any other fringe. This plan would have, however, furnished a very decisive solution of the difficulty with regard to the thickness, for by making the plate upon which the fuchsin was deposited the back mirror of an interferometer, and comparing the fringes due to the reflection from the bare plate with those due to the reflection from the top surface of the fuchsin, the shift of the central fringe could have been easily determined.

The photometric measurements were made by means of a Brace spectrophotometer. The methods of adjusting the instrument have been already given by Tuckerman.[1]

The extremely great absorption gave rise to some difficulties which do not occur in ordinary photometric work. For instance, a little of the red light (for which fuchsin is transparent) reached the eye as stray light by reflection in the telescope tubes and the comparison prism. This, too, in spite of the fact that the telescope tubes were well blackened and diaphragmed. This small amount of red light was more intense than the green light that was being measured, since the green light had been cut down a thousand times or more by the absorption of the film.

This was remedied by using a screen having absorption bands in the red and the blue, but which transmitted green quite freely. In this way green light alone entered the instrument when measurements in the green were being made.

Fig. 4 shows the general arrangement of the apparatus. A beam of sunlight is brought to a focus on each of the totally re-

[1] *Astrophysical Journal*, p. 145. Oct., 1902.

flecting prisms *p* and *p'* by means of a split lens *S*. Then prisms reflect the light on to the mirrors *m* and *m'*, and these direct the light on to the collimator slits. Part of the light from the right-hand collimator reaches the observing telescope by

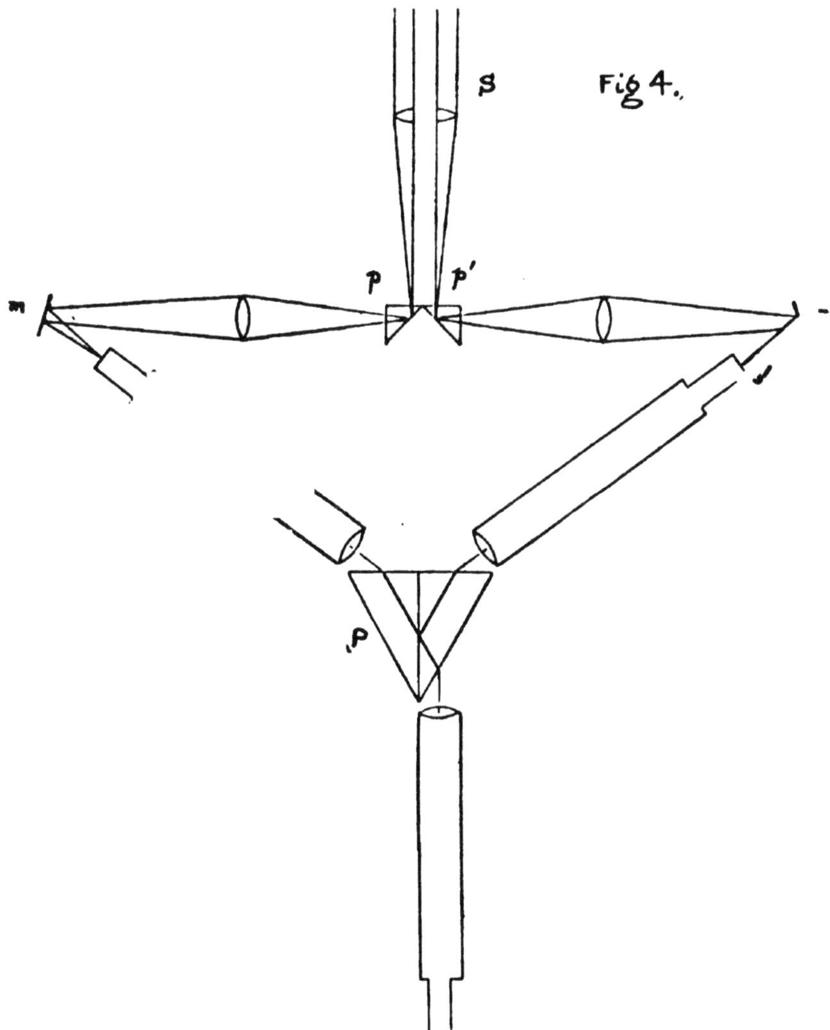

reflection from a silvered strip cemented in the comparison prism *P*. The rest passes above or below the strip and is lost. On the other hand, of the light which comes from the left-hand telescope, only that which passes above or below the silvered strip reaches the observing telescope, with the result that in the eye-piece may be seen a field illuminated from the left-hand collimator, with the exception of a central strip illuminated from the right-hand collimator; and when the intensity of the illumination from both these sources is alike, the field appears uniform. This uniformity may be brought about by varying the width of either of the slits *s* or *s′*. If the fuchsin film is placed before the slit *s* it will cause the center of the field to seem dark, and a match may again be produced by narrowing the slit *s′*, and if we know the original width of the slit, the ratio

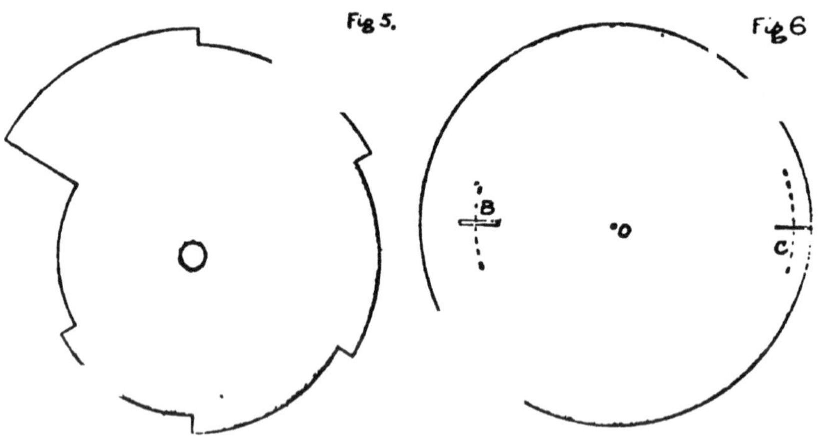

of the change in width to the original width gives very closely the proportion in which the introduction of the fuchsin has diminished the intensity of the light. If, instead of diminishing the intensity at *s′* by narrowing the slit, a revolving sector is used which cuts down the light by the proper amount, the error due to the lack of proportionality between intensity and slit-width does not enter in. However, it is found more convenient to use a sector cut into an arbitrary number of parts, as shown in fig. 5. and as with this we can, in general, only make an approximate match, the varying of the slit-width is used as a fine adjustment. This gives practically as good results as can be obtained with a variable sector. A sector of eighty notches was used except for the measurements within the absorption band, in which case a tin disc, as shown in fig. 6, was used, having a slit cut in the edge 0.7 mm. wide and another, about a cm. from the edge, 1.5 mm. wide. This disc was rotated very rapidly before the collimator slit by means of a small motor, and thus the intensity of the light incident upon the collimator slit was diminished by 99.856 or 99.682 per cent, according to which of the slits of the disc was used. Since the edges of the slits of the disc are parallel, care had to be taken to have the collimator slit at a definite distance from the center of the disc. This was done by using the disc close up against the collimator slit; and then, by holding a sharp edge on a line *B* or *C* scratched on the disc, the shadow of this edge would fall on the center of the collimator slit, when the center of the disc was at a distance *OC* or *OB* from the ray entering the collimator.

The use of this tin disc, which cut down the intensity of the light so greatly, made it possible to match very accurately, although the light had been reduced by the fuchsin to such an enormous degree. Since the slope of the absorption curve was very steep, it was necessary to have the collimator slit *s* quite narrow, not more than half a millimeter wide at most, in order that the readings should be made with sufficiently homogeneous light; otherwise the absorption as observed would have too low a value. With the slit *s* half a millimeter wide, *s'* would have been less than a five-thousandth part of a millimeter wide, when the adjustment was made using no sector, and accurate settings of the slit at this width would have been impossible. This feature of the photometer, together with the fact that a bare flame or direct sunlight can be used instead of a flame or sunlight behind ground or opal glass, which would cause a loss of 95 per cent to 98 per cent, and the absence of Nicol prisms, which would cause a further loss, makes it singularly well adapted for work on strongly absorbing substances.

The diminution in intensity caused by transmission through the fuchsin film is due to two factors, reflection from its surfaces and absorption within the film. To determine the absorption it is necessary either to eliminate the reflection by measuring the differential absorption of two films of different thickness or by making a separate determinative of the reflection and subtracting it. Preparations were made to measure the reflection, but this part of the work was not finished, though I expect to continue this work and measure the reflection. Not having the reflection measurements, I must be content with computing them. The effect of an error in the reflection upon the final values of the absorption coefficients will be greatest within the absorption bands, and the 10 per cent change in the reflection only changes the absorption coefficient 1 per cent.

For computing the reflection from an absorbing substance into air, Cauchy's theory leads to the following formula:

$$R = \frac{(\mu-1)^2+k^2}{(\mu+1)^2+k^2}$$

while the electromagnetic theory leads to

$$R = \frac{\mu^2(1+k^2)+1-2\mu}{\mu^2(1+k^2)+1+2\mu}$$

which may be seen to be identical with the previous formula when we remember that in Cauchy's theory the quantity k is identical with μk of the electromagnetic theory. For the reflection at the interface between an absorbing substance and a transparent substance whose indices of refraction are μ and μ' respectively, the electromagnetic formula becomes

$$\frac{(\mu-\mu')^2+\mu^2k^2}{(\mu+\mu')^2+\mu^2k^2}.$$

According to this formula, the reflection is the same on either side of the interface between the media, while with Cauchy's formula the reflection on the two sides will, in some instances, be different, which would lead to the loss by reflection being different, according to whether the light went through the glass plate and then through the fuchsin or through the fuchsin first, an effect which the writer has been unable to observe. The values of the reflexion as computed in this way are plotted in plate III.

If r_1, r_2, and r_3, represent the reflection at the fuchsin-air, fuchsin-glass, and air-glass surfaces respectively and I_0 the original intensity of the incident light, the effect of the reflection is to make the transmitted light have the value

$$(1-r_1)(1-r_2)(1-r_3)\, I_0,$$

neglecting the effect of multiple reflections which does not enter in within the absorption bands. For those colors for which the film is transparent, the effect of multiple reflection is easily allowed for to the degree of accuracy possible when working with strongly absorbing substances.

Knowing the thickness of the film, we may compute the absorption coefficient μk from the following formula:

$$\frac{I}{I_0(1-r_1)(1-r_2)(1-r_3)} = e^{-\frac{4\pi\mu k}{\lambda}}$$

in which $\frac{I}{I_0}$ is the ratio of the intensities as determined by the spectrophotometer, t is the thickness of the film, and λ the wave-length in vacuo. The photometric measurements were all made upon a film $192\mu\mu$ thick. The transmission of this film is plotted in plate II. Using these values, the curve given in plate I was computed. In the following table these values are compared with Pflüger's and Walter's results:

Values of μk

λ	PFLÜGER	WALTER	CARTMEL
589	0.79	0.792	.86*
527	1.22	1.419	1.[illegible]54*
486	0.98	1.168	1.07*
455	0.43	0.533	0.56*

Walter used Cauchy's formulas for the elliptic polarization which is associated with metallic reflection, and as this could only

* By interpolation.

be applied to those radiations which are strongly absorbed by the fuchsin, he obtained only the four values given above. In order to make a comparison with Walter's work, Pflüger measured the same four quantities directly, using the difference in transmission of two films of different thickness.

In conclusion, my best thanks are due Professor Brace for help and encouragement throughout the work, and for the excellent laboratory facilities afforded me at the University of Nebraska, where the work was done. I must also thank Professor B. E. Moore for various courtesies extended to me, and for suggestions in regard to the photometric measurements.

600
500
400

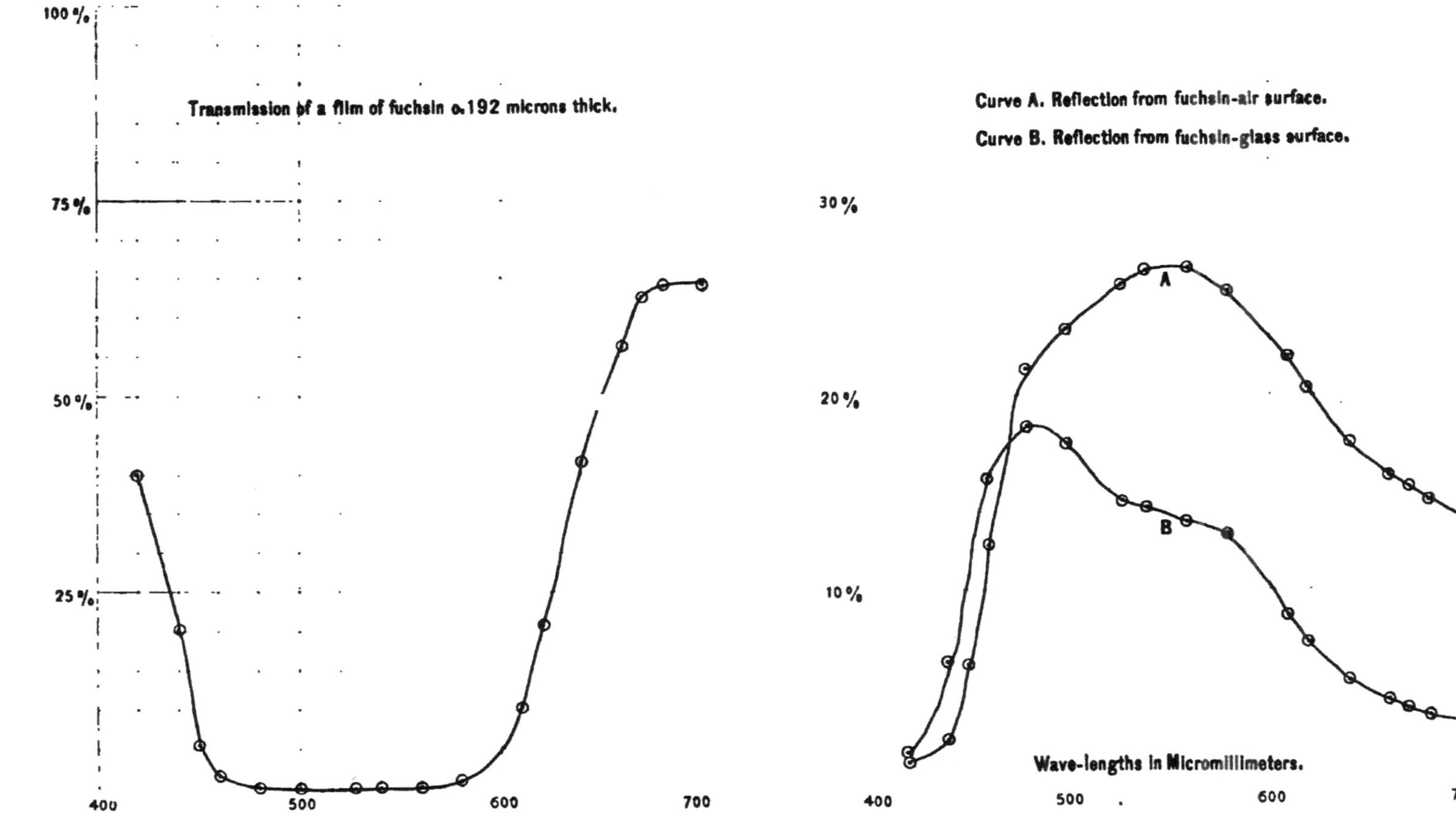

Transmission of a film of fuchsin o.192 microns thick.
100 %
75 %
50 %
25 %
400
500
600
700
Curve A. Reflection from fuchsin-air surface.
Curve B. Reflection from fuchsin-glass surface.
30 %
20 %
10 %
A
B
Wave-lengths in Micromillimeters.
400
500
600
700

III.—*Mallophaga from Birds of Costa Rica, Central America*

BY M. A. CARRIKER, JR.

INTRODUCTION

The specimens of Mallophaga, upon which the contents of this paper are based, were collected by the author during the summer of 1902, while engaged in a search for natural history specimens in different regions of Costa Rica, Central America. As will be seen from the text, the birds were collected in three principal localities, namely: Juan Vinas, on the Atlantic slope, at an altitude of approximately 3,000 ft.; the volcano Irazu, situated on the continental divide, the majority of the specimens being found between the altitudes of 8,000 and 10,000 ft., although some were collected at the summit, which attains an altitude of 11,198 ft.; Pozo Azul, on the Rio Grande de Pirris, about thirty miles from the Pacific coast, with an altitude of not more than 300 ft.

The birds collected are, with the exception of a few duplicates at the University of Nebraska, now in the possession of the Carnegie Museum, Pittsburg, Pa., and I am greatly indebted to Mr. W. E. Clyde Todd, curator of birds and mammals in that institution, for his kindness in giving me the correct determination of the bird hosts, the host list given in this paper having been arranged and corrected by him. I also extend thanks to Mr. C. F. Underwood, of San Jose, Costa Rica, for his kindness in giving me determinations for my specimens while in that country, and for the privilege of examining birds, collected by him, for Mallophagan parasites. Lastly, my thanks are due Professor Lawrence Bruner, of the University of Nebraska, under whose direction this paper was written, for his many valuable suggestions during the course of the work.

The types of all new forms described in this paper are in the collection of the author, with co-types of most of them in the collection of the Department of Entomology in the University of Nebraska.

A few words in regard to the distribution of Mallophaga collected would probably be interesting. It was found that birds belonging to genera not strictly tropical, and found principally in the higher altitudes, were much more often infested with the parasites than those strictly tropical and inhabiting the lower altitudes. This statement, however, is only true in a general sense, but when exceptions were encountered it was usually found that the parasites belonged either to the genus *Menopon* or *Colpocephalum,* and were present in great numbers, these two genera, especially the latter, having been taken in large numbers on strictly tropical birds.

The list of Mallophaga collected on *Tinamus robustus* is a very interesting one, including two new genera and several new species, all of which are very aberrant forms of the genera in which they have been placed. Other very interesting forms are *Colpocephalum extraneum* sp. nov. and *mirabile* sp. nov. collected on *Nyctidromus albicollis* and *Zarhynchus wagleri,* in that they possess a very marked mesothoracic suture and have the metathorax enormously developed posteriorly. It seems to me that many of the present genera need a thorough revision and that some of them should be split up into two or more genera or sub-genera, *Colpocephalum, Menopon,* and *Physostomum* especially needing it.

BIBLIOGRAPHY

Burmeister, H. *Handbuch der Entomologie.* Berlin, 1832–33.

Carriker, M. A., Jr. *Some New Mallophaga from Nebraska Birds.* Jour. N. Y. Ent. Soc., vol. X, pp. 216–229, 1902.

Denny, H. *Monographia Anoplurorum Britanniae.* London, 1842.

Geer, Charles Baron de. *Mémoires pour servir à l'histoire des insectes.* Stockholm, 1752–78.

Gervais, Paul. *Histoire naturelle des insectes apteres.* Paris, 1847.

Giebel, C. G. *Insecta Epizoa.* Leipsig, 1874.

Kellogg, V. L. *List of Biting Lice (Mallophaga) Taken from Birds and Mammals of North America.* Proc. U. S. Nat. Mus., vol. XVII, pp. 39–100.

—— *New Mallophaga I.* Contributions to Biology from the Hopkins Seaside Laboratory of the Leland Stanford Junior University, no. IV, 1896.

Kellogg, V. L. *New Mallophaga II.* Contributions to Biology from the Hopkins Seaside Laboratory of the Leland Stanford University, no. VIII.

Kellogg, V. L., and Chapman, Bertha L. *Mallophaga from Birds of the Hawaiian Islands.* Jour. N. Y. Ent. Soc., vol. X, pp. 151–69.

—— *Mallophaga from Birds of Pacific Coast of North America.* Jour. N. Y. Ent. Soc., vol. X, pp. 20–28, 1902.

—— *New Mallophaga III.* Contributions to Biology from the Hopkins Seaside Laboratory of the Leland Stanford Junior University, no. XIX, 1899.

Kellogg, V. L., and Kuwana, S. I. *Mallophaga from Alaskan Birds.* Proc. Acad. Nat. Sci. of Phil., pp. 151–59, 1900.

—— *Mallophaga from Birds of the Galapagos Islands.* Proc. Wash. Acad. of Sci., vol. IV, pp. 457–99, Sept., 1902.

Nitzsch, Chr. L. *Die Familien, und Gattungen der Thierinsecten (Insecta Epizoica).* Mag. der Ent. von Germar und Zinken, t. III, pp. 261–316, 1818.

Osborn, H. *Insects Affecting Domestic Animals.* U. S. Dept. Agri., Bull. 5, New Ser.

—— *Mallophaga Records and Descriptions.* Ohio Nat., vol. II, pp. 175–78.

—— *Mallophaga Records and Descriptions.* Ohio Nat., vol. II, pp. 201–04.

Piaget, E. *Les Pediculines,* Leide, 1880.

—— *Les Pediculines,* Supplement, Leide, 1885.

—— *Quelques nouvelles pediculines,* Tidschr. voor Ent. XXXI, pp. 147–64, 1888.

—— *Quelques nouvelles pediculines,* Tidschr. voor Ent. XXXIII, pp. 223–59, 1890.

Taschenberg, E. L. *Insektenkunde.* Bremen, 1879.

DESCRIPTIONS AND DISCUSSIONS

Docophorus bisignatus N.

Giebel, *Insecta Epizoa*, p. 106, pl. IX, fig. 9.
Piaget, *Les Pediculines*, Sup., p. 11, pl. II, fig. 1.

Numerous individuals of both sexes collected on ***Guara alba*** at Poza Azul, Costa Rica, June, 1902. This species is readily recognized by the bilobate, clypeal, signature.

Docophorus platystomus N.

Nitzsch, in Burmeister, *Handbuch d. Ent.* 1839. II, pp. 426.
Giebel, *Insecta Epizoa*, p. 69, pl. IX, fig. 1.
Denny, *Anoplurorum Brittaniae*, p. 108.
Piaget, *Les Pediculines*, p. 17, pl. I, fig. 1.
Osborn, *Notes on Mallophaga and Pediculidae*. **Canad.** Ent., 1884. XVI, p. 197.
—— *Insects Affecting Domestic Animals*, p. 216.

Numerous specimens collected on *Buteo borealis costaricensis*, volcano Irazu, April; on *Buteo abbreviatus*, Pozo Azul, June, 1902. While these specimens do not exactly agree with Piaget's plate and description in some particulars, they can be referred to that species.

Docophorus platystoma umbrosus var. nov.

MALE.—Body, length 2.22 mm., width 1.02 mm.; head, length .86 mm., width .81 mm. Much larger than Piaget's measurements for *platystomus*, especially the head, which is much narrower in proportion at the temples, with the clypeus very much narrower and deeply emarginate at the tip; the whole head is darker, also the lateral markings of thorax and abdomen; the pustules in the lateral fasciae are very clear and prominent, not obscured as in *platystomus*, while the great number of dorsal abdominal hairs of Piaget's specimens are wanting, there being but a scattering median row on each segment.

Two males collected on *Leucopternis semiplumbea*, at Pozo Azul, Costa Rica, June, 1902.

Docophorus transversifrons sp. nov., pl. I, fig. 1

MALE.—Body, length 1.76 mm., width .62 mm.; brownish golden; head large with broad truncate clypeus, and abdomen subclavate, with narrow, light smoky brown lateral bands and golden brown transverse bands.

Head, length .66 mm., width .58 mm.; deep golden brown, with long, broad, truncate clypeus and constricted, rounded temples; occiput convex; clypeal angles with two hairs; one marginal and one submarginal hair at clypeal suture; two submarginal hairs just before trabeculae; trabeculae rather small for the genus, pointed, golden; antennae short and stout, first joint thickest, second longest, third and fourth subequal, shortest, and all uniformly golden brown; eye convex, colorless, with a hair; temples evenly rounded, with two rather long hairs and several short bristles; narrow, dark brown, antennal bands, broken at clypeal suture, not reaching clypeal angles, and with bases swollen and bent inward at trabeculae; paler brown, internal, sinuated bands inside of antennal bands; clypeal signature large, broad pointed posteriorly, slightly emarginate at sides, scarcely reaching end of clypeus, and whole deep golden brown; mandibles heavy, chestnut; pale brown occipital bands running from sides of occiput to bases of antennal bands; temples narrowly edged with brown, and whole region brownish.

Prothorax small, quadrangular, with slightly diverging sides; posterior margin bluntly angulated; posterior angles with one stout hair, lateral bands of deep brown, pale at the margins and curving around on posterior border nearly to middle of segment; interior of segment pale golden brown; brown coxal bands visible. Metathorax short, transverse, with strongly convex, diverging sides, and flatly angulated posterior margins, with three hairs on each side; lateral submarginal bands of deep brown; whole segment golden brown. Legs rather short, stout, with well-developed tibiae and aborted tarsi; uniformly concolorous with body, with the exception of pale brown semiannulations at tips of femora and tibiae.

Abdomen short, subclavate, with lateral margins of segments slightly convex, and rounded, projecting, posterior angles furnished with two hairs in segments four to seven, four on each side of the eighth; one hair on posterior margin of segments three to seven just within the lateral bands; a median row of long slender hairs across the middle of segments two to six; narrow lateral bands of translucent brown, broken at the sutures and projecting into adjacent anterior segments; clear pustules at spiracles on segments two to seven; transverse bands of golden brown, separated by clear sutures and broken medially; apical segment clear, with brown tip; genitalia deep brown, short, stout, with inward curving points.

FEMALE.—Body, length 2.15 mm., width .83 mm.; head, length .71 mm., width .66 mm.; larger and lighter than male, with more oval abdomen; lateral bands narrower and transverse bands shorter and more widely separated; apical segment large, rounded, clear, with a median, transverse blotch.

Two males and one female collected on *Micrastur guerilla*, at Pozo Azul, Costa Rica, June, 1902. A rather unique form for a Raptorial bird, but one readily recognized by the long, broad, truncate clypeus.

Docophorus californiensis Kell.

Kellogg, *New Mallophaga* II, p. 483, pl. LXVI, fig. 6.

Numerous males and females collected on *Melanerpes formicivorus* and *Dryobates villosus jardinii*, volcano Irazu, April, 1902, and on *Chloronerpes yucatanensis* and *Melanerpes aurifrons hoffmani* at Juan Vinas, Costa Rica, March, 1902.

Docophorus bruneri sp. nov., pl. I., fig. 2

FEMALE.—Body, length 1.19 mm., width .42 mm.; clear umber brown, with large head having clypeus pale golden and temples deep brown; abdomen elliptical, with narrow, pitchy, lateral bands and median, transverse, brown bands.

Head, length .42 mm., width .42 mm.; very large in proportion to the body, with broad, truncate-conical clypeus, slightly swollen at clypeal angles; one hair in clypeal angles, one marginal and two submarginal ones on sides; narrow golden brown antennal bands; clypeal signature large, pale, pointed posteriorly, with dusky lateral lobes projecting posteriorly, and the whole separated from adjoining parts by a narrow clear line; trabeculae large, curving, bluntly pointed, concolorous with front of head: antennae long and stout, with first two segments longest, and equal, and last three shorter, and subequal, all with semiannulations of brown and ground color same as temples; eyes large, protruding, clear, with a very fine bristle; temples evenly rounded with two short pustulated hairs and two bristles; occiput convex, heavy, pitchy brown, occipital bands paler and narrower anteriorly, running forward past the posterior root of mandibles to bases of the antennal bands; mandibles large, chestnut; whole temple from occipital bands to margin deep, clear brown.

Prothorax small, short, quadrangular, with concave anterior margin, convex sides and posterior margin; posterior angles with one hair; narrow, dark brown, submarginal, lateral bands; brown coxal bands visible; metathorax larger than prothorax, pentagonal, with convex diverging sides and sharply angulated posteriorly; posterior angles with one long hair and two short ones; posterior margin with five long hairs on each side; dark brown lateral bands, paler at margins; meso-coxal bands visible. Legs stout, with swollen femora and tibiae, concolorous with thorax, with a darker spot at the tips of femora and inner base of tibiae.

Abdomen short, elliptical, with posterior angles acute and slightly projecting, furnished with one hair in fourth segment and two in segments five to seven; narrow pitchy brown lateral bands on segments one to seven, separated at the sutures and projecting into adjacent anterior segments; large clear pustules at the spiracles on segments one to seven; a long pustulated hair arising from the posterior margin of segments one to seven, just within the lateral bands; a second long pustulated hair on posterior margins of segments one, two, three, half way in toward middle of abdomen; a row of six long, pustulated hairs on the median posterior portion of segments four to six; eighth segment with two long hairs on each side; ninth segment very short, indented at tip; clear brown transverse bands, interrupted medially, on segments one to seven, and separated by clear sutures; segments eight and nine entirely brownish; median, ventral transverse bands, of dark brown, separated by broad clear sutures. except between segments five and six.

MALE.—Body, length 1.00 mm., width .42 mm.; head, length .37 mm., width .37 mm.; markings same as female; genital hooks short and stout, tip of abdomen clear.

Numerous individuals of both sexes collected on *Menacus candaei*, at Juan Vinas, Costa Rica, March, 1902. A species rendered striking by the large head, pale clypeus, and dark brown temples, and narrow lateral bands of abdomen.

Docophorus underwoodi sp. nov., pl. I, fig. 3

FEMALE.—Body, length 2.09 mm., width .89 mm.; clear, with bands and spots of deep, clear brown and pitchy.

Head, length .70 mm., width .62 mm.; conical, end of clypeus emarginate, clear, with one hair in rounded angles; sides concave, with a short hair at the suture and two submarginal ones back of suture; trabeculae large, bent, bluntly pointed and pale brown; antennae medium, first segment largest, colorless, second a trifle smaller, and last three smaller and subequal, and last four deep brown, with narrow clear bases; eye large, clear, with a hair; temples evenly rounded, narrowly margined with brown, with two pustulated and two non-pustulated hairs; occiput convex; clear brown antennal bands, broken at the sutures and not reaching clypeal angles; internal narrow bands of clear brown, following margin of ocular band, running backward half way to the mandibles, then bending inward a short distance, then straight back, and finally spreading out to bases of mandibles; area between antennal and internal bands a pale brown; clypeal signature long, narrow, dark at tip, emarginate, with posterior portion gradually tapering to a point at mandibles; mandibles large and heavy, chestnut; heavy, pitchy occipital bands running from sides of occiput to bases of antennal bands; a pitchy ocular band from eye to junction of antennal and occipital bands, a small, dark brown occipital signature.

Prothorax quadrilateral, with rounded anterior and posterior angles and flatly convex, slightly diverging sides; dark brown lateral, marginal bands and a pitchy line across postero-lateral portion of segments. Metathorax scarcely longer than prothorax, pentagonal, with convex, widely diverging sides and angulated posterior margin; posterior angles with one short hair, and posterior margin with a row of ten short, pustulated hairs on each side; deep clear brown lateral bands, widening posteriorly; a pitchy brown band around posterior border, slightly submarginal in median portion; meso-coxal bands visible. Legs stout, femora and tibiae swollen at the tips; clear, with bases and tips of femora annulated with pitchy; a pitchy band along posterior border of tibiae, and tips annulated with brown.

Abdomen broadly oval, with lateral margins of segments more or less convex and the rounded posterior angles protruding, with one hair in segments two to eight and one in anterior portion of segments five to seven; a row of hairs across the middle of segments one to seven, pustulated along posterior margin of transverse bands; narrow, pitchy, lateral bands on segments one to eight, separated by a small clear spot in posterior angles; clear brown, lateral transverse bands, extending inward one-third the width of abdomen, narrowing inwardly, with a large, round, clear area just within the lateral bands and a darker oval spot near the inner ends; a large, round, ventral, brown patch covering the apical portion of the abdomen; two median brown spots on the eighth segment and the tips of the ninth brownish.

MALE.—Body, length 1.61 mm., width .62 mm.; head, length .63 mm., width .55 mm.; very similar to female, except the shorter, rounder abdomen; genitalia short, stout, deep brown.

Numerous males and females collected on *Psilorhinus mexicanus,* at Juan Vinus, Costa Rica, March, 1902. This form is of the type of *D. rotundus* Piag. and *corvi* Osb., but is easily distinguished from *corvi* by the slightly fuscus color and dark brown. instead of pitchy markings, and from *rotundus* by the absence of lateral angles in the prothorax and by the double occipital bands; also female is much larger.

Docophorus communis N.

Nitzsch, Germar's *Mag. of Ent.,* 1818, vol. III, p. 290.
Kellogg, *New Mallophaga* II, 486, pl. LXVI, fig. 7.

Large numbers of this type of *Docophorus* collected on a great variety of hosts (see host list). I have made no attempt to separate them into varieties from lack of time and material, much as they need it. As Mr. Kellogg says, this group needs revision badly, but it is a difficult undertaking, with much time and material as a requisite.

Docophorus cancellosus sp. nov., pl. I, fig. 4

FEMALE.—Body, length 2.51 mm., width .83 mm.; head and thorax translucent, smoky brown; abdomen clear, with pitchy lateral, and deep brown, transverse bands; head conical with narrow emarginate clypeus, and squarish temples.

Head, length .74 mm., width .66 mm.; conical from trabeculae forward, with sinuate sides and narrow emarginate front; two short fine hairs before trabeculae, which are of medium size, bluntly pointed and golden brown; antennae rather small, first joint thickest, second joint longest, last three shortest and subequal, all uniformly golden brown; eye large, clear, prominent, with a short hair in the middle and one at posterior margin; temples squarish, with rounded angles, furnished with several short hairs; occiput slightly concavo-convex; heavy, deep brown antennal bands, widening and darkening to pitchy at bases, which bend inward almost to bases of mandibles; paler brown bands inside of antennal bands; clypeal signature very indistinct; mandible very heavy, chestnut; temples narrowly edged with pitchy brown and whole templar regions deep smoky brown inward to the broken, occipital bands, which do not reach the occipital margin; short faint ocular bands running inward from eyes to occipital bands; a small, brown, occipital signature.

Prothorax small, quadrilateral, narrowed in front, with concave interior margin, convex diverging sides, and concave posterior margin; posterior angles rounded, with a short hair; deep brown, lateral bands, curving around on posterior portion of segment; prominent brownish coxal bands; metathorax larger than prothorax, with convex, diverging sides and angulated posterior margin; posterior angles with one hair, and posterior margin with five hairs on each side, three near the angles and two farther in; broad lateral bands of brown and a broad, pale brown, transverse band in median portion; coxal bands prominent. Legs short and stout, deep smoky brown, with pitchy edging on anterior margins of femora and both margins of tibiae.

Abdomen rather long for the genus, elongate elliptical, with convex lateral margins to segments and projecting rounded posterior angles, furnished with one hair in segments four and five, two in six and seven, one in eighth, and one short one on each tip of the short, indented ninth; a long hair on posterior margins of segments two to six, just within the lateral bands; two short ones in the median portion of posterior margin of segments one to six, and four in the seventh; two in anterior angles of eighth; broad pitchy lateral bands in segments two to seven, separated by clear posterior angles and projecting into anterior adjacent segments; heavy, dark brown, transverse bands on segments one to six, separated by broad clear sutures; clear pustules at spiracles in segments two to seven, with pale spots running inward from them for a short distance; pale brown, median, ventral bands connect the ends of the dorsal bands and form a dark spot on segment six; tip of ninth pale brown.

MALE.—Body, length 1.93 mm., width .81 mm.; head, length .71 mm., width .65 mm.; abdomen much shorter than female and orbicular, with longer, narrower, transverse bands, narrow clear sutures, and the whole median portion obscured by a brown, ventral patch; ninth segment larger, protruding with rounded tip furnished with fourteen long hairs; genitalia short, compact, and very dark brown; tip of ninth segment with a crescent-shaped band.

Two males and two females collected on *Rhamphastos tocard* at Pozo Azul, Costa Rica, June, 1902. A very striking species, easily recognized by the dark conical head, with emarginate clypeus, and by the heavy bands of the abdomen.

Nirmus fuscus epustulatus var. nov.

FEMALE.—Body, length 2.15 mm., width .52 mm.; pale yellow golden, with the usual markings of *fuscus* in the form of heavy complete antennal bands, narrow templar bands, lateral bands on thorax and abdomen, and the heavy, median, transverse bands of the abdomen.

Head, length .58 mm., width .44 mm.

This variety of *fuscus* is recognized at a glance by the absence of clear pustules and fewer hairs on the transverse abdominal bands, there being but four on each segment, while *fuscus* has eight on segments one to five, and six on segments six and seven.

With the exception of the smaller size, there are no other appreciable differences.

Six females collected on *Accipiter bicolor*, at Juan Vinas, Costa Rica, March, 1902.

Nirmus curvilineatus Kell. and Kuw.

Kellogg and Kuwana, *Mallophaga from Birds of the Galapagos Islands.* Proc. Wash. Acad. of Sci., vol. IV, p. 470, pl. XXIX, fig. 4.

This species, described from specimens collected on *Nesopelia galapagoensis* and *Oceanites gracilis* in the Galapagos Islands, was taken in large numbers from several individuals of *Buteo borealis costaricensis*, on the volcano Irazu, April, 1902.

The taking of this species on a *Buteo* complicates still more the already confusing state of its distribution. It seems to me that the present host is more typical of the three, and it leaves room for the query as to whether Mr. Kellogg's specimens might not have straggled from *Buteo galapagoensis*. My specimens agree perfectly with Mr. Kellogg's description and plate in every detail.

Nirmus atopus Kell.

Kellogg, *New Mallophaga* III, p. 18, pl. II, fig. 4.

Numerous males and females collected on *Piaya cayana mehleri*, one male and female on *Myiarchus lawrencei nigricapillus*, and one male and female on *Stelgidopteryx ruficollis uropygialis*, at Juan Vinas, Costa Rica, March, 1902. I am inclined to think that the specimens taken from the last two hosts were stragglers, although there is no direct proof but, since it was described by Mr. Kellogg from a *Piaya* and I took it in such large numbers on a different variety of the same host species, it is probably confined to the Cuculidae.

Nirmus rhamphasti sp. nov., pl. II, fig. 1

MALE.—Body, length 1.40 mm., width .48 mm., short and robust, with pale testaceus head and thorax, heavy antennal bands broken at the suture; abdomen oval, clear, with clear brown lateral bands.

Head, length .44 mm., width .43 mm.; front sharply conical, sides slightly concave, clypeus narrow and emarginate, with one small hair at the suture; trabeculae small, pale golden; antennae short, of median thickness, pale golden, second segment longest; temples large, expanding laterally and posteriorly, with one hair; eye prominent, colorless, with a very short bristle; occiput convex; antennal band broad, smoky brown, curving inward slightly at base, inner margin sinuate, broken by clear clypeal suture, beyond which they are pale testaceus; clypeus and oval fossa clear, except a dusky submarginal band across the tip; mandibles heavy, deep chestnut; a small black ocular fleck and a short brownish ocular blotch; region between pale occipital bands clear; a rather large brownish occipital signature; whole head, except oval fossa and part between occipital bands, an even testaceous.

Prothorax quadrangular, sides flatly convex, without hairs; rather broad, brownish, lateral bands, extending around on posterior portion; coxal bands visible, interior of segment same color as head. Metathorax larger, pentagonal, posterior margin angulated on abdomen, with four slender hairs on each side and two in the rounded posterior angles; sides flatly convex, widely diverging; broad lateral brown bands, curving inward slightly in median portion of segment. Legs short and robust, pale golden, narrowly edged with testaceus on anterior margins of femora and tibiae.

Abdomen, oval, with round, projecting, posterior angles, bearing one hair in segments three and four, two in five to eight, and six on the posterior margin of the ninth; rather heavy, lateral bands of smoky brown in segments one to seven, separated by clear posterior angles, and the portion extending into the adjacent anterior segment, darker; eighth segment with pale brown spot in lateral portions and ninth with posterior half brownish; a median, longitudinal, dusky area, scarcely broken at the sutures and darker and wider on segments five and six; genital hooks very short, slender, and widely curving; a hair on the posterior margin of segments three to six, just within the lateral bands; four short hairs on the median portion of the posterior margins of segments two to six, and two on the first.

Two males collected on *Rhamphastos tocard,* at Pozo Azul, Costa Rica, June, 1902. This species resembles Piaget's *coniceps* more than any other form of the genus.

Nirmus hastiformis sp. nov., pl. II, fig. 2

FEMALE.—Body, length 1.73 mm., width .53 mm.; uniformly golden brown, with rounded forehead, encircled by the antennal bands, and spear shaped abdomen, having narrow lateral bands, scarcely darker than the heavy transverse bands.

Head, length .47 mm., width .45 mm.; front converging, slightly concave sides and evenly rounded clypeus, with a few very short, fine hairs; trabeculae small, pointed, pale golden; antennae of medium size, joints subequal, pale golden; temples slightly expanded laterally and posteriorly, with one long hair and several short bristles; occiput convex; eye small, obscured, with a fine, short bristle; mandibles heavy, bright chestnut; antennal bands, scarcely darker than the golden brown temples, encircle the forehead, but are broken by a clear, diagonal line at the clypeal suture; region in front of the mandible clear, as is the portion enclosed by the occipital bands, and also a small area at the bases of the antennae; a black ocular fleck, and a golden brown, spear-shaped occipital signature.

Prothorax quadrangular, broader than long, with sides flatly convex and posterior angles with one hair; lateral bands scarcely darker than the interior of the segment. Metathorax pentagonal, posterior margin obtusely angled on abdomen, with four slender hairs on each side, and two spines in the rounded posterior angles; sides slightly convex and diverging; lateral bands and interior portion of segment the same color as in the prothorax. Legs stout, though little swollen, clear golden brown.

Abdomen rather long, spear-shaped, with rounded, widely projecting, posterior angles, furnished with from one to three hairs in segments four to seven and one in the anterior and posterior angles, and four on the posterior margin of the eighth; narrow, deep golden brown lateral bands on segments one to seven, widely broken by clear posterior angles, and slightly projecting into the adjacent anterior segments; eighth segment very long, as long as the first, with nearly straight, converging sides and completely obscured by the transverse, golden brown band; ninth segment short, and almost transverse posteriorly, uniformly pale golden, with one short hair on each side of the tip; segments one to seven almost completely obscured by deep golden brown, transverse bands, darker medially and in posterior segments, and separated transversely by clear sutures; a clear pustule at the spiracles on segments two to seven and a single long hair on the posterior margins of segments two to six, just within the lateral bands.

A single female collected on *Trogon caligatus*, at Juan Vinas, Costa Rica, March, 1902. This species is easily recognized by the rounded front, with concave sides, completely encircled by the antennal bands; by the length and shape of the eighth segment of the abdomen and the presence of the broad, transverse abdominal bands.

Nirmus parabolocybe sp. nov., pl. II, fig. 3

Female.—Body, length 1.66 mm., width .40 mm., almost entirely obscured by blotches and bands of smoky brown, with narrow pitchy bands on head, thorax, and abdomen; slender, legs short and stout.

Head long, with narrowly parabolic front, bare, with a slight colorless protuberance at the tip of the clypeus; trabeculae small, colorless; antennae short, slender, second segment the longest, whole pale brownish; eye small, clear, with bristle; temples nearly square, narrowly margined with pitchy; flatly rounded angles, bearing one slender hair; occipital margin truncate; narrow, pitchy, submarginal, antennal bands ending at the clear oral fossa, and with the bases curving diagonally inward at the trabeculae for a distance of one-fourth the width of the head; short, deep brown ocular bands running transversely inward from the eye to base of antennal bands; mandibles small, chestnut; a large smoky brown occipital signature; whole interior of head except oral fossa evenly obscured with smoky brown.

Prothorax, small quadrilateral, anterior margin slightly convex, posterior transverse, sides convex; posterior angles with one weak hair; deep, almost pitchy brown, lateral bands, slightly submarginal, curving around on posterior portion of segment, but separated medially; coxal lines show faintly; interior of segment paler than the head; a small clear pustule in the posterior angles.

Metathorax much larger than prothorax, pentagonal, sharply angulated on abdomen; sides convex, diverging; posterior angles with two slender hairs and two more on each side of posterior margin; a pitchy spot in the anterior angles, a large pitchy blotch running inward from the posterior angles, narrowing and nearly meeting in center; remainder of segment same color as head.

Legs very short and stout; concolorous with head, margined anteriorly with pitchy and femora semiannulated at tips.

Abdomen long, subclavate, pointed abruptly by the last two segments; posterior angles rounded and protruding with a single weak hair in segments three to six, three in segment seven, and three on each side and two on posterior margin of eighth segment; ninth segment small, colorless, indented at tip; segments one to seven with deep, smoky brown lateral bands having a pitchy hue through the middle and extending anteriorly into adjacent segments; smoky brown transverse bands on segments one to eight, separated by broad clear sutures and having in segments two to seven, two median longitudinal rows of darker, quadrilateral blotches, joined on segments six and seven; large clear pustules on segments two to seven, just within lateral bands

Male.—Body, length 1.34 mm., width .32 mm.; head, length .32 mm., width .25 mm.; differs from the female only in size and in the color and shape of the eighth and ninth abdominal segments, the eighth being clear, except for a narrow brown band around the anterior margin, while the ninth is very small and obscured with brownish; genital hooks small, short, and blunt, being only the length of the eighth abdominal segment.

Numerous males and females collected on *Muscivora tyrannus* and *Tyrannus melancholicus*, at Juan Vinas, Costa Rica, March, 1902. This is quite a distinct form, though resembling somewhat in shape of head and abdominal markings *N. angustifrons* Car.

Nirmus marginellus N., pl. II, fig. 4

Giebel, *Insecta Epizoa,* p. 147.
Piaget, *Les Pediculines,* Sup., p. 21, pl. III, fig. 1.

FEMALE.—Body, length 2.13 mm., width .74 mm.; head, length .60 mm., width .60 mm.

MALE.—Body, length 1.90 mm., width .72 mm.; head, length .56 mm., width .56 mm. This species is of the *interrupto fasciati* type, and is distinguished by the shortness or rather by the width of the head and body, the head being an almost perfect equilateral triangle, while the abdomen is a perfect oval, a shape unusual among this group; the head is much darker than the body, being a translucent reddish brown, with heavy chestnut antennal bands; abdomen very pale with reddish fulvous lateral bands and very faint golden transverse bands.

Numerous males and females collected from *Momotus lessoni,* at Juan Vinas and Pozo Azul, Costa Rica, March and June, 1902.

From Giebel's description alone the species is unrecognizable, but Piaget has thoroughly established it from specimens collected on *Prionites momota.*

My specimens agree quite closely with his descriptions and plate with the exception of the head measurements of the male, my specimen being smaller than the female, while he gives the head measurements the same for the two sexes. It is possible that he may have made an error in this point.

Nirmus francisi sp. nov., pl. II, fig. 5.

FEMALE.—Body, length 2.19 mm., width .63 mm.; absolutely colorless, with bold markings of pitchy, and smoky brown on head, thorax, and abdomen.

Head, length .59 mm., width .52 mm.; sharply conical from trabeculae forward, with sides concave, and the narrow tip emarginate; three short, fine hairs along sides; trabeculae medium, colorless, equal to the last three, which are deep brown; antennae slender, first two joints longest, eye prominent, colorless, with a short bristle, and slightly obscured by a pitchy ocular blotch; temples almost square, with rounded sides and angles, colorless with one hair in the angle; occipital margin nearly truncate, occiput slightly convex; marginal, antennal bands, broken at the clypeal suture, run backwards from the sides of the clear oral fossa nearly to the trabeculae, then straight inward to the bases of the posterior roots of the mandibles; the portion in front of the clypeal suture is clear brown, back of that, brown, with a pitchy stripe along inner portion, and pitchy after leaving the margin; occipital bands of deep brown, fading outwardly, and not reaching occipital margin, join the posterior ends of antennal bands; mandibles chestnut; a smoky brown occipital signature.

Prothorax small, quadrilateral, with anterior and posterior margins straight, lateral margins convex, and one weak hair in posterior angles; pitchy, lateral, submarginal bands, curving part way around on posterior margin, and having a short projection into the anterior angles of the metathorax; brown coxal bands show through.

Metathorax larger than prothorax, pentagonal, sides convex and widely diverging, with posterior margin sharply angulated on abdomen; a series of six slender hairs on each side of posterior margin; heavy pitchy transverse bands in the region of posterior angles, with a projection into the first segment of abdomen; a narrow, submarginal, chestnut band along posterior margin. Legs short and stout, with pitchy spot at base, and semiannulation at tip of femora, and brownish spot at outer side of tip of tibiae.

Abdomen narrowly oval, with rounded, projecting, posterior angles, furnished with one hair in the third segment, and three in four to eight, with two others in anterior angles and four on posterior margin of eighth; ninth segment small, indented at tip, with two chestnut blotches in sides; segments one to eight with narrow, lateral, pitchy bands, separated by clear posterior angles, and projecting slightly into anterior adjacent segments; two median longitudinal rows of deep, smoky brown blotches in anterior portion of segments, with a latero-posterior projection which reaches the lateral margins of abdomen in segments five to eight; spots joined in the eighth segment and connected in remainder by a pale brown, ventral, transverse band, which is expanded into a continuous pale blotch on segments six to seven.

MALE.—Body, length 1.63 mm., width .57 mm.; head, length

.57 mm., width .48 mm.; abdomen shorter, nearer a perfect oval; ninth segment flatly rounded, with a fringe of long hairs; eighth segment with only a narrow brown band around anterior margin, lateral bands absent; genital hooks short, stout, and curving at tips.

Numerous males and females collected on *Zarhynchus wagleri* at Juan Vinas, Costa Rica, March, 1902. This is a very striking and beautiful form, something like the type of *N. ornatissimus* Giebl, a type found on many *Icteridae.*

Nirmus melanacocus sp. nov., pl. II, fig. 6

FEMALE.—Body, length 1.65 mm., width .46 mm., very pale testaceous, almost clear, except the head, which is translucent brownish, with pitchy antennal and templar bands; abdomen clavate, with pitchy lateral bands and brownish, median, transverse bands.

Head, length .38 mm., width .39 mm.; almost an equilateral triangle, with slightly convex sides and truncate occipital margin; clypeus narrowly truncate; trabeculae small, colorless; antennae of medium size, pale testaceus, second segment the longest, third and fourth equal; eye flatly convex, colorless, with a small bristle; angles of temples almost square, with one slender hair; heavy pitchy brown antennal bands not quite reaching the clear oral fossa, with the fossa bending inward slightly at the trabeculae; mandibles long and slender, chestnut; temples margined laterally with narrow pitchy bands, connected with the antennal bands by pale brown bands; whole head, except oral fossa and region enclosed between the pale occipital bands, an even clear brown; a large, deep brown occipital signature.

Prothorax small, quadrangular, with convex sides and a single weak hair in posterior angles; broad pitchy brown lateral bands, curving around on posterior margin; coxal bands visible. Metathorax larger, pentagonal, with convex, diverging sides, and sharply angulated posterior margin, having five weak hairs on each side; a pitchy brown spot in anterior angles, connected with lateral bands of prothorax; pitchy brown transverse blotches in region of posterior angles; coxal lines visible, interior of segment clear. Legs short and stout, narrowly margined with brownish on femora, concolorous with head.

Abdomen clavate, widest at sixth segment; posterior angles of segments rounded and protruding, with one hair in third, and two in segments four to seven; eighth segment with three hairs on each side; ninth small, clear, and indented at tip; segments one to seven with pitchy lateral bands, slightly paler along outer margins, separated by clear posterior angles and projecting into adjacent anterior segments: segments one to eight with broad, median, transverse bands of pale, smoky brown, separated by clear sutures except between segments six and seven; the band in the eighth segment longer and narrower, nearly reaching the lateral margins of segment.

A single female collected on *Piranga bidentata sanguinolenta* taken on the volcano Irazu, February, 1902.

This form resembles *N. ampullatus* Piag., and is distinguished from that species chiefly by the narrower clypeus, heavier antennal bands, and smaller legs.

Nirmus pseudophaeus sp. nov., pl. III, fig. 1

FEMALE.—Body, length 2.18 mm., width .60 mm.; almost exactly the shape of *N. fuscus*, pale translucent, smoky brown, much darker on head; antennal and temporal bands on head much as *fuscus*, but lateral bands of abdomen wanting.

Head, length .60 mm., width .44 mm.; shape and markings practically the same as in *fuscus*, except that they are a smoky brown instead of golden brown, while the interior is paler and also smoky brown; temples with one long hair.

Prothorax quadrilateral, with sides slightly sinuate and one short hair in the posterior angles; very pale, indistinct lateral bands. Metathorax larger, with slightly expanded anterior angles, sides rounding and diverging, posterior margin truncate, with a short, median, pointed projection on abdomen; posterior angles rounded with one slender hair: four more slender submarginal hairs along lateral portion of posterior margin; a pale brown, median, anterior blotch, about in the region of the mesothorax; a darker band across posterior portion of segment, more intense in region of posterior angles; posterior margin with a clear border. Legs stout, with femora narrowly edged with dusky anteriorly.

Abdomen oval, widest at fourth and fifth segments; posterior angles acute, but scarcely projecting, with one hair in third and fourth segments, two in five and six, three in seven, two on each side, and two longer pustulated ones on posterior margin of eighth, and a pustulated one in the anterior angles of the ninth; lateral bands absent; clear, pale, brown bands across the posterior portion of segments one to eight, darker, and extending to anterior margin of segment in the median portion, and separated transversely by broad clear sutures; ninth segment small, rounded, indented at the tip and pale brown; two long dorsal hairs in the median anterior portion, and a transverse row of six across the median portion of first segment; segments two to seven with a median transverse row of nine slender hairs, all with only very faint, small pustules or none; clear pustules in regions of spiracles, on segments two to seven.

A single female collected on *Pezopetes capitalis*, on the volcano Irazu, February, 1902.

The finding of this *fuscus*-like species on a *Tanagradae* is about as inexplicable as the taking of *N. curvilineatus* on *Nesopelia galapagoensis*, but there was absolutely no chance of straggling from a Raptor, since none were killed in that locality on that trip. The present species is easily recognized by the *fuscus*-like head, the absence of lateral abdominal bands, and the pale continuous transverse bands.

Nirmus brachythorax ptiliogonis var. nov.

> *Nirmus brachythorax* Gieb., *Insecta Epizoa*, p. 134.
> *Nirmus brachythorax* Piaget, *Les Pediculines*, p. 150, pl. XII, fig. 8.

FEMALE.—Body, length 1.61 mm., width .38 mm.

These specimens really come much nearer to Piaget's *N. brachythorax cedrorum* than to *brachythorax*, but since that form is a variety, I will make this another variety, but will compare it with the var. *cedrorum*.

No measurements are given for the variety, but this form is much larger than *N. brachythorax cedrorum*, having a broader head, occipital margin squarely truncate, temples more nearly square, and angles less rounded; abdomen paler, lateral bands pale fulvous instead of deep brown and median bands barely noticeable.

Male.—Body, length 1.33 mm., width .40 mm.; head, length .33 mm., width .33 mm.

Very abundant on *Ptiliogonys caudatus*, on the volcano Irazu, Costa Rica, February and April, 1902.

Nirmus caligineus sp. nov., pl. III, fig. 2

Female.—Body, length 1.83 mm., width .59 mm.; rather short and stout, pale golden brown, with darker, smoky brown head, smoky brown lateral and transverse bands on abdomen, separated by clear sutures.

Head, length .53 mm., width .51 mm.; sharply conical, with convex sides, bearing four fine, short hairs, and narrow truncate clypeus; trabeculae small, colorless; antennae of medium size, tinged with brownish, second joint the longest; temples broad, narrowly margined with pitchy brown, with evenly rounded angles, bearing one long, pustulated hair; eyes small, with a short bristle; occipital margin concave, occiput slightly convex; rather narrow, smoky, antennal bands ending at the clypeal suture; paler internal bands enclosing the oral fossa, running backward from the lateral angles of the clypeus, converging slightly in middle and spreading out in front of the mandibles; mandibles large, chestnut; narrow occipital bands, but slightly darker than the clear brown temples, running from the posterior roots of the mandibles to the occipital margins; bases of antennal bands connected with occipital bands by a very faint band; small, semi-clear areas at the bases of the antennae, between the antennal and internal bands, and between the occipital bands; a brownish, oval, occipital signature.

Prothorax small, quadrangular, with anterior margin concave, posterior flatly rounded, and sides flatly convex and diverging; posterior angles with one short hair; submarginal, deep brown, lateral bands curving around on the posterior margin; narrow brown lines cutting across from the middle to ends of lateral bands; brown coxal lines visible; interior of segment golden brown, margins smoky brown. Metathorax much larger than prothorax, pentagonal, posterior margin sharply angulated on abdomen, sides nearly straight, widely diverging and posterior angles broadly rounded, with two hairs; five hairs on each side of posterior margin; a pitchy brown spot in anterior angles; a deep brown lateral band, broadening rapidly posteriorly, with a pitchy projection from posterior angles towards middle of segment; meso-coxal lines visible; interior of segment golden brown. Legs short and stout, concolorous with body.

Abdomen elliptical, with rounded, projecting posterior angles bearing one hair in the third segment, two in the fourth, fifth, and seventh, four in sixth, two on each side and four on the posterior margin of eighth; narrow lateral bands of smoky brown on segments one to seven, broken by clear posterior angles and projecting into anterior adjacent segments; broad, brownish transverse bands on segments one to eight, darker on posterior segments, separated by broad, clear sutures and having two median longitudinal rows of larger quadrilateral spots on segments two to seven; eighth continuous; ninth segment small, deeply indented at tip, and with brown spots on each side; a long hair on the posterior margin of segments three to six; a median row of six hairs along posterior margin of segments one to six and two on segment seven.

MALE.—Body, length 1.49 mm., width .55 mm.; head, length .49 mm., width .46 mm.; slightly darker than female with the abdomen oval, ninth segment rounded, with about ten fine hairs on posterior margin.

One male and one female collected on *Merula grayi*, at Juan Vinas, Costa Rica, March, 1902. This form resembles Kellogg's *N. simplex*, but differs principally in the longer and narrower clypeus and the duskiness of the head and thorax.

Lipeurus longipes tinami, var. nov., pl. III, fig. 3

Lipeurus longipes Piaget, *Les Pediculines,* p. 329, pl. XXVIII, fig. 3.

MALE.—Body, length 1.90 mm., width .35 mm.; head, length .48 mm., width .40 mm.

This species is easily recognized by the conical head with truncate front and heavy antennal bands; by the short weak anterior legs and enormously developed middle and posterior pairs, the posterior ones being as long as the abdomen; pro- and metathorax quadrangular, with blackish lateral bands; abdomen slightly clavate, with deep brown lateral bands and pale brown transverse bands.

The present variety differs from *longipes* chiefly in the size and shape of the head, the head measurements for *longipes* being .44 mm. by .33 mm., while the body is practically the same; this form has the head in front of the antennae wider and shorter, with the front broader and perfectly truncate, while it is emarginate in *longipes.*

FEMALE (not seen by Piaget).—Body, length 2.15 mm., width .48 mm.; head, length .52 mm., width .44 mm.; whole head darker, especially the temples; antennae scarcely different from the male, except in the absence of the appendage on the third segment and the last three segments being subequal; anterior pair of legs and thorax the same, except a trifle larger; two posterior pairs of legs slightly smaller; abdomen much larger, slightly clavate, but not pointed so abruptly; lateral bands narrower and darker; in addition to the continuous pale transverse bands, which are about the same as in the male, are heavy, translucent, brownish, quadrilateral bands, extending inward from the lateral bands a trifle more than one-third the width of the abdomen; on the last segment this band is continuous; apical segment tapering gradually to the slightly indented tip, which is furnished with two long hairs on each side, two short ones at the tip, and two short ones on each side in the anterior portion; a large median ventral blotch of pale brown extends from the middle portion of segment five to middle of segment eight.

Two males and two females collected on *Tinamus robustus,* at Pozo Azul, Costa Rica, June, 1902.

Lipeurus longisetaceus Piag.

Piaget, *Les Pediculines,* Sup., p. 57, pl. VI, fig. 4.

MALE.—Body, length 1.87 mm., width .33 mm.; head, length .40 mm., width .25 mm.

FEMALE.—Body, length 2.41 mm., width .32 mm.; head, length .56 mm., width .25 mm.

This species, described by Piaget from specimens collected on *Tinamus solitarius,* is distinguished by its exceedingly slender form and nirmoid appearance; the head is conical with straight sides and rounded front and concave occiput; the clypeus is distinctly set off and pale reddish brown, with two slender fleshy appendages on the dorsal surface; the antennal bands are heavy, while the temples are margined with blackish; mandibles large; whole interior of the head clear; thorax and abdomen with heavy lateral bands, interior clear.

Two females and one male collected from *Tinamus robustus,* at Pozo Azul, Costa Rica, June, 1902.

Piaget gives such a clear description and figure that a drawing in this paper is unnecessary.

Lipeurus postemarginatus, sp. nov., pl. III., fig. 4

FEMALE.—Body, length 2.21 mm., width .59 mm.; clear, elongate; head broadly parabolic in front with clear brown temples, antennal bands encircling front and pitchy at base; abdomen spindle shaped, with pitchy brown lateral bands and clear brown transverse bands.

Head, length .51 mm., width .41 mm.; slightly conical, broadly parabolic in front, with five short hairs on each side, trabeculae small, clear; antennae rather short and slender, first segment thickest and subequal to second and fifth, third and fourth shorter, equal, the whole pale fulvous; one short hair in front of the trabeculae; eye prominent, colorless, with a short bristle; temples slightly expanded, evenly rounded, with one long hair and several short bristles and narrowly margined with deep smoky brown; occiput concave; whole temple uniformly clear brown inward to the clear occipital bands, and separated from the base of antennal bands by a transverse clear space; mandibles large, chestnut; the broad antennal bands completely encircle the front, broken at the clypeal suture, front part deep fulvous, sides darker, with the rounded bases pitchy; a paler internal band, parallel to the lateral portion of the antennal band; interior of head clear.

Prothorax small, quadrilateral, with sides convex and diverging; anterior margin slightly concave, posterior truncate; lateral bands of brown, widening posteriorly, with a pitchy blotch in region of anterior angles; a single hair in the posterior angles; coxal lines showing through.

Metathorax much larger than prothorax, quadrilateral, anterior and posterior margins truncate; sides convex, diverging, and slightly sinuate in region of mesothoracic suture; broad lateral bands of deep smoky brown, with a curving pitchy blotch in the median portion of the sides; a large, median, ventral dusk spot: two short hairs in the posterior angles and four very long, stout hairs on each side of the posterior margin, adjacent to the lateral angles. Legs rather small and weak, pale fulvous with a few short hairs.

Abdomen large, spindle shaped, widest at the fourth and fifth segments; posterior angles rounded, protruding, furnished with one short hair in segments one to four, one short and one long one in five to seven; eighth segment conical, with a conspicuous indentation at the tip, one long hair in the anterior angles and two large, pustulated hairs in the median portion; narrow pitchy brown, lateral bands broken by clear posterior angles; a clear pustule in the middle of segments two to seven, just within the lateral bands; clear, pale brown transverse bands, paler on first three segments, extend inward from the lateral bands nearly to the center of the abdomen, broken by broad clear sutures; pale brown ventral bands connect the ends of these heavier dorsal bands; a long hair arises from the posterior margin of the transverse bands just inside the lateral bands, on segments three to seven; a row of six short dorsal hairs across the middle of segment five, and two in the middle of segments six and seven; a heavy and continuous transverse band on segment nine, with another paler band encircling the posterior emargination.

Male.—Body, length 1.46 mm., width .38 mm.; head, length .38 mm., width .33 mm.; differs greatly in size from the female, shape of abdomen, size of legs, and antennae.

Head markings the same; first segment of antennae greatly swollen and lengthened, remaining segments subequal, but slightly larger than in female and third having a slight protuberance; two posterior pairs of legs larger than in female, the posterior pair being as long as the abdomen; abdomen short, thick, almost parallel sided and abruptly pointed by the last three segments, which are much aborted; transverse bands continuous and paler.

One male and one female collected from *Ortalis cinereiceps* at Juan Vinas, Costa Rica, March, 1902.

This form is of the type of *mesophelius* and *intermedius*, resembling most *intermedius*, from which it is distinguished by the much shorter metathorax, with truncate posterior margin, by fuller and rounder temples, and by paler abdomen.

Lipeurus assesor Gieb., pl. III, fig. 5

Giebel, *Insecta epizoa*, p. 207.
Piaget, *Les Pediculines*, p. 294, pl. XXIV, fig. 3.

Female.—Body, length 3.45 mm., width .63 mm.; head, length .76 mm., width .50 mm.

This species is easily distinguished by the long, almost straight sided head, with rounded front, and concave occiput; the peculiar curving marks on each side of the front, the black ocular blotches and dusky temples in strong contrast to the clear interior of the head; the thorax has heavy, lateral, blackish bands, while each segment of the abdomen is furnished with broad, black, lateral blotches and dusky transverse bands.

A single specimen collected from *Gypagus papa* at Pozo Azul, Costa Rica, June, 1902.

Although Piaget gives only a drawing of anterior half of head and last two segments of abdomen, his description is good and, together with Giebel's, leaves no doubt but what this specimen can be referred to Giebel's species.

Goniocotes eurysema sp. nov. pl. III, fig. 6

FEMALE.—Body, length 1.93 mm., width .85 mm.; clear fulvous, with head large, temples produced laterally and angulated behind; abdomen oval, with pale brown lateral bands and narrow, pitchy, submarginal bands.

Head, length .53 mm., width .76 mm.; broadly and rather flatly rounded in front, with slight depressions at the point where the antennal bands touch the margin, trabeculae absent; antennae short, slender, first two segments the longest and equal, last three shorter, equal; temples expanded laterally into a rounded protuberance, bearing two long hairs; posterior portion slightly extended and sharply angulated; occiput deeply re-entering, convex; a broad brownish band running around the front of the head; short, pitchy bands run from the base of the mandibles to the anterior margin at the depression; just within these bands are chestnut-colored protuberances, running a short distance backward from the frontal band; pitchy ocular blotches; a slightly dusky, narrow band runs around the temples from the eye to the posterior angles; heavy pitchy occipital bands run backward from the posterior roots of the mandibles to the sides of the occiput, thence spreading out into a narrow occipital margin; regions between ocular blotches and antennal bands, dusky brown.

Prothorax small, quadrilateral, with anterior and posterior margins slightly concave; sides nearly straight, diverging; posterior angles produced backward into a slender rounded protuberance, bearing one long hair; heavy, pitchy, submarginal bands run backward from the anterior angles to the anterior angles of metathorax; a portion of the coxal lines showing through; median portion of segment clear, region between lateral bands and margin dusky. Metathorax short, transverse, with lateral margins convex and posterior margin slightly angulated on abdomen; heavy, curving, pitchy bands running backward from anterior angles, extend half way across the first segment of the abdomen; lateral margins with two long hairs; a transverse chestnut band across the anterior portion, and another, broken in the center, across the posterior portion of the segment. Legs short and stout, typical of the genus, with pale fulvous edgings on the anterior faces of femora.

Abdomen oval, slightly clavate; posterior angles rounded, projecting, with from one to three short hairs; segments one to seven with rather broad, smoky fulvous, lateral bands, whose anterior portion extends inward, forming a rounded, backward-curving protuberance, almost obscured by the narrow, pitchy, transverse bands, widening and fading inwardly; eighth segment clear, rounded and indented, bearing one and two short hairs on each side of the tip.

MALE.—Body, length 1.38 mm., width .64 mm.; head, length .43 mm., width .62 mm.; differs from female in having long pitchy ocular blotches extending backward from the eye; a shorter, thicker abdomen, more abruptly rounded posteriorly, with the last segment smaller, less protruding, with a fulvous band around the flatly rounded posterior margin; genital hooks long, slender, and almost parallel.

Numerous males and females collected on *Odontophorus guttatus*, on the Volcano Irazu, Costa Rica, April, 1902. It is of the general type of *G. major* Piag., but it is easily distinguished from that species by the heavy occipital bands and the wide diverging prothorax.

ORNICHOLAX nov. gen.

Body short, compact, head large, and with the general appearance of Goniocotes; antennae small, without appendages, and similar in the two sexes; trabeculae large, triangular, movable; prothorax small, short; mesothorax large, broad as head, and separated from metathorax by a very distinct suture; metathorax much narrower than the mesothorax and plainly divided into two lobes by a longitudinal, clear suture; abdomen of both sexes with lateral bands, with but eight segments and with the seventh much aborted; legs very short and stout; dorsal surface of the thorax and abdomen thickly and deeply punctured.

Found as yet only on *Tinamus robustus*, but is probably common to the Crypturi.

Ornicholax robustus n. sp., pl. IX, figs. 1-1c

Two male specimens and one female taken on *Tinamus robustus*, at Pozo Azul, Costa Rica, June, 1902. This is a strikingly different form from anything so far described, enough so, indeed, to make it the type of a new genus.

Female.—Body, length 2.85 mm., width 1.35 mm. With the exception of the central portion of the abdomen, it is uniformly pale brown, with darker reddish-brown markings. Head, length .86 mm., width .90 mm., somewhat shield-shaped, rounded in front with four short hairs on each side; sides nearly straight; temples obtusely angled, with two longish hairs; occipital margin convex, with two stiff hairs on each side; occiput deeply concave; trabeculae nearly twice the length of the first segment of the antennae, triangular with a darker band along the lateral margin; antennae short, concolorous with head, first segment longest, second and fifth, and third and fourth equal; antennal bands extending in a curve from the anterior point of the trabeculae to the base of the mandibles; a narrow, transverse serrated band across the posterior portion of the head; a dark occipital signature between the occipital margin and the transverse occipital band; occipital bands extending from the base of the posterior root of the mandibles to the transverse occipital band; a narrow transparent lobe extends along the lateral borders of the temples.

Prothorax short, narrow, with lateral angles produced to a blunt protuberance, furnished with a short stiff spine; with the exception of the posterior median portion it is completely encircled by a broad, reddish-brown band.

Mesothorax broad, lateral portion expanded anteriorly; lateral angles blunt, furnished with two stout hairs; postero-lateral portion slightly concave, with two stout hairs towards the angle; the portion touching the prothorax broadly bordered, while two longitudinal curving bands extend across the segment from the ends of the lateral bands of the prothorax, enclosing a median clear spot; posterior margin transverse, slightly concave.

Metathorax scarcely wider than the prothorax, somewhat triangular in shape, but completely divided longitudinally by a clear suture; without hairs and with a broad, dark, slightly curving band extending across the anterior portion and continuous with the lateral bands of the abdomen.

Abdomen short, oval, with the first segment much wider than any of the others, and the seventh aborted (not extending to the lateral margin of the abdomen); lateral margins of the segments convex and the posterior angles of first and second segments with one short hair; the third, fourth, and fifth with two hairs, the sixth with three, and the eighth with a fringe of about twelve hairs on each side; tip of eighth segment slightly indented; a broad submarginal, darker band completely encircling the abdomen, with the enclosed portion much clearer than the remainder of the body; a stout, slightly pustulated hair on the posterior margin of segments one to five, just inside the lateral band; seventh segment with a fringe of about twelve short, weak hairs along the posterior margin. Legs short and stout; tarsi aborted; tibiae almost as large as femora; claws long and stout.

The male is slightly smaller than the female, measuring, length 2.41 mm., width 1.14 mm. Head, length .77 mm., width .83 mm.; the hairs of the temples, thorax, and abdomen longer and stouter than in the female; abdomen more nearly orbicular; seventh segment of the abdomen appearing as two lobes, one on each side of the eighth, which is narrower than in the female, with sides deeply emarginate and tip deeply concave; the fifth segment with three hairs in the posterior angle; the sixth with four, the seventh with six, and the eighth with three short hairs in each angle and four slightly longer, submarginal ones on each side of the middle; the genitalia are simple, curving slightly inward, and about half the length of the eighth segment.

KELLOGGIA nov. gen.

Body short, compact, and with the general appearance of Goniocotes, with the exception of the thorax; head of medium size, thorax small; antennae small, without appendages and similar in the two sexes; trabeculae entirely absent; whole thorax small, much smaller than the first segment of the abdomen; meso- and metathorax separated by a distinct suture; metathorax narrower than mesothorax, and completely divided into two lobes by a longitudinal suture; abdomen differing greatly at the tip in the sexes; female with seven segments, male with eight but with the seventh aborted; lateral bands present in both sexes; legs short and stout; dorsal surface of thorax and abdomen coarsely punctured.

Found as yet only on *Tinamus robustus.*

Kelloggia brevipes n. sp., pl. I, figs. 2-2c

Two adult females and five adult males taken on *Tinamus robustus* at Pozo Azul, Costa Rica, June, 1902.

FEMALE.—Body, length 2.22 mm., width 1.05 mm.; with the exception of the central portion of the abdomen, it is a uniform testaceous with darker smoky brown markings on head and thorax.

Head, length .73 mm., width .73 mm., triangular and about equilateral, narrowly rounded in front, without hairs; sides of head from antennae to temples perfectly straight, with two short marginal and one submarginal hair; temples rounded with two long stiff hairs; occiput deeply concave with three short stiff hairs on each side; antennae short, simple, and inserted near the front of the head, first joint largest, second and fifth equal, and third and fourth equal; no trabeculae; mandibles prominent, dark colored antennal bands running in a slight curve from the base of the mandibles to the frontal margin at the sides, thence [illegible] across the front of the forehead; occipital bands faint, running backward from the posterior root of the mandibles to [illegible] of the prominent occipital signature.

Prothorax short, thick, quadrangular, with an emargination [illegible] anterior angles and the whole segment deeply inserted [illegible] occipital margin; posterior angles drawn out to a blunt [illegible] with a short, stout bristle; margin encircled by a [illegible] band, narrower along the posterior margin; coxal lines [illegible] through.

Mesothorax much broader than prothorax; lateral portions slightly expanded forward and with an emargination just in [illegible] the lateral angles; have dark submarginal bands ex- [illegible] side of the segment, with inner [illegible] extending slightly into the metathorax; [illegible] with a long stout bristle and two stout hairs; the median posterior margin rounded; the postero-lateral margins concave, with two stout hairs; metathorax much narrower than mesothorax, extending completely over the first segment of the abdomen; divided into two distinct lobes by a longitudinal suture; posterior tips of lobes bluntly pointed.

Abdomen short, thick, narrowing gradually from first segment to the rounded point; first segment much broader than the others; segments two to six projecting under adjacent anterior segments at their lateral portions; segments one to six with broad lateral bands, seventh entirely colored, while the median portion is almost entirely uncolored; segments one to five with one long and one short hair in the posterior angles, segment six with three hairs and segment seven with four weak, slender hairs on each side of tip; a long, stout dorsal hair on each side of the posterior margin of segments one to five, nearly to the inner margin of the lateral bands; dorsal surface of thorax and abdomen strongly, though not closely, punctured.

Legs short and stout, concolorous with body.

The male differs slightly from the female, especially in the tip of the abdomen.

Measurements of male, length 1.76 mm., width .88 mm.; head, length .6 mm., width .65 mm.; the hairs of the thorax and abdomen are longer and stouter; sixth segment of abdomen very wide (longitudinally), almost as wide as the first; seventh segment much aborted, appearing as two lobes, one on each side of the eighth, which is deeply inserted into the sixth; segments four to seven with three hairs in the posterior angles; eighth with one very long and two short hairs at each angle, and the tips slightly indented.

Goniodes minutus sp. nov., pl. IV, figs. 1 and 2

MALE.—Body, length .96 mm., width .49 mm.; short and robust, head large, translucent golden brown throughout, with darker smoky brown frontal band, and lateral bands on thorax and abdomen.

Head, length .28 mm., width .39 mm.; quadrilateral, front broad and flatly rounded; sides nearly straight, slightly diverging with three short bristles; temples bluntly rounded, with one long stout bristle and a spine; occiput concave, with two short bristles on each side; a narrow dark brown band around front, with eight small lobate posterior projections; antennae of medium size, but first joint greatly swollen, ovoid, second much smaller, third emarginate on one side and pointed, with fourth arising from the emargination, fourth and fifth larger than third, subequal; mandibles rather small, chestnut.

Prothorax large, almost bell-shaped, projecting under the head, with anterior and posterior margins concave, posterior angles rounded, with a short spine; wide lateral bands of reddish brown. Metathorax about as long as prothorax and wider; with broadly rounded posterior margin; anterior angles rounded, with two stout bristles and a spine; three stout hairs in the lateromedian portion and two shorter ones on each side of median posterior border; wide lateral bands of reddish brown, coxal bands visible. Legs short and stout, concolorous with body.

Abdomen broadly oval, with lateral angles scarcely protruding, and furnished with a stout spine; three stout hairs in the angles of the sixth segment and three on each side of the tip of seventh; two short hairs on each side of the middle portion of the posterior border of segments one and two, and one hair on the posterior margin of segments one to three just within the heavy lateral bands of reddish brown; broad transverse bands of golden brown, separated by narrow, clear sutures, wider between segments four and five and five and six; segment seven entirely the color of lateral bands and deeply inserted into the sixth; genital hooks short, stout, and simple.

FEMALE.—Body, length 1.11 mm., width .52 mm.; head, length .25 mm., width .43 mm.; temples more projecting, occipital margin less concave; antennae simple, first, second, and fifth joints longest, subequal; third and fourth shorter, equal; abdomen narrower, apical segment protruding, rounded at tip; markings and color same as in male.

Numerous males and females collected on *Tinamus robustus* at Pozo Azul, Costa Rica, June, 1902. A very distinct species, having little resemblance to any species heretofore described.

Goniodes laticeps Piag., pl. IV, fig. 3

Piaget, *Les Pediculines*, p. 259, pl. XXI, fig. 6.

This striking form is easily recognized by the broad head-shield and the peculiar large posterior tibiae, edged with chestnut and fringed with fine hairs on both sides. The female was not seen by Piaget, and a drawing of it is given here.

Numerous males and females collected on *Tinamus robustus*, at Pozo Azul, Costa Rica, June, 1902.

Goniodes aberrans sp. nov., pl. IV, figs. 4 and 5

MALE.—Body, length 2.00 mm., width .68 mm.; abdomen spindle shaped, pointed posteriorly, with heavy transverse bands of smoky golden brown; head deeply constricted back of the antennae, and temples enormously developed posteriorly into a slender, almost pointed process.

Head, length .57 mm., width .70 mm.; front narrow, flatly rounded, with two median, submarginal hairs; antennae placed at the very front of the head, with first joint greatly swollen, second smaller, truncated-conical, third slender, with fourth arising near its base, fourth and fifth slender subequal; a marked constriction behind the antennae, at which point the head is scarcely wider than the length of the first segment of the antennae; a long, stout bristle just in front of the constriction and a shorter one on the first segment of the antennae; behind the constriction the sides of the head diverge widely with convex margins to the bluntly pointed temporal angles, which are furnished with two long, stout bristles; whole posterior margin of head deeply and regularly concave, with two short marginal hairs, and one submarginal on each side of the occiput; a deep chestnut, occipital border, curving forwards at the ends; narrow chestnut antennal bands curving inward from sides of front to bases of mandibles; mandibles small, chestnut; whole head and antennae an even, slightly smoky, golden brown.

Prothorax large, flatly dome-shaped, with the whole border anterior to the posterior angles forming a regular, almost half-circle; posterior angles almost right angles, furnished with one stout bristle; posterior margin flatly convex; narrow, deeply submarginal, lateral bands of dark brown; whole of lateral regions slightly darker than median portion, which is smoky golden brown. Metathorax wider and shorter than prothorax, with rounded anterior angles and transverse posterior margin, slightly angulated medially on abdomen; three long stout bristles in the region of the slightly obtuse posterior angles, and one shorter one farther in towards the middle; narrow, semicircular bands start from the median, anterior portion and pass outward and backward across the segment and half way across first abdominal segment; whole segment uniformly smoky, golden brown. Legs extremely short, femora swollen, but concealed under body, only the short parallel sided tibiae projecting.

Abdomen spindle shaped, and pointed posteriorly, with posterior angles scarcely projecting and furnished with two stout hairs in segments one to five, sixth with four, and seventh with three on each side of tip; six hairs along the posterior margin of segments one to five; (the ventral portion of the abdomen is furnished with hairs very similar to the dorsal portion, and the body is so thin and translucent that care must be taken not to confuse the ventral with the dorsal hairs) all the segments with heavy, smoky, golden brown transverse bands, darker in posterior portion and separated transversely by rather broad, clear sutures; seventh (apical) segment deeply inserted into the sixth, with pointed tip and sides slightly concave; genitalia long, with posterior half slender and tapering to a point, with the slender portion projecting from abdomen.

FEMALE.—Body, length 2.11 mm., width .73 mm.; head, length .63 mm., width .79 mm.; body about the same shape as male; head without the lateral constrictions, front more rounded; antennae shorter, simple, and length of joints much as in *Nirmus*; a large, fleshy ovipositor with numerous stout hairs along sides and on tip protrudes from the tip of the abdomen.

Two females and one male collected on *Tinamus robustus*, at Pozo Azul, Costa Rica, June, 1902.

This is a very distinct and striking form of *Goniodes*, and I am a little doubtful as to whether it really belongs there. The spindle shaped abdomen, with continuous transverse bands and the peculiar sexual organs are very aberrant for this genus.

Goniodes longipes Piag.

Piaget, *Les Pediculines*, p. 253, pl. XX, fig. 7.

A single female specimen of a *Goniodes* was collected on *Odontophorus guttatus*, volcano Irazu, Costa Rica, April, 1902, in company with several individuals of an undescribed species of the same genus. There can be no doubt about its being this form, for it agrees exactly with Piaget's description and plate, although the specimens from which he described it were taken on quite a different host.

Laemobothrium delogramma sp. nov., pl. IV, fig. 6.

FEMALE.—Body, length 10.00 mm., width 2.65 mm.; deep smoky brown, with pitchy markings on head and thorax, pitchy edgings on legs and pitchy lateral bands on abdomen; abdomen with a series of dorsal hairs on each segment arising from clear pustules.

Head, length 1.84 mm., width 1.65 mm.; slightly conical, truncate clypeus and sides straight, interrupted by the antennal swellings; palpi projecting by last two segments; temple produced to a blunt point posteriorly; occiput transverse with numerous short bristle-like hairs, two larger ones at the angles; four long and numerous short hairs on the surface of the antennal swellings; one hair at the eye; a fringe of short hairs on the sides of the temples; temples with three rather long hairs, two of which are pustulated, and several short ones; a pitchy brown blotch running back from the angles of clypeus to the antennal swelling and half way around its inner border; a second band curving inward and backward from the posterior portion of the antennal swelling, to the occipital margin; antennal swellings obscured with brown; mandibles large, tipped with black, temples margined with deep brown; a medium oval ventral spot of brown, the unmarked portion of the head clear, pale brown.

Prothorax shield-shaped, with prominent lateral angles, antero-lateral portion emarginate, and postero-lateral margin slightly angulated in median portion; anterior margin truncate; lateral angles with a long hair and several short bristles; sides with numerous short hairs; a pitchy spot in anterior angles, from which run pitchy bands, parallel with the lateral margins of segment and joining anterior angles of metathorax; a narrow cross band in the anterior portion; a median longitudinal ventral patch, darker in anterior portion; postero-lateral margins narrowly edged with pitchy brown. Metathorax about as long as prothorax, with slightly concave anterior, and truncate posterior, margins; sides very slightly convex, diverging, bordered with numerous fine hairs; two pustulated hairs in the middle of posterior margin and one on each side in the middle portion of the segment, just with the lateral bands; whole anterior portion of segment pitchy brown; lateral pitchy bands, narrower anteriorly, join the abdominal bands; a median brownish patch divided longitudinally by a clear narrow line running the entire length of segment; regions between median blotch and lateral bands, clear pale brown. Legs long and stout, with anterior pair smaller, with femora and tibiae all margined with pitchy brown; interior portion clear, obscured in femora by a central patch of brown; femora with a row of pustulated hairs along anterior margin; a few on tibiae.

Abdomen large, spindle-shaped, with lateral angles scarcely visible and sides fringed with numerous fine hairs of different lengths, longer posteriorly; posterior margins of segments almost straight with a transverse series of pustulated hairs of different lengths, the shorter hairs arising from smaller pustules; heavy lateral pitchy bands from segment one to the middle of the ninth, unbroken at the suture, and having in segments two to eight three clear pustules, one in the anterior and two in the posterior portion of the segment, from which arise rather long hairs; clear pustules at the spiracles on segments three to eight; inner border of lateral bands rather uneven, fading into a wide, clear, submarginal longitudinal band, on which is a longitudinal row of very fine, short hairs, together with many scattering ones of the same size; median, transverse, deep brown bands on segments one to eight, separated transversely by narrow clear sutures, and longitudinally by a clear line in segments one and two, and interrupted by clear spots on remainder; sides of transverse bands uneven and separated from lateral bands by the clear submarginal area; tip of ninth segment encircled by a brown band; a median longitudinal deep brown blotch, narrowing posteriorly; a row of about six short hairs in median portion of segment.

MALE.—Body, length 8.47 mm., width 2.32 mm.; head, length 1.67 mm., width 1.57 mm.; darker than the female, with narrower submarginal clear bands on abdomen, and with the tip having two brown lateral blotches instead of one median blotch.

Numerous males and females collected on specimens of *Gypagus papa* at Pozo Azul, Costa Rica, June, 1902.

This species also much resembles Piaget's *Lae. titan,* more so than my *oligothrix,* but can easily be separated from it by the presence of median clear lines and spots on the abdomen, by a wider, submarginal clear area, by the absence of a narrow darker transverse band across the anterior margin of the segments, by the presence of clear pustules in the lateral bands, by the lighter ground color of the thorax, and different bands, and by the more slender tibiae.

Laemobothrium oligothrix sp. nov., pl. IV, fig. 7

FEMALE.—Body, length 9.61 mm., width 2.71 mm.; deep smoky brown throughout, with pitchy markings on head and thorax and broad pitchy lateral bands on abdomen.

Head, length 1.79 mm., width .76 mm.; sharply conical with the straight sides interrupted by the swellings of the antennal fossae, and the clypeus squarely truncate; palpi small, projecting by the last two segments; temples produced posteriorly, bluntly pointed; occiput transverse; clypeus, antennal fossae, and lateral portion of temples with numerous fine hairs; temples with three rather weak hairs, one of which is pustulated; whole head an even, clear brown, with an irregular pitchy band on each side, starting just behind the lateral angles of the clypeus, running backward along inner edge of antennal fossa, separating behind it, with one branch curving outward to margin at eye, thence backward to widest portion of temple, and the other branch passing backward and inward to the occiput, with a break at the submarginal, pitchy, occipital band; mandibles large and heavy, points deep brown; a crescent-shaped patch between the lateral bands of the head, at the point where they nearest approach each other.

Prothorax somewhat shield-shaped, with antero-lateral portion emarginate with some short hairs, postero-lateral portion rounded, with a few short hairs; lateral wings clear, pale brown, whole median portion of segment deep brown; pitchy bands run from the lateral angles to the ends of the lateral bands of the metathorax, around on the margin at the lateral emargination and from the anterior portion of the emargination diagonally backward to the center of the segment, a longitudinal median ventral band of darker brown. Metathorax about as long as prothorax, with straight diverging sides and truncate posterior margin; sides with a few short hairs; heavy pitchy lateral bands, curving around on anterior portion of segment, continuous with lateral abdominal bands, and with branches running diagonally inward at the posterior portion of the segment and connected by a paler band; interior of segment uniformly deep smoky brown. Legs long and stout, concolorous with body, especially the two posterior pairs; anterior femora swollen and orbicular, posterior two lengthened; anterior margin and a portion of posterior margin of femora and both margins of tibiae heavily edged with black, first joint of tarsi swollen, second long and slender; a series of short pustulated hairs along the inner margin of the black edging of the two posterior pairs of femora; other short hairs on margin of femora and tibiae; femora deeper brown than tibiae and darker at base and in median portions.

Abdomen large, elongate oval, with posterior, lateral angles scarcely visible; sides fringed with short slender hairs, heavier and more abundant at posterior portion; posterior margins of segments almost truncate, with a row of eight to ten short hairs on segments two to seven, lateral margins with heavy pitchy bands, narrowing posteriorly and ending at the middle of the last segment; short, pitchy bands run diagonally backward from lateral band in the first segment, corresponding to the bands in metathorax; interior of abdomen an even, deep brown, separated from lateral bands by a narrow pale area; heavier, pitchy brown, narrow, transverse bands at the anterior margin of segments two to eight, ending at the pale lateral band; a darker median area in segment seven; segment eight translucent, with a deep brown blotch in anterior portion and a clear brown, crescent-shaped band around the tip.

MALE.—Body, length 8.29 mm., width 2.17 mm.; head, length 1.65 mm., width 1.56 mm.; similar to female, except in size, and the tip of abdomen, which is slightly swollen laterally and flatly indented at the tip, with a narrow band around anterior border, a dusky spot in each side of the median portion, and the same band around tip as in female.

Numerous males and females collected from specimens of *Buteo borealis costaricensis,* shot on the volcano Irazu, April, 1902. This form is quite close to Piaget's *Lac. titan,* but is disguished from that species by the clear temples, the absence of a dark band across the middle of the antennal swelling, by the heavier lateral abdominal bands, internal metathoracic bands, and absence of clear pustules on abdomen, and by the difference in the shape of the legs and fewer hairs on margins of abdomen.

Physostomum jimlnezi sp. nov., pl. V, fig. 1

FEMALE.—Body, length 3.06 mm., width .97 mm.; smoky fulvus throughout, abdomen with darker lateral bands and median transverse bands; legs very dusky, femora margined with blackish.

Head, length .71 mm., width .57 mm.; conical, with flatly rounded, bare front; palettes projecting by apical segment; sides of head slightly sinuated with one weak hair at anterior margin of antennal fossae; a slight ocular notch with one very fine bristle; temples produced backward, ending in a blunt spine, very slightly turned outward, with three weak hairs; occipital margin reentering, occiput flatly convex; a pale fulvous band across the forehead at the base of the palettes; a pale indistinct band along the lateral margins of the head, disappearing at the antennal fossae; antennal fossae of unusual size, with inner portion bordered with blackish but outer bounded only by a very faint line; a black ocular fleck; a narrow, fulvous, submarginal occipital band: a slightly darker occipital patch; whole head smoky fulvous.

Prothorax unusually large with clear lateral wings; anterior and posterior margins concave; lateral angles very obtuse; antero-lateral margins slightly concave, postero-lateral portions straight and diverging; median portion of segment dusky; pale internal bands running back from the lateral margins, near the anterior angles, to metathorax. Metathorax scarcely larger than prothorax, with anterior and posterior margins truncate; sides slightly sinuate, diverging; a faint clear line at the meso-metathoracic suture; anterior region of segment, and broad lateral bands of brownish fulvous, continuous with lateral bands of abdomen.

Anterior legs short and weak, same color as head; two posterior pairs long and stout, with femora margined before and behind with blackish; the whole femora and tibia deep smoky brown.

Abdomen narrowly oval, with scarcely protruding acute angles, furnished with one weak hair; posterior margin of first three segments flatly angulated; four and five transverse and six concave; broad, marginal, unbroken, lateral bands on segments one to seven, with inner border emarginate on segments four to seven; eighth segment clear, vulva convex, fringed with fine hairs; in the middle of the broad lateral band is to be found the peculiar chain shaped band of chitin common to the genus; broad, median, transverse bands of brownish fulvous, separated by narrow, clear sutures, on segments one to seven.

Three females collected from *Amazilia tzacatl*, at Juan Viñas in March and three females from *Selasphorus flammula* on the volcan Irazu, Costa Rica, February, 1902.

This species much resembles [illegible] *cephalum* Kell., but is easily distinguished from that species by the large clear prothorax and the absence of numerous long hairs on the head and thorax. It seems to be a common parasite of several species of hummingbirds since it was taken on quite different species in very different localities.

Physostomum doratophorum sp. nov., pl. V, fig. 4

FEMALE.—Body, length 2.41 mm., width 1.10 mm.; short and broad, with sides of head deeply emarginate, abdomen broadly oval, whole body pale golden, darker on head, with clear brown markings on head, thorax, and abdomen.

Head, length .58 mm., width .61 mm.; front broad, slightly concave, with rounded lateral angles; sides deeply emarginate slightly forward of the middle, and just at the base of the palettes, which are very large, filling the emargination and extending around on the sides in front; palpi projecting by almost entire length, long and stout, apical segment largest, globular; sides of head behind bases of palpi convex and diverging; a slight ocular notch; temples small, rounded, with two short weak hairs; occipital margin concave, but the wide occiput strongly convex, with a narrow darker border; a narrow brown band across the forehead just in front of the palettes; anterior margin of palettes bordered by a narrow dark brown band; antennal fossae of medium size, encircled by only narrow lines, and interior slightly darker than general color of the head; a black ocular fleck; a narrow, brown, marginal band, encircling the lateral emarginations at the base of the palettes and palpi.

Prothorax of medium size, shield-shaped, with anterior margin concave, anterior portion of sides slightly concave, and whole margin back of lateral angles nearly evenly rounded; five short, slender hairs on each side of the rounded lateral portion; narrow, brown, marginal bands along the portion of the sides anterior to the lateral angles; narrow, pale, internal bands curving backwards from the anterior angles to the anterior angles of the metathorax; two median, longitudinal bands enclosing a spear-head shaped area. Metathorax about as long as prothorax, sides broadly diverging, slightly sinuate, there being a slight lateral constriction at the suture of the meso- and metathorax and a narrow clear dorsal line; posterior margin flatly rounded; posterior angles rounded and slightly protruding; narrow brown bands from the anterior angles along the lateral margin, broken at the suture, back nearly to the posterior angles, then curving inward and extending half way across the first segment of the abdomen, where they join the internal chitin bands; heavy, pale, internal bands run backward from the anterior angles into the abdomen; the region of the mesothorax and posterior angles deep golden, about the color of the head. Legs long and stout, femora slightly swollen, tibia much swollen at the tips and second segment of tarsi unusually large, being almost globular; femora margined anteriorly and posteriorly, tibiae posteriorly, and tarsi anteriorly with brown.

Abdomen short and broadly oval, almost clear, with the exception of the transverse bands; posterior angles scarcely protruding, without hairs; posterior markings transverse; eighth segment flatly rounded behind, with a row of short fine hairs; deeply submarginal, chain-like, chitin bands, extending from the prothorax, run almost straight back as far as the middle of the sixth segment; between these bands are faint golden transverse bands, broken by clear sutures; outside, in each segment are spots of the same color as the transverse bands; a few short dorsal hairs on each segment inside and outside of the internal bands.

Four females collected from *Selasphorus flammula*, on the volcano Irazu, April, 1902. I have placed this species in the genus *Physostomum*, but do not think that it rightly belongs there. However, since Mr. Kellogg has placed a very closely related species (*promineus*) in this same genus, I will follow him in the matter at present, though I believe that further collecting of Mallophaga from hummingbirds will produce additional species of this type, upon which a new genus can be safely established.

Physostomum leptosomum sp. nov., pl. V, fig. 2

Female.—Body, length 3.13 mm., width .82 mm.; head and thorax clear pale brown, with brown markings, abdomen brownish golden with deep smoky brown, lateral bands.

Head, length .80 mm., width .61 mm.; slightly conical, with front broad and evenly rounded; sides straight with two short fine hairs in front of the antennal fossae; ocular notch small, with three short, fine hairs; temple produced posteriorly and bluntly pointed, with two rather long and one short hair; occipital margin deeply reentering, occiput very slightly concave; the protruding palettes faintly tinged with brown; a brownish band across the forehead, joining the bases of the palettes and bearing a row of fine, dorsal hairs; antennal fossae clear, encircled by a narrow brown line, except on the inner side where the line expands into a brown band; a brownish, semicircular band curves from the lateral margin of the head at the base of the palettes upward to the transverse band, and between this and the antennal fossae are large irregular shaped brown blotches; a narrow, submarginal, occipital band, blackish in the median portion.

Prothorax the same width as the head at the templar angles, quadrilateral, with anterior and posterior margins concave, the anterior half of the lateral margin straight and the posterior half evenly rounded; lateral angles very obtuse (scarcely noticeable), with a short spine and hair, two more hairs on lateral margin behind the lateral angles; broad lateral bands of clear brown and a median, longitudinal, ventral patch of paler brown. Metathorax scarcely larger than the prothorax, with anterior portion rounded and covered by prothorax; sides very slightly concave, posterior border truncate and same width as first segment of abdomen; one slender hair in the posterior angles and a spine in the anterior angles; heavy, brown, lateral bands, passing around on the anterior margin; brown bands pass diagonally forward from the posterior angles to the median anterior portion, broad and pale at base, narrowing and darkening anteriorly; internal portion of segment dusky brown. Legs of medium size, pale fulvous, lighter than body.

Abdomen rather broad, sides parallel, with the tip abruptly rounded by a portion of the seventh and the eighth segments; segments of nearly equal width throughout, with transverse posterior margins, except the sixth, which is slightly concave; one short hair in the region of the posterior angles of segments one to seven; eighth with a fringe of short hairs along the round margin; deep brown, lateral bands on segments one to seven and anterior portion of eighth, heavier along the inner portion and broken by sharply diagonal, clear lines at the sutures; interior of the abdomen an even golden brown, with a slightly darker median patch covering a portion of segment four, all of segments five and six, and a part of seven.

MALE.—Body, length 2.43 mm., width .69 mm.; head, length .69 mm., width .55 mm.; differs from the female only in smaller size and in last segment of abdomen, which is without the dark lateral bands in anterior portion, i, longer, more flatly rounded, and slightly sinuate on the sides; a dusky, median, transverse band in the anterior portion; genital hooks are slender, widely diverging in the median portion, with ends straight and converging, and extending the width of the sixth and seventh segments.

Two females collected from *Myiarchus lawrencei nigricapillus*, and two males from *Myiozetetes cayanensis*, Juan Vinas, Costa Rica, March, 1902. This form has a superficial resemblance to *P. sucinaceum* Kell., but differs in the shape of the head, markings of thorax, presence of palettes, and size.

Physostomum angulatum Kell.

Kellogg, *New Mallophaga* II, p. 515, pl. LXX, fig. 5.

One female of this well-marked species collected on *Tanagra cana*, Juan Vinas, Costa Rica, March, 1902.

This is another strange instance of distribution, but there can be no doubt as to the identification of the specimen since it agrees perfectly with Mr. Kellogg's plate and description of *angulatum*.

Physostomum australe Kell.

Kellogg, *New Mallophaga* II, p. 516, pl. LXX, fig. 4.

One female collected on *Tanagra cana*, at Juan Vinas, Costa Rica, March, 1902.

The finding of this species on *Tanagra cana* is as unaccountable as that of *P. angulatum*, but the identification is absolutely certain.

Physostomum subangulatum sp. nov., pl. V, fig. 3

FEMALE.—Body, length 4.37 mm., width 1.06 mm.; almost uncolored, with a faint tinge of golden on abdomen, with narrow, pitchy, submarginal, lateral bands on abdomen and thorax and pitchy and deep brown markings on the head.

Head, length 1.05 mm., width .80 mm.; conical, with rather narrow, evenly rounded, hare front; large projecting palettes connected by a clear band across clypeal suture; sides of head slightly sinuate in region of antennal fossae; ocular notch shallow with three short bristles; eyes small, nearly obscured by a black fleck; temples rounded, clear, with three weak hairs, produced backward to a point; occipital margin deeply reentering and evenly concave, with a narrow black submarginal border; a brown occipital signature; antennal fossae small, inner border obscured by a pitchy band fading to brown; region between antennal fossae and lateral margins, and a short space in front, clear brown; a dark brown marginal blotch at the ocular notch; pitchy brown antennal bands run forward from the occipital band, passing along the inner margin of the antennal fossae, thence forward to the palpi, with two round, darker spots on them, between the fossae and palpi; a short, curving band connects their ends with the margins at the base of the palettes; a narrow, curving, longitudinal, brown band runs along the inner borders of the anterior portion of the antennal bands.

Prothorax hexagonal, lateral angles rounded, with a bristle and short hair; antero-lateral margin straight, postero-lateral margin curving, with one weak hair; anterior and posterior margins deeply concave; pitchy lateral bands, marginal in front of lateral angles, submarginal behind, and joining anterior angles of metathorax; two narrow, median, longitudinal, pitchy lines, and two diagonal lines in each side of the posterior portion of segment. Metathorax larger than the prothorax, quadrilateral, with sinuated, diverging lateral margins and truncate posterior margin; posterior angles with one hair and a bristle; strong, pitchy, submarginal bands running backward from the anterior angles to the posterior margin of segment; a second band starting at the anterior angles follows the margin backward to median portion of segment, then curving inward cuts the submarginal band and passes forward to the posterior border of the prothorax; the enclosed portion between the marginal and submarginal bands pale brown. Legs long, rather slender, and almost colorless.

Abdomen nearly parallel sided, with acute, scarcely projecting posterior angles, having one weak hair in segments one to four, and two in segments five to seven; eighth segment evenly rounded, with two short hairs on each side; vulva convex, fringed with fine hairs; whole abdomen clear, with only faint tinge of golden, excepting the heavy, pitchy, submarginal, lateral bands, extending from the end of the metathoracic bands to the middle of the eighth segment, where they are marginal and narrow.

Four females collected on *Tanagra cana,* at Juán Vinas, Costa Rica, March, 1902. This species is a curious combination of *P. angulatum* and *australe* Kell., having the head resembling *australe* and the thoracic and abdominal markings resembling *angulatum.*

It seems quite remarkable that three closely related species of this genus should be found upon the same host, although they were collected on different individuals. As Mr. Kellogg says, this genus is a peculiar one and must be thoroughly revised as soon as sufficient material is accumulated.

Physostomum picturatum Car.

Carriker, *New Mallophaga from Nebraska Birds.* Jour. N. Y. Ent. Soc., vol. X, no. 4.

One female collected on *Helminthophila peregrina,* Juan Vinas, Costa Rica, March, 1902.

The finding of this species on this host in Costa Rica is not so surprising as might at first appear, since it was described from specimens collected on *Helminthophila celata.*

Physostomum pallens Kell.

Kellogg, *New Mallophaga* III, p. 49, pl. IV, fig. 7.

This species described from specimens collected on *Protonotaria citrea* was taken on *Compsothlypis pitiayumi,* a closely related genus, on the volcano Irazu, Costa Rica, April, 1902.

But a single female was taken.

Colpocephalum gypagi sp. nov., pl. VI, fig. 2

Female.—Body, length 1.96 mm., width .71 mm.; head smoky brown, with four large pitchy areas; body clear with lateral pitchy spots and paler transverse bands on segments four to seven of the abdomen; femora and tibiae heavily margined anteriorly with deep brown.

Head, length .35 mm., width .53 mm.; flatly rounded in front, with conspicuous ocular emarginations and with the palpi and antennae protruding by apical segment; front with six very fine hairs; three more, slightly longer, on sides in front of the ocular emargination; ocular fringe rather sparse for the genus; temples slightly angulated in the rear, with three rather long hairs, two marginal and one submarginal; occipital margin deeply concave, with six short hairs; very large, pitchy ocular and occipital blotches, the ocular blotches connected with the occipital blotches by dark brown bands and the occipital blotches connected by a similar band.

Prothorax small, oval, with lateral angles produced to a blunt spine, and furnished with a short spine and a long hair; postero-lateral margins with two long hairs; posterior margin transverse, with six short hairs; deep brown chitin bands curving across lateral angles, which are brownish, fading inward; transverse band pale brown. Metathorax larger than the prothorax, quadrilateral, with widely diverging sides and rounded posterior margin; anterior and posterior angles pale golden, remainder of segment clear; region of posterior angles with numerous short, dorsal hairs.

Abdomen large, widest at third segment, thence constricted and tapering sharply to the pointed tip; lateral angles projecting but little, with several short, weak hairs, and lateral margins with several short hairs also; lateral regions of segments one and two with some short, dorsal hairs; posterior margins of segments one to eight with a fringe of fine hairs; first three segments almost clear, except a brownish spot in lateral portions of segment three; segments four to seven with pitchy blotches in the median portion of lateral margins, and pale brownish transverse bands, separated by clear sutures; segments eight and nine uniformly brown except the tip of the ninth, which is clear; tip of ninth with two longish hairs and a fringe of fine hairs, also a few short hairs along the lateral margins; on the ventral surface is a transverse row of stout hairs across the posterior margin of the eighth segment, and a short row of stiff bristles in the lateral portions; also a fringe of stout marginal hairs along the anterior portion of the eighth, curving upward around the sides of the segment. Legs of medium length and stout, especially the anterior pair; smoky brown the same as the head, with heavy, deep brown anterior borders on the femora and tibiae.

A single female collected from *Gypagus papa*, at Pozo Azul, Costa Rica, June, 1902. This form is of the same type as *setosum* Piag., from which it is distinguished by the absence of transverse bands on the first two segments of the abdomen, by the much narrower lateral bands on segments four to seven, by the presence of transverse bands on segments four to eight, by the presence of a continuous fringe of hairs on the posterior margin of all the abdominal segments, and by the much shorter hairs on the posterior angles of the abdomen.

Colpocephalum osborni Kell. var. **costaricense** var nov.

A large number of males and females collected from *Buteo borealis costaricensis*, on the volcano Irazu, Costa Rica, April, 1902. While these specimens resemble quite closely Kellogg's *osborni*, there are sufficient important differences to give them a varietal rank. This form is larger, measuring: female, length 1.70 mm., width .56 mm.; head, length .31 mm., width .47 mm.; male, length 1.57 mm., width .49 mm.; head, length .31 mm., width .46 mm.; the pitchy lateral spots are absent from the first and second segments of the abdomen; there is a dusky transverse band across the metathorax; the pitchy spots of the head are not so closely united, while the marginal bands of the legs are paler.

With these exceptions it agrees with Mr. Kellogg's description of the species.

Colpocephalum extraneum sp. nov., pl. VI, fig. 3

FEMALE.—Body, length 2.15 mm., width .76 mm.; angulated before and behind, legs long and stout, meso- and metathorax divided, metathorax extraordinarily long and shield shaped, and abdomen constricted posteriorly.

Head, length .45 mm., width .65 mm.; front very flatly rounded, sides sinuate, there being a depression at the point where the palpi project and at the ocular emargination; front with six short hairs, sides with four; palpi projecting by the long apical segment; an antennae projecting by almost all of last two segments; ocular fringe very thick and long; temples expanded broadly, roundly angulated before and behind, with four long hairs on the lateral margin; occipital margin concavo-convex, with a narrow pitchy border; pitchy ocular blotches; mandibles rather small, points pitchy; brownish bands running from end of ocular blotch to frontal margin; a brownish band, curving backward and broadening medially, connects the anterior portion of ocular blotches; whole temples clear brown.

Prothorax hexagonal, with lateral angles produced to a blunt point, furnished with three spines; anterior and antero-lateral margins straight; postero-lateral slightly concave and posterior convex; posterior angles with one hair, a median ventral blotch in anterior portion of segment; narrow brown edging to whole segment; dark brown coxal bands in the form of a flattened semi-circle across the postero-lateral portion of segments.

Mesothorax distinctly divided from metathorax, with convex lateral and truncate posterior margins; a pitchy brown band around sides and anterior angles, broken medially, with short bands running backward from their ends. Metathorax very wide and long, at least three-fourths as long as abdomen; sides straight, diverging; posterior margin elliptical, the region posterior to the lateral angles being longer than that anterior to them; lateral angles very obtuse, with three spines; sides of posterior margin with three long hairs; lateral margins with a narrow pitchy band curving across lateral angles; dark brown bands running inward from posterior portion of lateral margins, almost meeting medially; a lunate ventral patch in median anterior portion; two brownish patches in the posterior portion; coxal outlines, pitchy, showing through (not shown in the plate). Legs long and stout; femora much swollen, especially posterior pair, and all with anterior margins brownish; tibiae swollen apically, with anterior and posterior edgings of brown; second joint of tarsi long; the whole concolorous with body, and having numerous short marginal hairs; a patch of short hairs on the dorsal surface of the posterior femora.

Abdomen short, scarcely wider than metathorax and constricted sharply in the posterior portion; segments subequal in length, with rounded lateral margins; posterior angles with three short spiny hairs; narrow pitchy lateral bands on segments one to eight; deep umber brown transverse bands on segments one to eight, extending inward about one-third the width of the abdomen and scarcely broken at sutures; ninth segment pale clear brown with fringe of fine hairs on the flatly rounded posterior margin; one long hair in posterior angles of the eighth segment and two on each side of ninth, a row of fine, pustulated hairs on posterior margin of transverse bands, except on first segment; median portion of abdomen pale, clear brown.

A single female collected on *Nyctidromus albicollis*, at Pozo Azul, Costa Rica, June, 1902.

Colpocephalum luroris sp. nov., pl. VI, fig. 4

FEMALE.—Body, length 2.03 mm., width .76 mm.; clear tawny brown, with the large abdomen completely obscured by continuous transverse bands.

Head, length .43 mm., width .54 mm.; front very much flattened, sides sinuate; ocular emargination large but shallow and ocular fringe strong; eye large, obscured, protruding from the emargination; temples expanded laterally, angulated behind, rounded before, with three long hairs; occipital margin concavo-convex, with two hairs; front with two hairs in median portion and three at the lateral angles; palpi projecting by half of apical segment; two hairs before the ocular emargination; two long pustulated hairs just within the ocular blotch; an irregular, brownish blotch along interior margin of ocular emargination, and short curving bands from bases of mandibles to lateral angles of frontal margin, a pale brown occipital blotch; whole head clear tawny brown.

Prothorax hexagonal, with expanded lateral angles, furnished with one bristle; coxal bands very plain; a narrow transverse chitin band; posterior margin convex with four hairs. Metathorax larger than prothorax, pentagonal, with posterior and anterior margins truncate and sides slightly concave, strongly diverging; a single spine in posterior angles; posterior margin with a row of short, stout hairs; anterior angles and lateral margins edged with dark brown; brown bands (starting at lateral margins) cut across posterior angles into the abdomen. Legs long and stout, especially the front femora, which are much swollen; concolorous with body.

Abdomen large, oval, posterior angles projecting slightly, with two or three spines; a long hair in segments seven and eight, and two long ones on each side of the ninth, with a fringe of fine hairs between; posterior margins of segments one to seven with a row of short hairs, heavier, tawny, lateral bands on segments one to eight, broken at the sutures; whole interior of abdomen an even tawny brown.

A single female collected on *Zarhynchus wagleri,* Juan Vinas, Costa Rica, March, 1902.

Colpocephalum mirabile sp. nov., pl. VI, fig. 5

FEMALE.—Body, length 1.61 mm., width .66 mm.; clear with bands and markings of brown and pitchy; metathorax enormously developed, wider than head and abdomen and two-thirds the length of abdomen.

Head, length .34 mm., width .56 mm.; clear, with front almost evenly rounded except a slight depression at the projecting palpi; front with two long and six short hairs, one just before and three just behind palpi; a longer one pointing backward from anterior portion of ocular emargination; ocular fringe long and thick; eye very large, clear, with a black fleck; temples expanded and nearly evenly rounded, with four long hairs, occipital margin deeply reentering, occiput straight; mandibles medium with chestnut tips; short, brown, curving bands from bases of mandibles to frontal margin; curving pitchy ocular blotches; short, longitudinal crescent-shaped, brown bands inside of ocular blotches.

Prothorax large, hexagonal, with anterior margins straight, bluntly rounded lateral angles, concave postero-lateral margins, and rounded posterior margin; lateral angles with three spines; pitchy spots in the anterior angles; narrow, broken, pitchy brown edgings to the lateral margins; conspicuous pitchy coxal bands, running from the middle of the antero-lateral margins to median portion of segment; a faint median ventral spot.

Mesothorax distinctly set off from metathorax, with rounded lateral posterior margins; mesocoxal bands of pitchy brown very distinct, running from anterior angles of metathorax around the lateral margins of mesothorax, into the posterior angles of prothorax; from the ends of lateral bands to middle of segment, then bending sharply back into mesothorax. Metathorax very large (.46 mm. × .62 mm.), clear, with straight diverging sides, very obtusely rounded lateral angles and truncate posterior margin; lateral angles with four spines and a very long hair; the posterior portion of lateral margin slightly angulated in the median portion, with one spine; deep brown, dorsal bands, forming a figure 8 across the middle of the segment, with short appendages at the ends on the anterior side and long narrow bands curving backward from the posterior portion of ends, into the abdomen as far as the fourth segment. Legs long and stout, clear, with brown spots at tips of femora and tibiae; numerous short, marginal hairs.

Abdomen short, almost parallel sided, with segments five to seven much shorter than remainder, two segments, one, eight, and nine the longest, subequal; sides slightly convex, with a short hair in posterior angles of segments one to eight and a long one in segments one and two, and seven and eight; posterior margins of segments one to seven concave, especially five to seven, eight truncate and nine evenly rounded, with two long hairs on each side, a double fringe of very fine hairs at tip, and submarginal row of short, stout hairs around the whole posterior margin; a number of short spines on the lateral portion of the posterior margin of segments one to five; short hairs along the posterior margins of segments three to seven, thicker in the portion just inside the lateral bands; irregular pitchy brown spots in the lateral portion of segments one to eight, not as wide as segment, a narrow longitudinal clear line separates these spots from the interior of abdomen; smoky brown transverse bands in segments four to seven, separated by narrow, clear sutures; irregular, lateral, brown spots in eighth segment; interior of segments one to three, and nine, a pale translucent brownish. The abdomen has the appearance of having a flattened lateral area, with the whole central portion convex.

MALE.—Body, length 1.54 mm., width .51 mm.; head, length .34 mm., width .50 mm.; clearer than female, with much smaller metathorax; abdomen a perfect oval, with lateral spots of regular size and shape in all the segments except the ninth; posterior margins of all the segments with a row of hairs; transverse bands on segments three to six separated by wider sutures; genital hooks very large and long, reaching from third to posterior margin of the eighth segment, with the anterior two-thirds a single heavy shaft, widening posteriorly and separating into a perfect trident.

One female and three males collected on *Zarhynchus wagleri*, at Juan Vinas, Costa Rica, March, 1902.

Nitzschia bruneri sp. nov.

This form is very easily distinguished from *pulicaris* by the exceedingly short metathorax (length .34 mm., width .71 mm.), by the very slender hind femora and tibiae of the female, and by the paleness of the transverse abdominal bands. Measurements: Female, length 2.50 mm., width .95 mm.; head, length .50 mm., width .66 mm.; male, length 1.96 mm., width .75 mm.; head, length .45 mm., width .61 mm.

While working over specimens of *Nitzschia* from Costa Rica, I again went over the material collected from *Aeronautes melanolencus* in Sioux county, Neb., and which I had named: *pulicaris,* var. *tibialis.* I now find that some errors were made at that time and that several important points were overlooked, which clearly separate this form from *pulicaris,* and I accordingly give it full specific rank as *Nitzschia bruneri.*

Nitzschia bruneri, var. **meridionalis** var. nov.

FEMALE.—Length 2.73 mm., width .99 mm.; head, length .52 mm., width .76 mm.; male, length 2.18 mm., width .74 mm.; head, length .51 mm., width .70 mm.

The variety is distinguished from the species by the darker color, being a translucent brown instead of golden, by the absence of a marginal band on the lateral portion of the mesothorax, by much darker thoracic and lateral abdominal markings, by more and longer hairs at the posterior angles of the abdomen, by shorter and more spine-like hairs along posterior borders of segments, by much smaller posterior tibiae in the male, and, finally, by a difference in size.

Numerous males and females collected from *Chaetura griseiventris,* at Pozo Azul, Costa Rica, June, 1902. While these specimens closely resemble *bruneri,* they can be scarcely called that, and have accordingly been given varietal rank.

Menopon tridens costaricense var. nov.

FEMALE.—Body, length 1.48 mm., width .61 mm.; head, length .32 mm., width .49 mm.; whole body uniformly translucent fulvous, with black tips to the mandibles, black ocular flecks, narrow blackish occipital margin, while the peculiar, characteristic, occipital process is deep brown; occipital margin with four hairs, while the posterior margin of the pro- and metathorax and the abdominal segments is furnished with a row of stout hairs; just inside the lateral bands is a longitudinal area covered with short, fine hairs. The rotundity of the abdomen is also a prominent character.

A single female collected from *Porzana cinereiceps,* at Juan Viñas, Costa Rica, March, 1902. Unlike the varieties described by Mr. Kellogg, this form has the lateral bands of the abdomen uncolored, as Piaget gives for *tridens,* but while it agrees with the species in this respect, it differs radically in others.

Menopon ortalidis sp. nov., pl. VII, fig. 1

MALE.—Body, length 1.81 mm., width .71 mm.; very pale throughout, with only a faint tawny tinge; markings of head and thorax and lateral spots of the abdomen light smoky brown.

Head, length .37 mm., width .54 mm.; front, beyond ocular emargination, almost evenly rounded, with two short, fine hairs near the middle, two more just back of the projecting palpi, and two longer ones arising just in front of the ocular emargination; the ocular emargination shallow with a strong fringe of hairs; the temples slightly drawn out latero-posteriorly, with five long, stout hairs, four of which are pustulated; occipital margin concave, bare, with a slight, marginal, dusky band on each side of the occiput; the mandibles small, cinereous, and placed near the front of the head; palpi long and stout, projecting by nearly all of the last two segments; just behind the base of each palpus is a dark cinereous, cone-shaped protuberance, between which are two short dorsal hairs; the eye is large, clear, with a short hair and partly obscured by a large black fleck on the anterior side; the antennae rather stout, apical joint much the largest, nearly globular, and projecting by a trifle more than half its length; a dark band along the inner border of the ocular depression; pale smoky, occipital bands, curving from each side of the occiput to the anterior margin of the ocular depression; three short, dorsal hairs along each occipital band.

Prothorax very large, nearly as broad as the head and almost hexagonal in shape; lateral lobes expanded, dusky, and lateral angles bluntly rounded, with one short spine; the postero-lateral margins with six long hairs, and the posterior margin with four shorter ones; narrow, dusky bands start from the median portion of the antero-lateral margins, curve gently backward nearly to the posterior margin, then bend inward and unite, the whole enclosing a nearly quadrilateral space; fainter narrow bands run diagonally backward from the anterior corners of this quadrilateral, nearly to the center of the segment, joining a faint ventral spot; a still fainter band connects these diagonal bands transversely. Metathorax about the same size as the prothorax, with straight, diverging sides, and flatly rounded posterior mar-

[illegible]

[illegible] at Juan Vinas. [illegible]

[illegible]fasciatum [illegible]

[illegible] XXV. [illegible]

[illegible] mm. [illegible] mm.: head

[illegible] mm.: head. length

[illegible] fit any one of [illegible] that my specimens [illegible], and that what de- [illegible] with this specimen, there [illegible] another species.

This form is very readily distinguished by the broad head and the clavate abdomen, with heavy, chestnut, transverse bands and narrow, pitchy, lateral bands.

Several males and females collected on *Gypagus papa,* at Pozo Azul, Costa Rica, June, 1902.

Menopon macrocybe sp. nov., pl. VII, fig. 2

FEMALE.—Body, length 1.34 mm., width .48 mm.; head wider than the abdomen; abdomen almost parallel sided, with heavy, transverse, smoky brown bands.

Head, length .36 mm., width .57 mm.; very large, somewhat quadrangular, front almost truncate, sides convex and diverging, interrupted by the shallow ocular emarginations; two long and one short hair in front of ocular emargination; a short sparse ocular fringe; temples rounded, with four long hairs and several short ones; occipital margin deeply reentering, occiput transverse, with two long hairs; mandibles rather large, well toward the front of the head; dark brown antennal bands run diagonally backward from the clypeal angles to the bases of the mandibles, then straight back to the large pitchy ocular blotch; a black ocular fleck; whole head evenly clear, pale, brown.

Prothorax large, lateral angles bluntly pointed, anterior and latero-anterior sides straight; whole margin back of lateral angles evenly rounded, with one hair on each side; whole segment clear brown, darker in lateral portions, with a transverse band. Metathorax scarcely larger than prothorax, with straight, widely diverging sides and rounded posterior margin; six long hairs on each side of posterior margin, interior of segment clear brown, with lateral portion deep smoky brown. Legs large and stout, clear pale brown, almost the same color as head, with anterior margin of femora and both margins of tibiae edged with darker brown; second joint of tarsi very large in posterior pair of legs.

Abdomen short, almost parallel sided, abruptly rounded by the large apical segment; segments subequal in length, with lateral angles scarcely visible, furnished with one long hair and one short one in segments one to eight; ninth with one long hair on [illegible] hairs; a row of [illegible] along the posterior margin [illegible] with heavy con- [illegible] darker in the [illegible] and separated trans- [illegible] eighth and ninth.

[illegible] at Juan Vinas, [illegible] by the broad head [illegible]

[illegible]

[illegible] mm.; head, [illegible] in brown- [illegible] each side of the occi- [illegible] a black spot at their [illegible] margin instead of six; two [illegible] of the prothorax instead of [illegible] instead of slightly [illegible] and paler; in [illegible] the row along the [illegible] hairs transversely [illegible]

[illegible] head length [illegible] segment of the same [illegible]

[illegible] three females [illegible] the volcano [illegible]

[illegible]

[illegible] head having [illegible] long as broad [illegible] angles a [illegible] slightly flattened arc; whole occipital margin evenly concave with a narrow, pitchy submarginal band, and two median marginal hairs, two short hairs on the front, one at the slightly projecting labial palpi, two long ones and a short one in front of the ocular fleck; three long marginal ones and several short ones on the pointed temples; a pitchy brown submarginal band around front, broadening at bases into brownish areas covering the whole sides of the head except the paler apical portions of the temples; a large black ocular fleck; a short dark longitudinal band runs forward from bases of mandibles almost to margin of head; two curving bands in occipital region, joining at anterior ends and then extending laterally to the dark portion of the head.

Prothorax broad, with produced, bluntly pointed lateral angles, with anterior portion flatly rounded, and very obtuse posterior angles, making nearly straight postero-lateral margins and flatly rounded posterior margin; lateral angles with one long hair and a spine; posterior angles with a long hair, and posterior margin with six slender hairs; whole segment narrowly edged with chestnut; narrow, lateral, deeply submarginal, and a narrow median transverse band of chestnut; whole segment deep smoky brown. Metathorax scarcely larger than prothorax, with straight, widely diverging sides and flatly rounded posterior margin having row of about ten fine hairs and one in posterior angles; a lateral emargination at the mesothoracic suture; anterior and lateral margins edged with deep chestnut; segment deep smoky brown paler in mesothoracic region. Legs large and stout, posterior pair largest, with tibiae very large, longer than femora and edged with pitchy.

Abdomen broadly oval, with broad, flatly rounded tip; posterior angles rather sharp, projecting, with one long and one short hair in segments one to eight; segments nine with several long, and a fringe of fine hairs on the flatly rounded posterior margin, posterior margins of segments one to eight with a row of about twelve to sixteen short hairs; broad continuous, transverse bands of deep smoky brown on all the segments, darkening to pitchy in the lateral portions of segments one to eight and separated transversely by clear sutures except between segments eight and nine.

MALE.—Body, length .91 mm., width .45 mm.; head, length .24 mm., width .45 mm.; slightly paler than the female, with narrower abdominal bands; apical segments of abdomen same shape as in female; genitalia long, very slender and widely separated with tips curving inward slightly.

One male and one female collected on *Tityra personata,* at Juan Vinas, Costa Rica, March, 1902. This species approaches *M. maestum* Kell., but differs greatly in size, shape, and intensity of markings.

Menopon distinctum Kell.

Kellogg, *New Mallophaga* III, p. 126, pl. VIII, fig. 7.

One male and one female of this well-marked species, described from *Myiarchus cinerascens,* were collected from *Myiarchus lawrencei nigricapillus,* at Juan Vinas, Costa Rica, March, 1902.

Menopon stenodesmum sp. nov., pl. VIII, fig. 2

FEMALE.—Body, length 1.54 mm., width .60 mm.; clear pale testaceus, with brown and pitchy markings on head and thorax, pitchy brown, lateral bands and clear brown, median transverse bands on abdomen.

Head, length .33 mm., width .45 mm.; front rounded, with a depression at the projecting palpi and another at the ocular emargination, four hairs on front, between palpi; three short ones behind palpi and two longer ones just in front of the ocular emargination; ocular fringe heavy; temples expanding, clear, with four long hairs and a couple of short ones; occipital margin reentering, occiput very slightly convex, with a narrow, pitchy border; eye large, clear, with a black fleck; an elongated, pitchy, ocular blotch; mandibles large, brown; antennal bands run back from frontal margin at palpi, past the bases of the mandibles and along the inner border of the ocular blotches for half their length, then bend abruptly inward and join, forming a backward curving band of deep brown across the middle of the head; a large brown occipital signature; antennae project slightly.

Prothorax almost hexagonal, lateral angles produced, blunt, with two spines; posterior angles very obtuse with one long hair, lateral margin flatly rounded with four short hairs; lateral regions brownish; a median ventral brown spot, with pitchy lines running backward to its posterior angles from the antero-lateral margins. Metathorax much larger than prothorax, clear, bands of brown and pitchy; mesothoracic suture plainly marked; sides, back of suture, straight, diverging, posterior angles rounded, with one hair and three spines; posterior margin flatly rounded with numerous fine hairs; a pitchy band around anterior and lateral margin of mesothorax, broken in median portion, with narrow bands running slightly diagonally backward from the ends to the middle of segment; a ventral brown blotch at the junction of the pro- and mesothorax and a larger wedge-shaped one in the median portion of metathorax; brownish bands run straight backward from margin at mesothoracic suture across the lateral portion of segment and half way across first segment of abdomen, then bend abruptly inward from anterior portion of lateral bands, fading into the median blotch; some brown coxal lines visible, in addition to above. Legs of medium length, clear, with slightly swollen femora and tibiae except the posterior tibiae, which are slender and parallel sided; tibiae brownish at tips.

Abdomen elliptical, clear, with rounded, projecting, posterior angles, furnished in segments one to seven with three spiny bristles, in eighth with five; ninth segment large, clear, rounded posteriorly, with three long hairs on each side and a fringe of shorter ones between; rather narrow, pitchy lateral bands in segments one to eight; posterior margins of segments with a row of fine hairs; median transverse bands as follows: a crescent-shaped one open behind, extending across portions of first and second segments; narrow, straight bands, separated by wide, clear sutures, on segments three to six; and a large, somewhat quadrilateral blotch extending from posterior portion of sixth into the anterior portion of the ninth.

One female and one male collected on *Empidonax atriceps*, on the volcano Irazu, April, 1902, and one female on *Tanagra palmerum melanoptera*, Juan Vinas, Costa Rica, March, 1902.

This form resembles *Col. quadrimaculatum* Car. more than

any other, but is distinguished from that species by the larger size, slenderer posterior femora and tibiae, and darker, narrower lateral bands of abdomen.

Menopon thoracicum Gieb., pl. VII, fig. 3

Female.—Body, length 1.4 mm., width .54 mm.; pale fulvous, with narrow, dusky, occipital margin; two blackish ocular flecks; fuscus markings on head and thorax and deep fuscus lateral bands on abdomen.

Head, length .28 mm., width .40 mm.; front rounded, with four short hairs between the projecting palpi and two longer ones in front of the ocular emargination; the emargination distinct, rather shallow, and with ocular fringe; temples moderately expanded, rounded, with four long, slightly pustulated, hairs; occipital margin concave, transverse in center, with two short marginal hairs, and the whole narrowly margined with blackish; a large, faint, occipital signature, with pale bands curving from its anterior corners to the base of the mandibles, which are small, with dark points; ocular bands indistinct, filling the ocular depression, with a black fleck in the center and another on the large clear eye; a brown spot on the margin just in front of the palpi.

Prothorax with slightly produced, blunt anterior angles, bearing three spines, the posterior angles bear one long hair, and the flatly rounded posterior margin four hairs; the chitin bars quite distinct, in the form of slightly flattened semicircles in the region of the anterior angles, a pale transverse line; metathorax with quite a prominent suture setting off the mesothorax, which has an angulated posterior margin and heavy bands on the anterior angles; sides of metathorax straight, widely diverging, and with narrow marginal bands; posterior angles obtuse, dusky, and with one long hair and three spines; posterior margin flatly rounded, with a complete row of hairs; pale brown coxal bands showing through; legs concolorous with body, bearing a few short hairs and with indistinct marginal markings on tibiae.

Abdomen rather large, elliptical, lateral angles serrate, armed with one long hair and several short bristles; posterior margins of segments with a row of longish, slender hairs; ninth segment large, rounded posteriorly, with two long hairs on each side and a double fringe of very fine hairs on the tip; lateral bands broad, deep, smoky brownish, ending with the eighth segment; very dim, brownish, transverse bands on segments three to eight, separated from lateral bands by a clear place, and from each other by clear sutures; ninth segment dusky in lateral portion, tip clear.

MALE.—Body, length 1.00 mm., width .37 mm.; head, length .28 mm., width .37 mm., the head of the male being but slightly different from the female, while the abdomen is much smaller and slightly darker in color; genital hooks large but simple, resembling more the common form of the *Colpocephali.*

Numerous specimens of a *Menopon* were collected from *Catharus gracilirostris, Chlorophonia callophrys,* and *Piranga bidenta sanguinolenta* on the volcano Irazu, Costa Rica, April, 1902. These specimens can be referred, without doubt, to *thoracicum.* Giebel's description is, for the most part, quite comprehensive, and every point which he mentions agrees with this form. A detailed description, together with plate, is given in order to thoroughly establish the species.

Menopon thoracicum var. majus var. nov.

The female measures: length 1.72 mm., width .73 mm.; head, length .38 mm., width .56 mm.

One female was collected from *Merula grayi* and one female from *Tanagra cana,* at Juan Vinas, Costa Rica, March, 1902, which agree very closely with the species, except that they are paler and much larger.

Menopon thoracicum, var. fuscum var. nov.

This variety measures practically the same as the species, the variation being the pale brown color of the entire body, instead of light golden, with markings of head and thorax and abdomen a deep clear brown; in the species the transverse bands of the abdomen are very faint or wanting, while here they are very noticeable, being pale on segments one and two but darkening backward and almost uniting on the last four segments to form a continuous, brownish patch filling almost the entire space between the lateral bands, and broken by only very faint pale transverse lines at the sutures.

Numerous specimens collected on *Ramphocelus passerinii*, Juan Vinas, Costa Rica, March, 1902.

Menopon difficile sp. nov., pl. VIII, fig. 1

FEMALE.—Body, length 1.86 mm., width .70 mm.; clear smoky brown, with broad, slightly darker transverse abdominal bands; narrow, pitchy, thoracic, lateral abdominal and ocular bands.

Head, length .32 mm., width .53 mm.; lunate, with deeply concave occipital border and small ocular emarginations; palpi projecting by the apical joint and antennae concealed; front bare; three long and two very fine hairs in front of the ocular emargination; ocular fringe meager; temples flattened and rounded with three long and two shorter hairs; occipital margin with a narrow blackish border and six slender hairs; mandibles small, with darkened tips; a faint submarginal band encircling front of head beyond ocular emargination with a branch joining the anterior portion of the heavy, blackish, ocular bands; a large quadrilateral occipital spot with curving bands connecting its anterior angles with the bases of the mandibles; eye large, rounded, with a black fleck.

Prothorax broad, pentagonal, with broadly rounded lateral angles, furnished with two spines; whole front from lateral angles forward flatly convex; sides almost straight, converging, with two long hairs and a spine; posterior angles obtuse, with one long hair; posterior margin truncate with six short hairs; whole segment deep smoky brown, with pitchy diagonal coxal lines, and brown transverse chitin bands; a median, ventral brown blotch. Metathorax scarcely larger than prothorax, with sides straight and broadly diverging; a slight lateral constriction at the mesothoracic suture; posterior angles acute, with one hair and one spine; posterior margin flatly rounded, with about sixteen short hairs; anterior angles with a short pitchy band and curving pitchy meso-coxal bands; dark brown meta-coxal bands, curving backward from mesothoracic sutures, across lateral portion of segment, and extending into first abdominal segment; whole segment deep smoky brown. Legs stout, with swollen femora and rather long tibiae and tarsi, concolorous with body, with some short hairs.

Abdomen large, subelliptical, with one hair and several spines in segments one to eight; ninth segment rounded, with a fringe of hairs in the median portion; posterior margin of segments with a row of fine hairs; narrow, pitchy, lateral bands, broken at the angles; heavy continuous transverse bands of deep smoky brown, separated by clearer sutures on segments one to eight; ninth segment clear, with a brown posterior band and a brownish spot in lateral portions.

MALE.—Body, length 1.52 mm., width .57 mm.; head, length .29 mm., width .49 mm.; very similar to female.

Numerous males and females collected on *Buarremon brunneinuchus*, on the volcano Irazu, Costa Rica, February, 1902. This species resembles, in a general way, Piaget's *M. extraneum*, but differs greatly in markings of thorax, size of legs, and other details.

Menopon palloris sp. nov., pl. VIII, fig. 3

FEMALE.—Body, length 1.54 mm., width .53 mm.; pale, clear golden, with a slight smoky tinge; no conspicuous markings, temples bluntly angulated anteriorly and very short, abdomen very hairy.

Head, length .34 mm., width .50 mm.; front broad, flatly rounded, with four short hairs; sides sinuate, slightly diverging, with two long and two short hairs; anterior margin of temples almost transverse; a prominent ocular fringe, temples with anterior and posterior angles (similar to *Nitzschia*) bluntly rounded, and four long pustulated hairs along the lateral margins; occipital margin concave, occiput slightly convex, with two short hairs; a pale ocular blotch and a black fleck; mandibles small, chestnut at tips; a pale band over palpi, which project slightly; a narrow, brown, occipital margin.

Prothorax hexagonal, lateral angles slightly produced, bluntly rounded, with three spines; anterior, antero- and postero-lateral margins concave; posterior margin flatly rounded, with six hairs; one faint coxal band visible on each side. Metathorax larger than prothorax, with almost straight diverging sides and very flatly convex posterior margin, set with a row of fine hairs; mesothoracic suture plainly visible; posterior angles with a long hair and a spine; a pale, lateral, marginal band, interrupted at the mesothoracic suture. Legs long and stout, with swollen femora and tibiae, but short tarsi; a few short hairs on margin.

Abdomen subclavate, with lateral margins of segments convex, and posterior angles projecting, with one long hair and several stout bristles in segments one to seven; eighth with two long hairs and two bristles, and two long hairs on posterior margin; ninth evenly rounded at tip, with two long hairs on each side and a few short bristles; posterior margins of segments with a series of fine short hairs; two other transverse rows of very fine short hairs across segments three to seven and a single row across two and eight; whole abdomen a uniform, translucent golden, with a slight tawny tinge.

The male is slightly smaller, with abdomen somewhat constricted posteriorly.

A single male and female collected on *Stelgidopteryx ruficollis uropygialis,* at Juan Vinas, Costa Rica, March, 1902. Of the type of *M. rusticum* Piag. and *dissimile* Kell., but differs in the shape of the head, and the markings of thorax and abdomen.

Menopon laticorpus sp. nov., pl. VII, fig. 5

Female.—Body, length 1.40 mm., width .70 mm.; clear brown, with numerous markings and bands of deep smoky brown; abdomen very large and broad, oval; head broad and very short.

Head, length .27 mm., width .58 mm.; front flatly rounded, with rather prominent, though not deep, ocular emarginations; two rather long hairs in front; labial palpi very long and stout, projecting by fourth and part of third segments; two short hairs and one long one just back of the palpi; two short hairs pointing backward from front of ocular emargination; ocular fringe very sparse; temples short, produced laterally, and evenly rounded with two long pustulated hairs and several shorter ones, occipital margin concave, with four hairs, and a narrow, marginal, pitchy border; a pitchy brown submarginal band almost entirely around the front of the head; mandible small, pointed; two broad brown bands starting from the frontal band at the palpi, extend backward past the bases of the mandibles, around the inside of the pitchy ocular blotches, and backward to the occipital margin, their posterior portions spreading out laterally along the temple; lateral portion of temples clear; region outside of the pitchy, curving, ocular blotches, deep brown; a large black ocular fleck, nearly obscuring the large clear eye; a large deep brown blotch with a clear circular center nearly fills the median occipital region.

Prothorax large, with lateral angles produced and broadly rounded, with two spines; whole margin posterior to lateral angles evenly rounded, with five long hairs on each side; lateral wings deep clear brown; a pitchy band runs inward from lateral angles for a short distance along the antero-lateral margin, then curves backward nearly across the segment; narrow pitchy bands run diagonally backward from the anterior angles to the median portion of segment; a narrow, sinuate, transverse band; a triangular, median brown blotch. Metathorax short, broad, with nearly straight, widely diverging sides, and flatly rounded posterior margin, set with numerous short hairs; posterior angles with two longish hairs; pitchy bands around anterior angles, curving backward across segment; heavier pitchy bands running diagonally backward and inward, from anterior angles to middle of segment; narrower pitchy bands curving backward and inward from lateral margins, across the segment as far as the middle of segment two of abdomen; a deep smoky brown band across posterior portion of segment. Legs long and stout, with swollen femora, and tibiae slightly enlarged at tips, concolorous with body, and furnished with numerous short marginal hairs.

Abdomen very large, a perfect oval, with lateral, posterior angles acute, but scarcely projecting, and furnished with one long hair and a bristle in segments one to eight; ninth large, rounded behind, with three long hairs on each side and a fringe of fine hairs between; posterior margin of segments one to eight with a row of short hairs; segments one to eight with continuous transverse bands of deep clear brown across the posterior portion of the segments; a large band in ninth, not reaching lateral margins.

MALE.—Body, length .86 mm., width .41 mm.; head, length .25 mm., width .51 mm.; differs little from female except in much smaller abdomen, and slightly smaller head, the legs being nearly as large; genitalia long, very slender, widely separated, with slightly curving tips.

A male and female collected on *Thamnophilus doliatus*, at Juan Vinas, Costa Rica, March, 1902. This is of the same general type as *M. maestum* Kell. and *tityrus* sp. nov.

LIST OF HOSTS WITH PARASITES

Tinamus robustus
  Docophorus sp.? (juv.)
  Lipeurus longipes tinami var. nov.
  Lipeurus longisetaceus Piag.
  Ornicholax robustus sp. nov.
  Kelloggia brevipes sp. nov.
  Goniodes minutus sp. nov.
  Goniodes laticeps Piag.
  Goniodes aberrans sp. nov.

Porzana cinereiceps
  Menopon tridens costaricense var. nov.

Odontophorus guttatus
  Goniocotes eurysema sp. nov.
  Goniodes longipes Piag.

Odontophorus leucolaemus
  Menopon praecursor meredionale var. nov.

Ortalis cinereiceps
  Lipeurus postemarginatus sp. nov.
  Menopon ortalidis sp. nov.

Guara alba
  Docophorus bisignatus N.
  Lipeurus sp.? (juv.)

Gypagus papa
  Lipeurus assesor Gieb.
  Laemobothrium delogramma sp. nov.
  Colpocephalum gypagi sp. nov.
  Menopon fasciatum Rud.

Accipiter bicolor
  Nirmus fuscus epustulatus var. nov.

Buteo borealis costaricensis
  Decophorus platystomus N.
  Nirmus curvilineatus Kell. and Kuw.
  Laemobothrium oligothrix sp. nov.
  Colpocephalum osborni australe var. nov.

Buteo abbreviatus
Docophorus platystomus N.
Buteo platypterus
Menopon macrocybe sp. nov.
Leucopternus semiplumbea
Docophorus platystomus umbrosus var. nov.
Micraster guerilla
Docophorus ultimus sp. nov.
Docophorus transversifrons sp. nov.
Piaya cayana mehleri
Nirmus atopus Kell.
Rhamphastos tocard
Docophorus cancellosus sp. nov.
Nirmus rhamphasti sp. nov.
Chloronerpes yucatanensis
Docophorus californiensis Kell.
Melanerpes formicivorus
Docophorus californiensis Kell.
Melanerpes aurifrons hoffmanii
Docophorus californiensis Kell.
Menopon praecursor meredionale var. nov.
Dryobates villosus jardini
Docophorus californiensis Kell.
Trogon caligatus
Nirmus hastiformis sp. nov.
Nyctidromus albicollis
Colpocephalum extraneum sp. nov.
Chaetura griseiventris
Nitzschia bruneri meridionalis var. nov.
Amizillis tzacatl
Physostomum jiminezi sp. nov.
Selasphorus flammula
Physostomum jiminezi sp. nov.
Physostomum doratophorum sp. nov.
Thamnophilus doliatus
Menopon laticorpus sp. nov.
Tityra personata
Menopon tityrus sp. nov.
Manacus candaei
Docophorus bruneri sp. nov.
Muscivora tyrannus
Nirmus parabolocybe sp. nov.
Tyrannus melancholicus
Nirmus parabolocybe sp. nov.
Myiozetetes cayanensis
Physostomum leptosomum sp. nov.
Myiarchus lawrencei nigricapillus
Nirmus atopus Kell.
Physostomum leptosomum sp. nov.
Menopon distinctum Kell.
Empidonax atriceps
Menopon stenodesmum sp. nov.
Momotus lessoni
Nirmus marginellus N.

Psilorhinus mexicanus
Docophorus underwoodii sp. nov.

Zarhynchus wagleri
Nirmus francisi sp. nov.
Colpocephalum luroris sp. nov.
Colpocephalum mirabile sp. nov.

Junco vulcani
Docophorus communis N.

Acanthadops bairdi
Docophorus communis N.

Chlorophonia callophrys
Docophorus communis N.
Menopon thoracicum Gieb.

Calospiza guttata chrysophrys
Docophorus communis N.

Tanagra cana
Physostomum angulatum Kell.
Physostomum australe Kell.
Physostomum subangulatum sp. nov.
Menopon thoracicum majus var. nov.

Tanagra palmarum melanoptera
Menopon stenodesmum sp. nov.

Piranga bidentata sanguinolenta
Nirmus melanacocus sp. nov.
Menopon thoracicum Gieb.

Ramphocelus passerini
Menopon thoracicum fuscum var. nov.

Pselliophorus tibialis
Docophorus communis N.

Pezopetes capitalis
Docophorus communis N.
Nirmus pseudophaeus sp. nov.

Buarremon brunneinuchus
Menopon difficile sp. nov.

Stelgidopteryx rubicollis uropygialis
Nirmus atopus Kell.
Menopon pallidoris sp. nov.

Ptiliogonys caudatus
Docophorus communis N.
Nirmus brachythorax ptiliogonis var. nov.

Helminthophila peregrina
Physostomum picturatum Car.

Compsothlypis pitiayumi
Docophorus communis N.
Physostomum pallens Kell.

Wilsonia pusilla
Docophorus communis N.

Catharus gracilirostris
Menopon thoracicum Gieb.

Merula grayi
Nirmus caligineus sp. nov.
Menopon thoracicum majus var. nov.

Merula nigrescens
Docophorus communis N.

INDEX

COLPOCEPHALUM
extraneum, 51
gypagi, 49
luroris, 52
mirabile, 53
osborni costaricense, 50
DOCOPHORUS
bisignatus, 4
bruneri, 6
californiensis, 6
cancellosus, 10
communis, 9
platystomus, 4
platystomus umbrosus, 4
transversifrons, 5
underwoodi, 8
GONIOCOTES
eurysema, 28
GONIODES
aberrans, 35
laticeps, 35
longipes, 37
minutus, 33
KELLOGGIA
brevipes, 32
LAEMOBOTHRIUM
delogramma, 37
oligothryx, 39
LIPEURUS
assesor, 27
longisetaceus, 25
longipes tinimi, 24
postemarginatus, 25
MENOPON
difficile, 66
distinctum, 62
fasciatum, 58
laticorpus, 68
macrocybe, 59
ortalidis, 57
palloris, 67
praecursor meridionale, 60
stenodesmum, 62
thoracicum, 64
thoracicum majus, 65
thoracicum fuscum, 65
tityrus, 60
tridens costaricensis, 56
NIRMUS
atopus, 12
brachythorax ptiliogonis, 21
caligineus, 22
curvilineatus, 12
francisi, 17
fuscus epustulatus, 11
hastaformis, 14
marginellus, 17
melanococus, 19
parabolocybe, 15
pseudophaeus, 20
rhamphasti, 13
NITZSCHIA
bruneri, 55
bruneri meridionalis, 56
ORNICHOLAX
robustus, 29
PHYSOSTOMUM
angulatum, 46
australe, 46
doratophorum, 43
jiminezi, 41
leptosomum, 44
pallens, 48
picturatum, 48
subangulatum, 46

EXPLANATION OF PLATES

PLATE I

1. *Docophorus transversifrons* sp. nov.
2. *Docophorus bruneri* sp. nov.
3. *Docophorus underwoodii* sp. nov.
4. *Docophorus cancellosus* sp. nov.

PLATE II

1. *Nirmus rhamphasti* sp. nov., male.
2. *Nirmus hastaformis* sp. nov., female.
3. *Nirmus parabolocybe* sp. nov., female.
4. *Nirmus marginellus* sp. nov., female.
5. *Nirmus francisi* sp. nov., female.
6. *Nirmus melanococus* sp. nov., female.

PLATE III

1. *Nirmus pseudophaeus* sp. nov., female.
2. *Nirmus caligeneus* sp. nov., female.
3. *Lipeurus longipes tinami* var. nov., male.
4. *Lipeurus postemarginatus* sp. nov., female.
5. *Lipeurus assesor* Giebel, female.
6. *Goniocotes eurysema* sp. nov., female.

PLATE IV

1. *Goniodes minutus* sp. nov., male.
2. *Goniodes minutus*, head of female.
3. *Goniodes laticeps* Piaget, female.
4. *Goniodes aberrans* sp. nov., male.
5. *Goniodes aberrans*, head of female.
6. *Laemobothrium delogramma* sp. nov., female.
7. *Laemobothrium oligothrix* sp. nov., female.

PLATE V

1. *Physostomum jiminezi* sp. nov., female.
2. *Physostomum leptosomum* sp. nov., female.
3. *Physostomum subangulatum* sp. nov., female.
4. *Physostomum doratophorum* sp. nov., female.

PLATE VI

2. *Colpocephalum gypagi* sp. nov., female.
3. *Colpocephalum extraneum* sp. nov., female.
4. *Colpocephalum luroris* sp. nov., female.
5. *Colpocephalum mirabile* sp. nov., female.

PLATE VII

1. *Menopon ortalidis* sp. nov., male.
2. *Menopon macrocybe* sp. nov., female.
3. *Menopon thoracicum* Giebel, female.
4. *Menopon tityrus* sp. nov., female.
5. *Menopon laticorpus* sp. nov., female.

PLATE VIII

1. *Menopon difficile* sp. nov., female.
2. *Menopon stenodesmum* sp. nov., female.
3. *Menopon palloris* sp. nov., female.
4. *Menopon fasciatum* Rud., female.

PLATE IX

1–1*c*. *Ornicholax robustus* sp. nov.
2–2*c*. *Kelloggia brevipes* sp. nov.

PLATE I

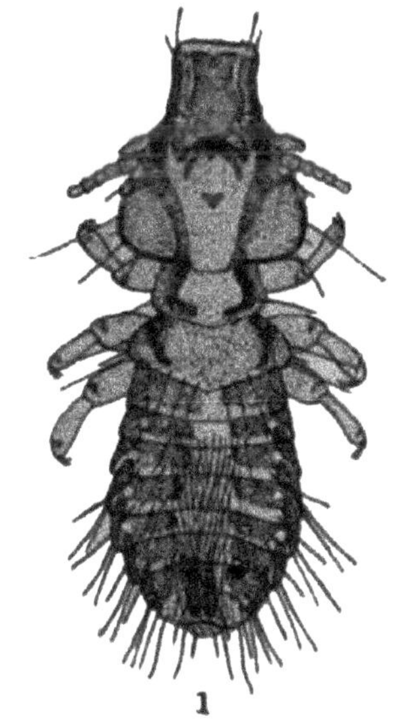

1

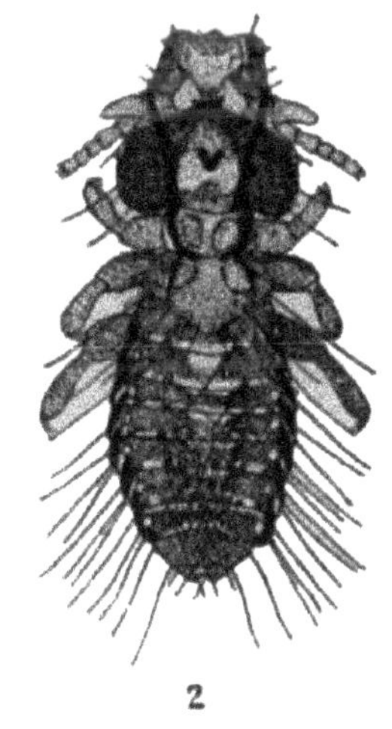

2

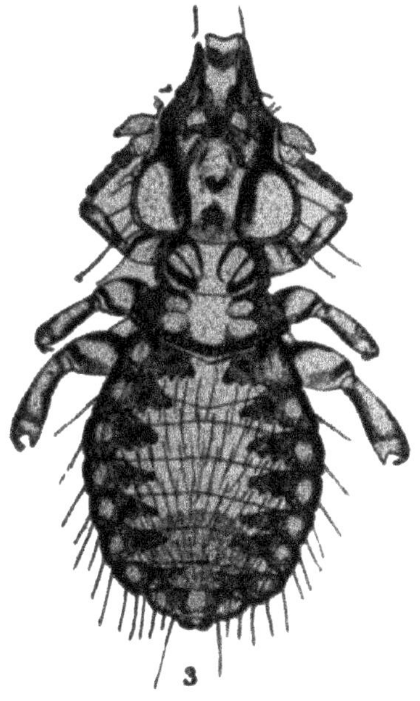

3

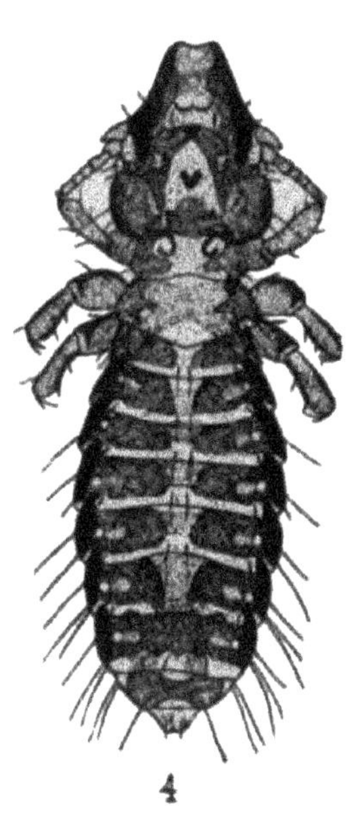

4

PLATE II

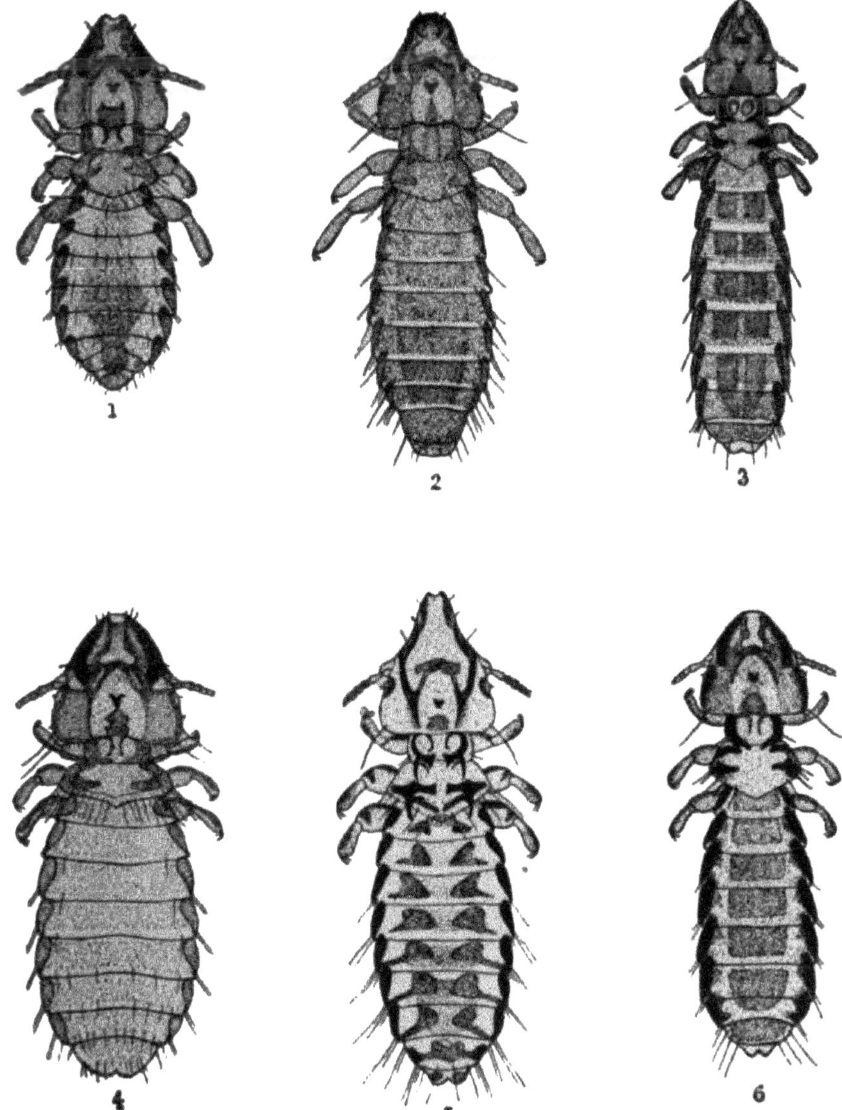

PLATE III

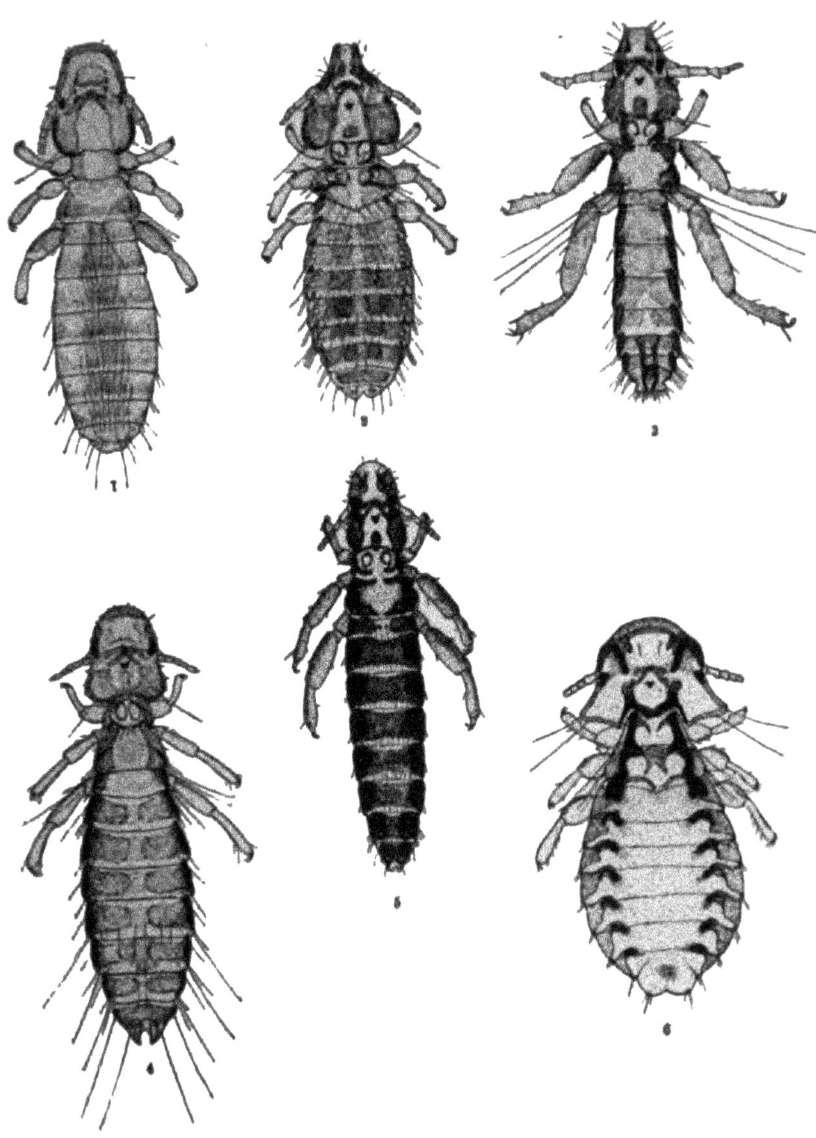

PLATE IV

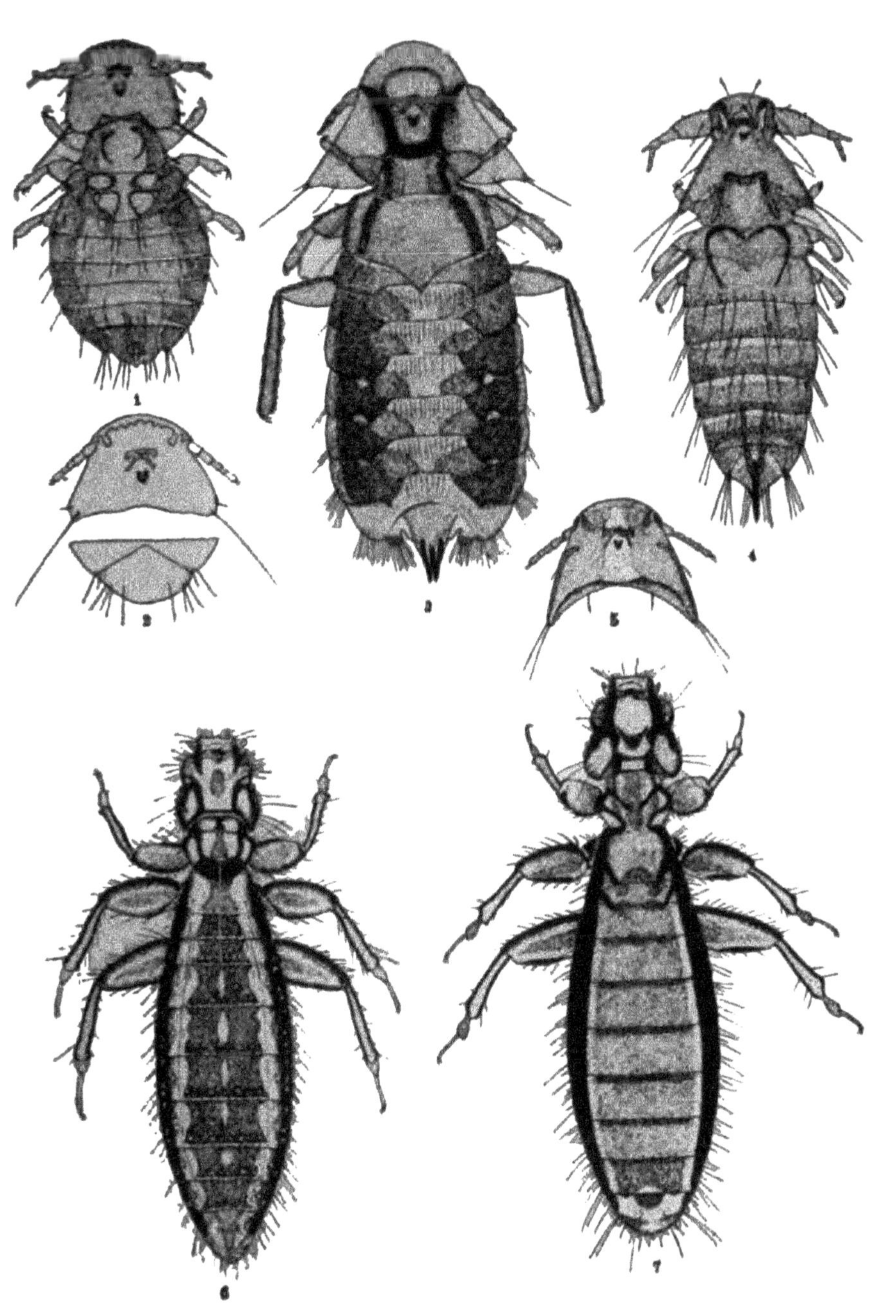

PLATE V

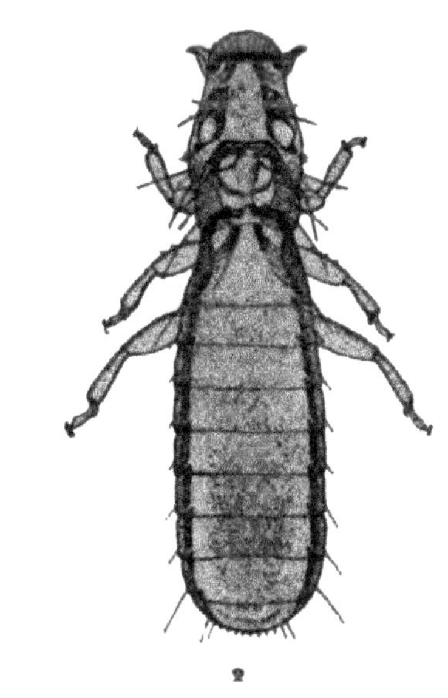

2

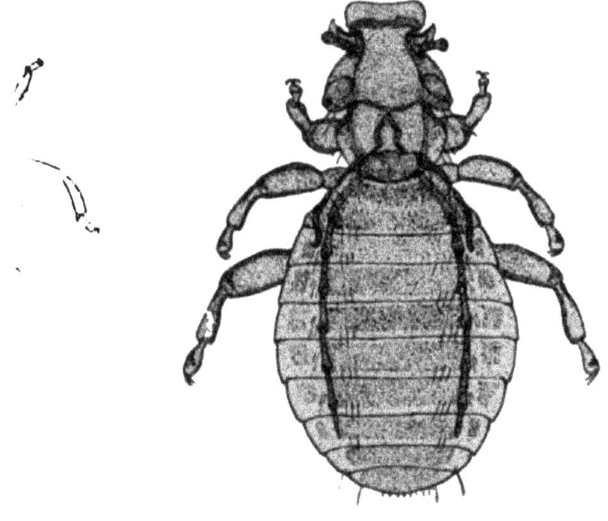

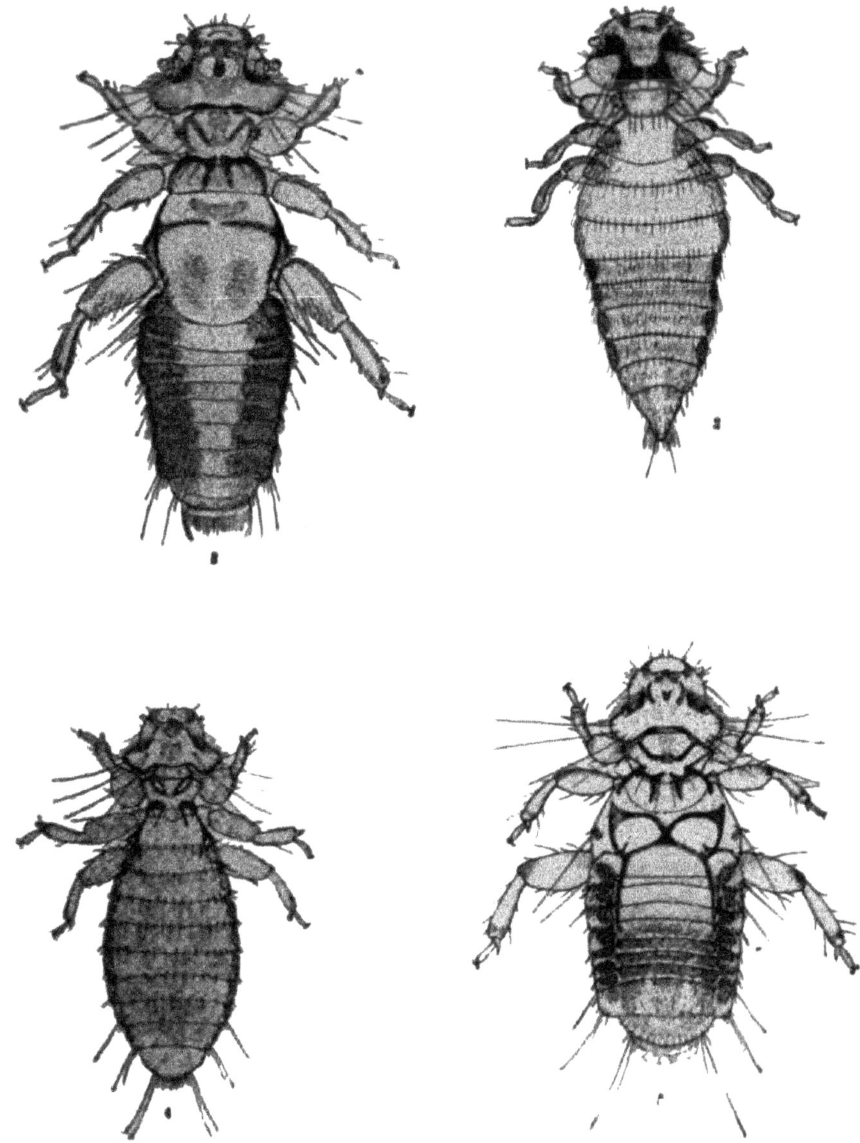

PLATE VII

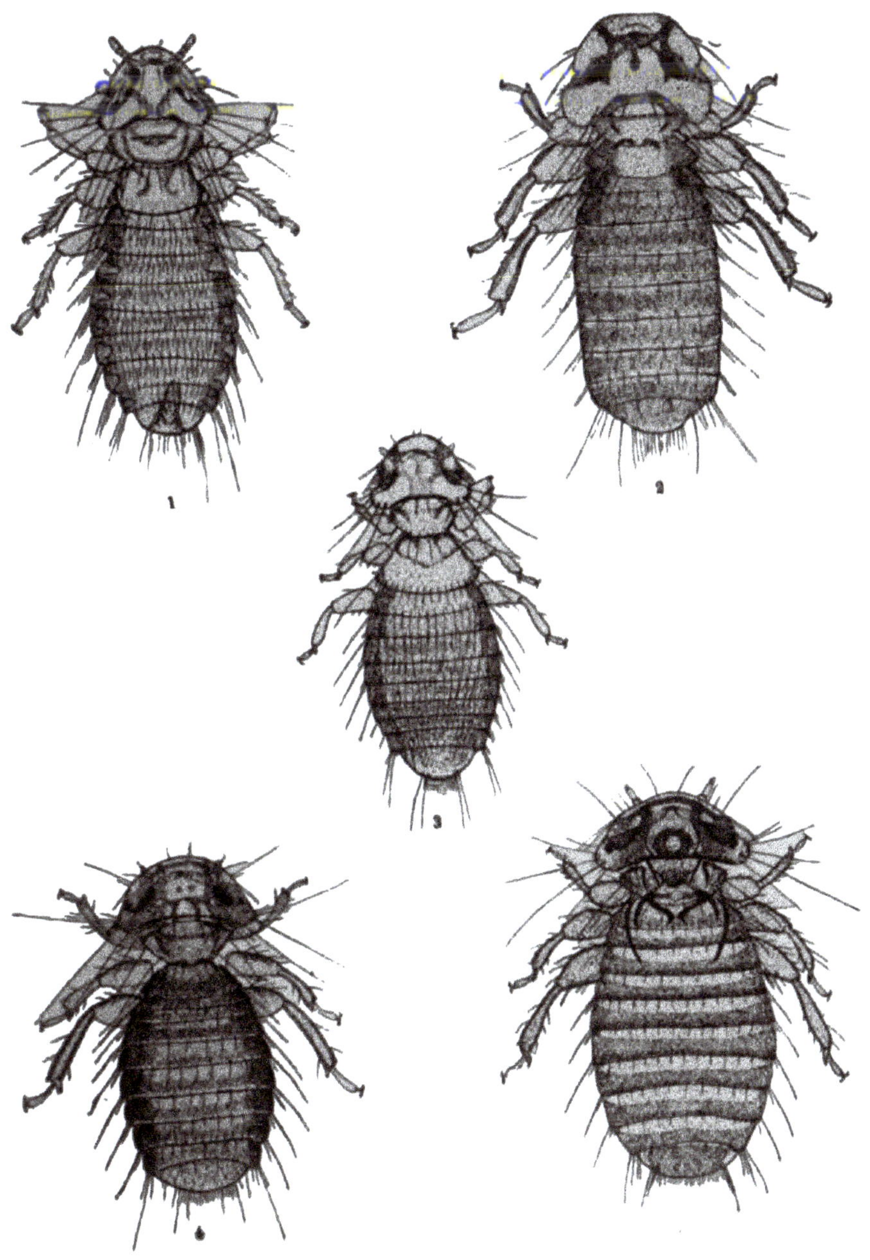

PLATE VIII

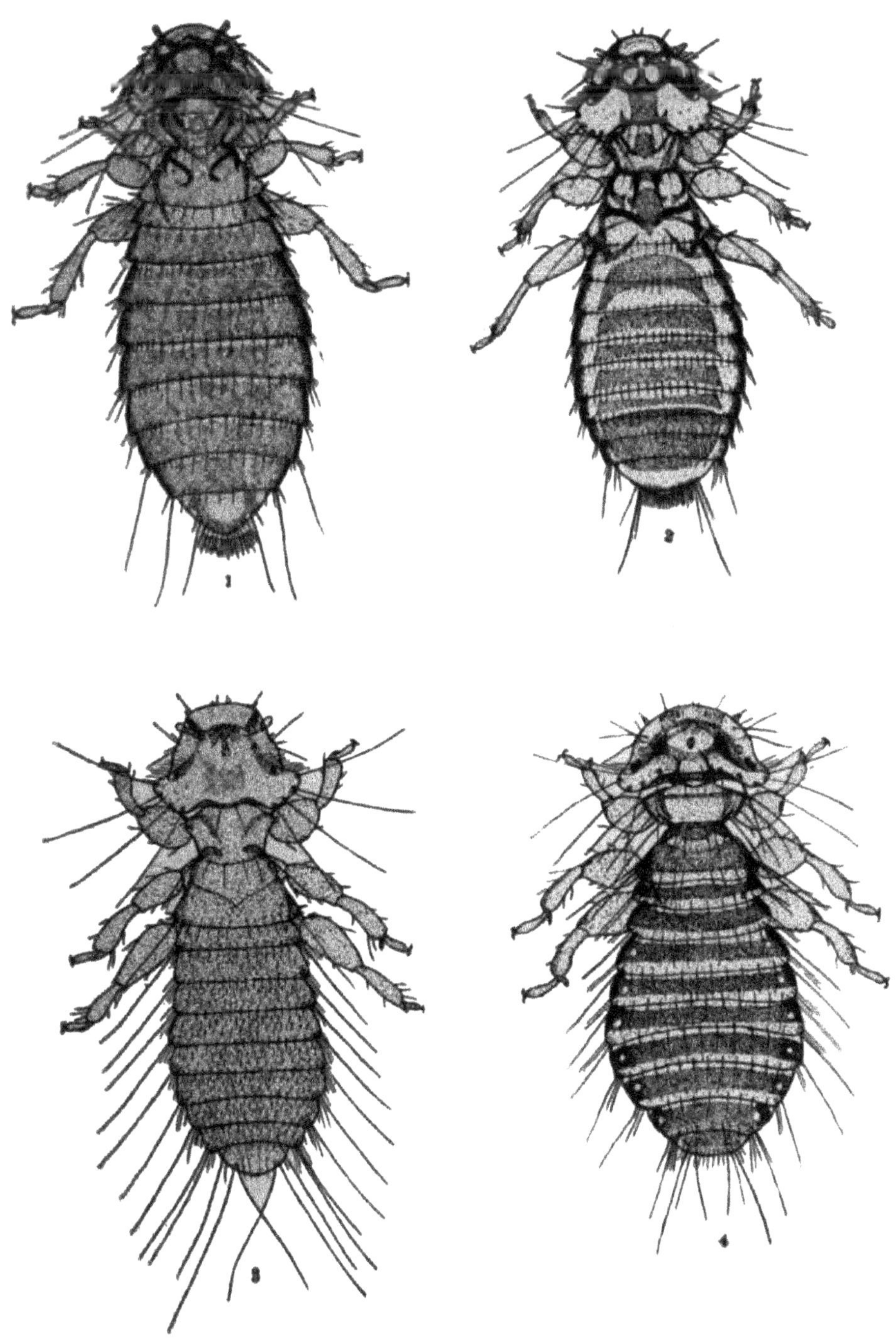

PLATE IX

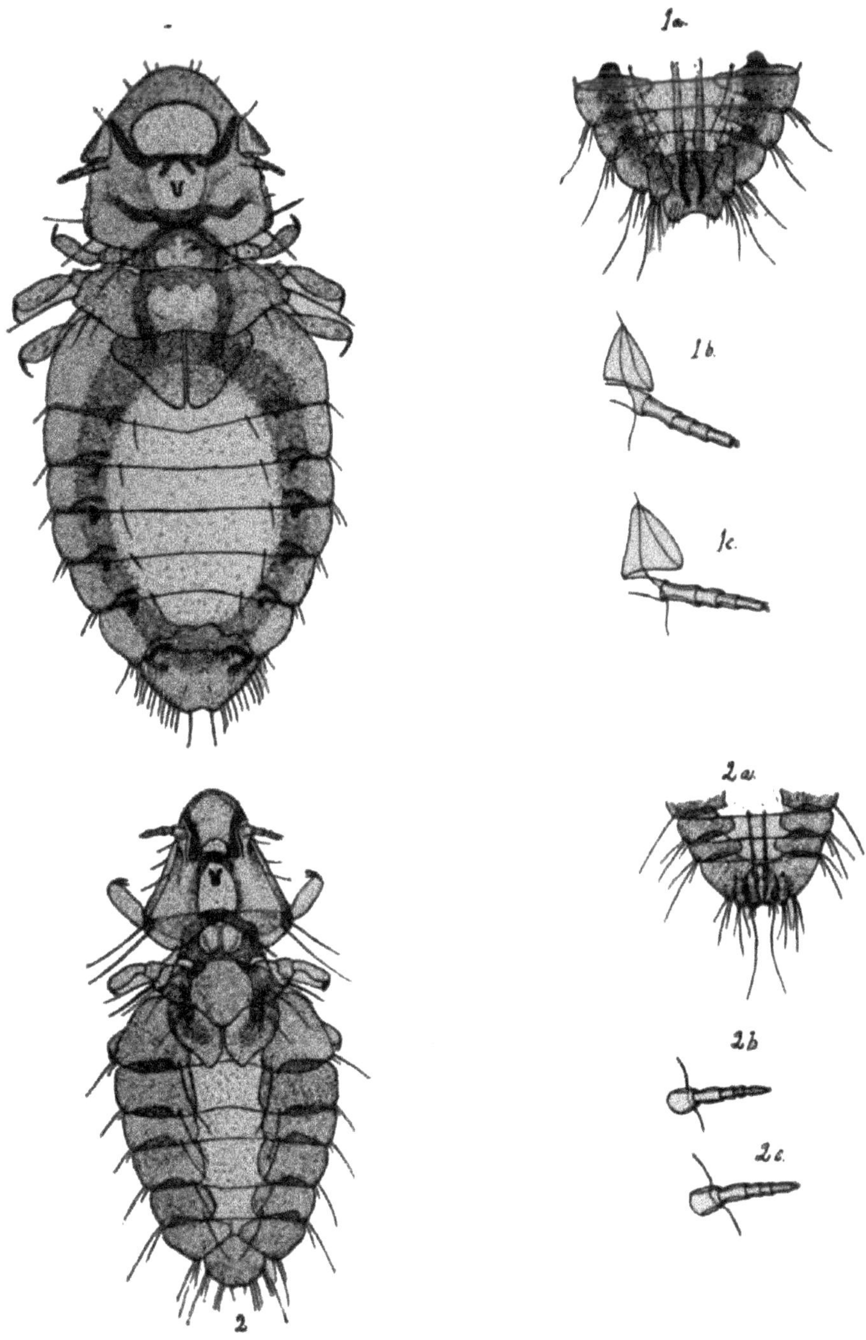

Volumes I and II of UNIVERSITY STUDIES are each complete in four numbers.

Index and title-page for each volume is published separately.

A list of the papers printed in the first two volumes may be had on application.

Single numbers (excepting vol. I, no. 1, and vol. II, no. 3) may be had for $1.00 each.

A few copies of volumes I and II complete in numbers are still to be had.

All communications regarding purchase or exchange should be addressed to

THE UNIVERSITY OF NEBRASKA LIBRARY

LINCOLN, NEB., U. S. A.

JACOB NORTH & CO., PRINTERS, LINCOLN.

Vol. III. July, 1903

University Studies

Published by the University of Nebraska

CONTENTS

I. George Sand and Her French Style
Prosser Hall Frye. 199

II. Notes on Certain Negative Verb Contractions in the Present
Louise Pound 223

III. On the Variation and Functional Relation of Certain Sentence-Constants in Standard Literature
R. E. Moritz 229

IV. On the Errors in the Methods of Measuring the Rotary Polarization of Absorbing Substances
Fred J. Bates 255

V. The Magnetic Rotary Dispersion of Solutions of Anomalous Dispersing Substances
Fred J. Bates 265

LINCOLN, NEBRASKA

Entered at the post-office in Lincoln, Nebraska, as second-class matter, as University Bulletin, Series 8, No. 11

UNIVERSITY STUDIES

VOL. III *JULY, 1903* NO. 3

I.—*George Sand and Her French Style*[1]

BY PROSSER HALL FRYE

Though it is to be feared that the influence which Matthew Arnold[2] speaks of as exerted by George Sand upon his own youth is exceptional and that as a matter of fact she has never been particularly popular with English readers, yet she certainly ought in justice to be so, for more than any other great French novelist she wrote in the English way. The English judge writing by its spontaneity rather than by its finish. They have hardly been able to understand, at least until very recently, much less to sympathize with the feeling of those French writers, who in assuming the name of artists, have tried to indicate something of the slow, self-conscious elaboration of their processes. To the Englishman writing is a gift, not an art; and he has never been tempted to confound the two. This is the reason that style and construction have counted for so relatively little in the English novel. Even so great a novelist as Thackeray has no composition to speak of; while the fact that a person with George Meredith's viciousness of expression should have won his reputation as an author, illustrates the native English indifference to grace of manner. And yet, to be just, Mr. Meredith does not

[1] George Sand: the novels, *Histoire de ma vie*, *Correspondance*, etc. See also, for original impressions, Caro: *George Sand;* Flaubert: *Lettres à George Sand;* des Goncourt: *Journal;* Sainte Beuve: *Portraits contemporains*, note to *George Sand;* Heine: *Lutezia;* Matthew Arnold: *George Sand;* etc.

[2] *George Sand.*

in this particular suit much better with the English ideal than he does with the French; for the former in its regard for spontaneity does at least imply a respect for naturalness.

The fact is, the English have formed their written upon the model of their spoken style. They seem, as it were, to assume that their literature is written offhand, and must be judged, even a little indulged, it may be, with this circumstance in mind; as though it were to be expected of an author, not that he should necessarily give long time and thought to his expression, but that he should write quickly and fluently, above all naturally—in short, as though his best possession were the pen of the ready writer. What he has accomplished, then, is to be criticised in accordance with these conditions, not as aiming at perfection, at the expense of unlimited pains and patience, at any cost! On the contrary, the main requirement made of himself by the French writer is that he attain this perfection, which the former has left as unattainable or inconvenient or impertinent—a perfection absolute and final, which he has always before his eyes as the goal of his aspirations and towards which he strives relentlessly. Time and labor are no object; only that when the work leaves his pen-cramped hand it shall be the best that can be made out of words, the very best without reserve or abatement. Ease, or at least the appearance of ease, may be desirable; not, however, because it is the main purpose of writing to write easily, but because it is a property of elegance that whatever is done, no matter with what difficulty, should be done too well to show the effort. But diffuseness, approximation, confusion, and the like unavoidable accompaniments of conversationalism and improvisation are forever unpardonable equally with the appearance of stress and strain. While the English write prose with something of the carelessness of talk, the French write prose with the same care that we give to poetry.

It is impossible to describe this state of mind better than Maupassant has done in speaking of an author who stands in every respect in the most striking contrast with George Sand and who represents most characteristically the literary tendencies and ideals, if not the actual performance, of his countrymen—Gustave Flaubert.

"Haunted by this absolute belief that there exists but one way of expressing a thing, one word to name it, one adjective to qualify it, one verb to animate it, he [Flaubert] would devote himself to superhuman efforts to discover for every phrase that word, that epithet, that verb. In this way he believed in a mysterious harmony of expressions, and when a word otherwise suitable seemed to him to lack euphony, he would go on searching for another with invincible patience, sure that he had not yet found the true, the unique word.

"For him writing was a redoubtable undertaking, full of torment, peril, and weariness. He would seat himself at his table in fear and love of that dear distracting business. . . . .

"Then he would begin to write slowly, stopping again and again, beginning over and over, erasing, interlining, filling the margins, criss-crossing, spoiling twenty pages for one he finished, and groaning with the effort of thought like a woodsawyer.

"Sometimes, tossing his pen into a great oriental pewter tray which he kept full of carefully cut goose quills, he would seize his sheet of paper, raise it to the level of his eyes, and leaning on his elbow, begin to declaim in a loud rasping voice, listening the while to the rhythm of his prose, pausing to catch a fugitive reverberation, combining the tones, separating the assonances, and disposing commas cunningly like resting places on a long road.

. . . .

"A thousand preoccupations would beset him at once, but this desperate certainty always remained fixed in his mind: 'Among all these phrases, forms, and turns of expression there is but one phrase, one form, one turn of expression to represent what I want to say.'

"And, red in the face, with swollen cheeks and neck, his muscles tense like a straining athlete's, he would struggle frantically with idea and expression, coupling them in spite of themselves, holding them indissolubly together by the force of his will, grasping the thought and subjugating it little by little with superhuman effort and fatigue, and caging it up, like a captive beast, in a solid and exact form."[1]

How excessive, but at the same time how indicative in its excess of the writer's scrupulousness! And while the passion of perfection may not be so virulent with every one of his nation as it was with Flaubert, yet was there ever Englishman, however exceptional, who wrote like this? It is necessary only to compare these remarks with our traditions of Scott's unfaltering pen and Shakespeare's unblotted page in order to recognize how different the spirit of French and English prose.

This difference of style as between the two nations may be referred, at least in effect, to a variety of causes, the most influential of which are probably these three.

In the first place the Englishman has never made so wide a divorce between thinking and writing as has the Frenchman. The former has temperamentally given thought such a decided preeminence over the presentation of thought that he has hardly considered the two as separate at all; but when he has had anything to write, has been content simply to think it out in words, and let it go at that. He has always managed to say what he wanted to say, if he has talked long enough; and writing is a sort of soliloquy in which no one can interrupt him. Consider how Browning conducts a poem, like a monologue upon which his readers are licensed for the nonce to eavesdrop, quite welcome to whatever, if anything, they can manage to pick up. One can, to be sure, put down his book, or throw it away; but his attitude under such circumstances is one of haughty indifference—he writes no better. The Frenchman, on the contrary, while thinking, considers that he is in privacy and may be as informal as he likes. In expressing himself, however, he remembers that he is in the presence of others, whom he is eager to please and impress—he feels that he must strike and maintain his pose. It is now an affair of manners, and manners maketh the Frenchman. He looks upon his thought as one thing, the presentation of his thought as quite another. And so when he

[1] Maupassant, Introduction to *Lettres de Gustave Flaubert à George Sand*. Compare also his introduction to *Pierre et Jean*.

comes to write, it is the result rather than the process that he aims to give, and then crystallized in polished sentences which shall have something of the finality of a formula, and forestall posterity. When he has once said a thing it is said forever. From this peculiarity of his mind results the importance taken in his literature by epigram.

Beside this intellectual difference between the two nations there exists also a difference of language which, though it may be sprung from the former, must be spoken of separately. French words, partly through the influence of the Academy, have comparatively little of that indistinctness or blur of outline, that sort of emotional penumbra which is so noticeable with English words and to which English poetry owes in great part its haunting suggestiveness. But they are defined and outlined, stamped clean to the very edges, covering the ideas upon which they are set with a nicety and exactitude that make French, for all its narrow vocabulary, an ideal instrument of thought, particularly analytic thought. About most English words there is something vague, floating, elusive—something left over to be accounted for after they are applied to the ideas which they symbolize. And this fringe of meaning, which scatters such an iridescent halo about English poetry, makes it necessary in English prose, where such diffraction is an embarrassment, to qualify, limit, and extenuate in order to define the thought with accuracy.

But these two conditions, far as they go, are not enough in themselves to explain all the phenomena we have been observing and have still to observe. It is necessary to take account also of a total difference as between the conceptions of genius held by the two peoples. Genius to the Frenchman means essentially an infinite capacity for taking pains—an intelligence capable of discerning the nature of the end proposed, of holding it steadily in view, and of applying cunningly and patiently every means at hand to its attainment. Characteristically, the ends of French genius are always rational, attainable by the eminently reasonable man—the man, it may almost be said, of common ideas and uncommon energies. To every race genius is the apotheosis it makes of its own best faculty; and intelligence is the Frenchman's best faculty, as imagination is the Englishman's. "Our literature," declares Nisard[1] in his well-known characterization of the French spirit, "is, as it were, the living image of this government of the faculties by reason. . . . This is the spectacle offered us by our masterpieces—they display nothing but a higher reason, sufficiently reinforced by the love of truth to dominate the imagination and the senses and to draw admirable assistance whence ordinarily come the greatest dangers." From this eminently practical point of view there is nothing absurd in Flaubert's sitting down with the avowed intention of producing a classic—and succeeding in doing so. While by the very fact his opinion concerning the spirit of the literature, which he knew well enough to produce a masterpiece in it by malice prepense, takes on a representative character.

"Talent," he declares for his part, and to appreciate the force of the word the reader must remember that it is one maker of *chefs-d'œuvre* coaching another,[2] "Talent is only long patience. Everything which one desires to express must be looked at with sufficient attention and during a sufficiently long time to discover in it some aspect which no one has as yet seen or described. In everything there is still some spot unexplored. . . . The smallest object contains something unknown. Find it. To describe a fire that flames, and a tree on a plain, look, keep looking, at that flame and that tree till in your eyes they have lost all resemblance to any other tree or any other fire.

"This is the way to become original."

To the Englishman, on the contrary, genius signifies something more, at least something other than the free play of intelligence. It implies inspiration, as he calls it—the revelation that seems to come down like a sudden light upon life, laying bare its very secrets, transmuting it with new meaning, and possessing the writer, like one beside himself, with an enthusiasm, a power, an eloquence beyond his own. And this capricious, heady, lawless spirit, this emotional transport and exaltation

[1] *Histoire de la littérature française.*

[2] Introduction to Maupassant's *Pierre et Jean*, translated by Hugh Craig.

which visits the author without warning and relieves him of the labor of preparation, it is doubtful whether the Frenchman has ever yet quite succeeded in appreciating; whether Shakespeare does not still appear to him under the image of Voltaire's drunken god adream; as the English have never learned, probably never can learn thoroughly to admire the pale, refrigerated shimmer of Racine. "We are very much mistaken," cries Zola,[1] "when we think that the characteristic of a good style is a sublime confusion with just a dash of madness in it; in reality the merit of a style depends upon its logic and clearness." The Frenchman, in short, tends always to subjectivize his emotion and possess it, thereby making his literature objective, while the Englishman tends to objectivize his and to allow it to possess him, thereby making his literature largely subjective.

And yet this difference, which is just the difference between art and genius, is the one critically differential of the two literatures. Language is at best an inadequate medium, no matter how well handled. And one in accordance with his temperament will prefer the relatively imperfect embodiment of a lofty ideal; and another, the well-rounded embodiment of a relatively low ideal. The former produces a literature of aspiration, in which the whole structure of language is bent and strained by the stress of meaning forced upon it, a romantic literature, strong in poetry and weak in prose, like English. The latter produces a finished and finite literature, neat, elegant, and limited, strong in prose and weak in poetry, a classic literature, like French. For the exuberance of life always tends to shatter and demolish form; and it is only by painful labor, by clipping and paring and pruning that a fresh and modern existence can be forced into vessels and moulds. This is probably something of what Flaubert meant by his celebrated and oft-quoted remark, "The idea springs from the form,"[2] a saying so hard for the Englishman, and yet almost a shibboleth to his own disciples. At all events the remark has this much truth: in Goethe's words, *"die Kunst ist nur Gestaltung,"* art is only form; and in deter-

[1] *Le Roman expérimental.*
[2] The Goncourts' *Journal.*

mining his form, in finding what he can or can not put into language without splitting it, the artist does at least determine what his idea shall be. To this general effect George Sand writes to Flaubert:[1] "It seems to me that your school does n't pay enough attention to the inwardness of things and is too much inclined to rest satisfied with their superficies. As a consequence of searching for form you neglect the profundities and address yourself only to the litterati." Ay; but he knew that he could not render the profundities without doing violence to the shape and figure of his work, and that he would not do. As Mr. Henry James says, "He had no faith in the power of the moral to offer a surface."[2]

For these causes, principal among others, English literature is distinguished from French by its preference, at least in effect, for improvisation and inspiration. And it is for this reason, because these are so exactly the characteristics of her writing, that George Sand deserves the attention of the English reader. "No writer," asserts Mr. James, "has produced such great effects with an equal absence of premeditation."[3] Her spontaneity, ease, and fluency; her individuality, sensibility, and inventiveness are the positive virtues which most please the English sense; while the vices of their reverse—her diffuseness, confusion, and haziness, her irregularity, extravagance, and wilfulness, in fine, her lack of discipline—are all defects which the English least notice or most readily excuse. She had no art in the strict sense: but she had inspiration, its virtues and vices, its qualities and defects.

The essential truth of this judgment of George Sand has never been disputed by her countrymen or indeed by herself. "She knows," writes Balzac,[4] "and said of herself just what I think, without saying it to her, namely, that she has neither force of conception, nor gift of constructing plots, nor faculty of reaching the true, nor the art of pathos, but—without knowing the French language—she has *style;* and that is true." But

---

[1] *Correspondance*, *Lettre* CVXLIX.

[2] *Essays in London and Elsewhere:* Gustave Flaubert.

[3] *French Poets and Novelists:* George Sand.

[4] *Correspondance*, March 2, 1838.

in spite of the charm of her writing, almost irresistible in the wooing of the soft slow sentences,[1] the inevitable weaknesses of the facility which stood her in place of literary method have been observed over and over, particularly where they are most noticeable, in her construction. Her lack of fundamental plan, of architectural design, has impaired a work that otherwise would have in perfection, as it now has in bulk, few peers. Sentences she could write, and chapters, exquisite in touch and feeling,—few better; but alas! for all their delicacy, fragments. When it comes to building up piece by piece a single whole, an entire fabric with the subdual of many parts to the perfect harmony of one great purpose,—there her weakness, the weakness of facility, is manifest. "*Le génie*," she says herself, "*vient du cœur et ne réside pas dans la forme*;"[2] and it was her misfortune to take her own statement too literally—so literally, indeed, that in Flaubert's sense she had no form at all.

For to Flaubert form meant something more comprehensive than style.

"While attaching great importance to observation and analysis, he attached an even greater importance to composition and style. In his opinion it was these two qualities in especial which made a book imperishable. By composition he understood that obstinate labor which consists in expressing only the essence of the successive acts of a life, in choosing only the characteristic traits, and in grouping and combining them so that they shall concur perfectly to the effect intended."[3]

It was not merely his language, then, for which Flaubert was so anxiously concerned in his obstinate wrestlings with expression—it was as well the figure, the shape, the whole concrete plastic embodiment—the *Gestaltung*—under which he should exhibit his conception, at once the emanation and the incorporation of the idea as surely as the pose of a statue is decisive of the final impression produced, to which the style was to add its particular evocation of sentiment like the music of an opera. This

[1] Compare Taine's essay, *George Sand*, for an appreciation of her style.

[2] *Histoire de ma vie*.

[3] Maupassant's *Étude*, prefixed to *Lettres de Gustave Flaubert à George Sand*.

was his conception of form, a complete organic whole, a creation in all its parts fatally answerable to the thought of its creator. In the words of Stevenson, who suffered under much the same infliction of literary conscience as Flaubert:[1]

"For the welter of impression, all forcible but all discreet, which life presents, it [art] substitutes a certain artificial series of impressions, all indeed most feebly represented, but all aiming at the same effect, all eloquent of the same idea, all chiming together like consonant notes in music or like the graduated tints in a good picture. From all its chapters, from all its pages, from all its sentences, the well-written novel echoes and reechoes its one creative and controlling thought; to this must every incident and character contribute; the style must have been pitched in unison with this; and if there is anywhere a word that looks another way, the book would be stronger, clearer, and (I had almost said) fuller without it. Life is monstrous, infinite, illogical, abrupt, and poignant; a work of art in comparison is neat, finite, self-contained, rational, flowing, and emasculate."

It is hardly surprising that of form in this consummate interpretation, as the deliberate artist understands it, George Sand should show small sense. With her quick, sensitive, and rather shallow nature she was by no means so likely to distinguish herself through the manifestation of intellect and will in literature as through the manifestation of sentiment and emotion—not so much in composition as in style. For these, as nearly as they can be discriminated, would seem to be the particular powers of the two.[2] A Greek tragedy imposed, not by its emotional and sentimental surface-play, but by its deep purposefulness, its severe determinism; and so to a lesser degree the drama of Racine, and to some extent all genuinely characteristic French work as compared with English; while a poem of Shelley's or Tennyson's, on the contrary, pleases by the prismatic shimmer of sentiment with which it is overlaid. The one is typically the affair of composition, the other of style. And toward the latter extreme George Sand's writing naturally gravitates in spite of the

[1] *A Humble Remonstrance.*

[2] Compare Pater's *Essay on Style.*

general tradition to which it belongs. It is full of color and feeling, it is splendidly romantic; but when one comes to consider it as a whole, to look toward its end and reflect upon its tendency, one is struck by its ineptitude to its purpose.

In this respect her work corresponds very closely with the account she herself gives of her own intellectual condition:[1]

"Wisdom," she remarks very justly, "consists perhaps in classifying one's impressions, in keeping them from encroaching upon one another, and in isolating, if necessary, the particular impression one wishes to receive. In this way arise the great works of genius." And of herself: "In order to put an end to my lack of mental discipline I have prescribed myself a regular life and a daily task—and then two-thirds of the time I lose myself in dreaming or reading or writing something very different from that in which I ought to be absorbed. Had it not been for this intellectual dissipation I should have acquired some sort of an education, for I comprehend readily enough—indeed, if anything, I get to the bottom of things a little too readily; I should have forced my memory to classify its ideas. To understand and to know has been my constant aspiration: but of what I have wished to realize I have realized nothing. My will has never governed my thought. . . . The external has always acted upon me more than I have acted upon it. I have become a mirror from which my own image is obliterated, so completely is it filled with a confused reflection of figures and objects."

These characteristic mental traits of hers show themselves in her writing in several ways. For the careful and consistent reader, one of the most painful experiences is prepared by the frequency with which she falls away in the latter part of her novels from the high standard of her beginnings,—and that not merely in her early work, when she was learning her trade, but in the work of after periods as well, when she had served a long apprenticeship to her art. It is sad to notice, for instance, that M. Faguet speaks of the first volume of a story like the *Beaux Messieurs de Bois Doré* as a *chef-d'œuvre* and then drops the remainder of it into the oblivion of silence as though in mercy

[1]*Impressions et souvenirs.*

of its defects.[1] And it is sadder still to find for oneself a book of such fair promise, which might have been completed faultlessly within the limit of three hundred pages, running on into a wreck of diminishing climaxes and crises and feeble afterthoughts, until it expires tardily of sheer exhaustion, without the needed apology for being so long a-dying, at more than twice its natural age,—spoiled for no other apparent reason than that the writer wrote too easily to stop when she had finished. Of her might be said what Dryden says of Fletcher: "He is a true Englishman—he knows not when to give over." It is hardly exaggeration to advise one wishing to read George Sand's best work to read only the first halves of her novels.

And yet the difficulty were not to be so escaped. This fault of saying too much, this plethora of words occurs again and again over smaller areas than an entire book. With the inveteracy of disease it infects the whole system. The author is not willing to make the reader a suggestion, to drop him a hint, to risk herself to his perspicacity. She must needs explain—often more for her own sake than for his, it would appear—until there is left over event and motive hardly a single shadow for him to penetrate, but everything lies exposed in an even glare of revelation, like the monotonous landscape of our great western prairie, without concealment or mystery. There are no skeletons in George Sand's closet; she has got them all out into the middle of the floor. And her dialogue is as prolix as her analyses. Her characters seem possessed with her own fondness for explication, and invariably talk matters out to a finish, however trivial, so that the reader is constantly outrunning the writer with a sense at the end of disillusion and disappointment. This circumstance is partly accountable for the feeling of commonplaceness which frequently torments one in his George Sand, even in what he is conscious on reflection are the rarest *aperçus*. The development of her thought is so slow, so gradual, so far foreseen that her utterances are stamped with none of that surprise which we have come to consider as the hallmark of a profound saying. One is so long prepared that, when the announcement

---

[1] *Étude sur le XIXe siècle.*

finally comes, it falls flat on his tired ear like an assertion of the obvious.

Perhaps this faultiness, behind which lies always her too ready fluency, may be explained, or at least illustrated, by her manner of work. It is well known nowadays, when the personal habits of authors are more studied than their books, that she wrote at night for certain fixed hours with the regularity of a day-laborer. "She works every night from one to four, and then sets to work again during the day for a couple of hours—and . . . it makes no difference if she's disturbed. . . . Imagine that you have a faucet open in the house; some one comes in, you close it. . . . That's the way with Mme. Sand."[1] The story goes of her that if she happened to finish the novel on which she was employed an hour or even less before her time was up for the night, she would calmly set the manuscript away, the ink still damp on the page, and placidly begin another, composing rapidly as she went until the clock released her.[2] Whether rightly or wrongly one misses something here—the fond lingering over the old work, the patient review and minute revision, the reluctance to part with the child of the brain which makes every *finis* to the author a lover's parting and which is so charactersitic of the French writers of the century.

It is another story that is told of Flaubert:[3]

"When he read to his friends the tale entitled, *Un Cœur simple,* several remarks and criticisms were passed on a passage of ten lines, in which the old maïd ends by confounding her parrot with the Holy Ghost. The idea seemed too subtle for the mind of a peasant. Flaubert listened, reflected, recognized the justice of the observation—but was seized with agony. 'You're right,' he said, 'only—I should have to alter my phrase.'

"That very evening, however, he set to work. He spent the night in changing ten words; he blackened and canceled twenty sheets of paper, and finally left things as they were, unable to construct another phrase whose harmony would satisfy him.

"In the beginning of the same tale the final word of a para-

[1] *Journal des Goncourt*, March 30, 1862.

[2] *Ibid.*, Sept. 14, 1863.

[3] Maupassant's *Étude.*

graph serving as the subject of the following, might give rise to an amphibology. This distraction was pointed out to him; he recognized it and attempted to change the sense, but could not recover the sonority which he wished for, and, discouraged, exclaimed: 'So much the worse for the sense; rhythm before everything!'"

Can there be a more significant contrast than that between these two pictures: Flaubert, the great, rough, positive Norman hesitating irresolutely over a novel for seven years, unable either to perfect or relinquish it; and George Sand, the woman, feeble and timorous, one might suppose, resolutely laying aside one piece of work and taking up another in order to fill out half an hour of scheduled time? By comparison there is something very like gradeur—the grandeur of renunciation, perhaps—in this ability of hers to put away the past when she was done with it, to leave her work to its deserts without just one more backward look, just one more correction, and to pass on confidently to the next duty without worrying over what was gone. "*Consuelo,*" she writes in reply to a letter of Flaubert's, "*la Comtesse de Rudolstadt,* what in the world is that? Can it be something of mine? I have forgotten every last treacherous word of it. Do you read it? Does it really amuse you? In that case I will reread it one of these days, and if you like me I shall like myself."[1] It shows at least a self-detachment, a sobriety and moderation not always evident in French literary workmanship of a modern school with its long brooding of the thought—often serving little better purpose than to addle the eggs—and its slow coagulation of the phrase, such as we have come to associate even with Balzac, who would never let his copy go, as Gautier tells us, till it was wrung from him by his implacable taskmaster, the publisher.[2]

But for all this excess of care we might well wish that George Sand had, without going too far, shown a little more concern for what she had done, a little more for what she was about to do, were it reasonable to suppose that all her errors were due to her habits of work and could have been retrieved by revision.

[1]*Correspondance.*

[2]*Portraits contemporains.*

Much, however, of her defective construction must be charged to another cause. A certain indefiniteness of conception, a failure to decide the end from the beginning and write up to it—in short, a powerlessness to fix and realize the idea of a book, is equally a condition of her structural frailty. "Descriptions and paintings are no proof that one knows how to write; they prove only that one has strong sensations. What is expected of the writer is the expression of general ideas, and by that he is judged."[1] For after all our talk about concreteness and what not, does not every great novel rest finally upon an idea, which the story serves as a specific instance to illustrate? It is difficult perhaps to determine but it is surely legitimate to ask whether the masters have not invariably seen in their fables something wider than the single incident recorded, something standing to that incident in the relation of a general principle to a particular case. "It is not enough to have seen, to have observed; it is essential besides that something general in the case of science, something universally human in the case of art, should be, as it were, engaged in our very observation."[2] Certain it is, at all events, that we can not think of a novel in any sense great which does not result for us in some conclusion, much more comprehensive than the case in point, in regard to human life and conduct. It may not be expressible in other terms than those particular ones in which the author has rendered it; it may not lend itself to intellectual formularization at all; but there it is in the reader's mind as the residuum of his reading—the book simple, concrete, and special; the idea complex, abstract, and universal. And it is hardly reasonable to suppose that the writer could have got it thither unless he wrote with it constantly before his eyes. So true is this that the idea a book leaves with us becomes its criterion. "When such a philosophical theme," so Taine insists, "meets a person capable of carrying it to the end and expressing it completely, the novel is of the first order."[3] While Lessing makes a similar distinction from the complementary

[1] Brunetière: *Le Roman expérimental.*

[2] Brunetière: *La Raportage dans le roman.*

[3] Essay on George Sand.

point of view: "To create for a purpose, to imitate for a purpose is what distinguishes the genius from the little artist who creates only to create and imitates only to imitate, who is quite content himself with the minor satisfactions of technique, makes this technique his sole aim, and requires that we also shall be content with just that same sort of minor satisfaction which arises from his artistic but purposeless exercise of his technique."[1] And to the same effect, were it not otiose to do so, it would be possible to cite the criticism of every age which has had a great literature;[2] while a lack of sense for this "*sorte de lieu commun moral*" is an almost infallible sign of critical and literary decadence. For life is a moral affair; and if literature succeeds in its purpose of representing life, its perusal, like experience, will result in the attachment of correct values to human action, not because it is the business of literature to inculcate morals, but because it is the business of literature to represent life, and life is a moral affair. The mere stylist like Gautier is felt to be less than first rate, in spite of the seduction of his manner, simply because he has no great ideas of human life to commemorate.

But this is very different from expecting a novel to be written for the promotion of social or religious doctrine or for the exploitation of theories or hypotheses of any kind. To attempt to use literature for such a purpose or to require of it the solution of philosophical problems is evidence of a strange perversion on the part of writer or critic. Philosophies are at best fluctuating and transitory; they change from generation to generation. The consequences of human action are alone of eternal interest to the human kind. And he who builds beyond the moment must build not upon the former but upon the latter. Nor do such ideas as a rule or as an exception afford a just measure for the evaluation of human life. On the contrary, they tend to force life and its expression into narrow, ready-made equations, true enough for the day but by so much the falser for the mor-

[1] *Hamburgische Dramaturgie*, Stück 34.

[2] For instance, Johnson: *Rambler*, No. 4; Addison: *Spectator*, No. 70; Dryden: *Grounds of Criticism in Tragedy;* etc. Indeed the idea has a clear literary pedigree back to Aristotle.

row; in other words, to reduce it temporarily to order by the summary process of straight-jacketing it. One attempting the representation, or better the interpretation of life, ought to bring to its study no preconceived ideas. All such ideas should, where they enter literature at all, be strictly distinguished as foreign to its purpose; that is, as extra-literary. They may not always be impertinent or uninteresting, but they are subordinate and inessential; and where they rise into prominence and importance above the life of the book, they are so,—both impertinent and uninteresting from the point of view of literature. And yet one finds persons enough to read a novel for nothing more than its historical background, or its treatment of a political issue or some other vexed question. In spite of the modern popularization of literature,—perhaps its vulgarization,—one has not ceased to recommend Scott for the historical information to be got out of him; or George Eliot for her curious cases of moral casuistry; or Mrs. Ward for her religious disputations:—clearly literary impertinences in any case and not the vitality that gives these writers their strength.

The best training for a novelist is not a system but an experience—a first-hand knowledge of men and their ways acquired from the give and take of existence, where the hard facts, by dint of battering the consciousness, finally gain recognition. If there is one thing, though but one, for which we are indebted to *naturalism,* it is the conviction that literature and science are in thus far alike—that both proceed not from speculation but from observation. This is the open school in which the novelist learns his lesson, not in the cloisters of a creed. Here he learns of human responsibility, of the consequences of human action, of the fatality of the human will; here he learns "what life and death is;" and here finally he gets his ideas of the world direct from the world itself, not in set formulae or generalized prescriptions, but embedded in the tissue of individual examples by which he conveys them to others. Literature can never be studied from any mirrored image, not even from literature itself, without distortion or conventionality. Some arrangements of facts he must make, no doubt; but these are not the classifica-

tion of them was confused and uncertain. And the result is much the same with other novels of hers of this and other periods, which are not strictly *Tendenz* perhaps but may be fairly classed together with the preceding as extra-literary, since they were not written under purely literary inspiration nor with purely literary motives, and since—the most important test—who reads them reads them primarily for something over and above their literary interest—for the side-light generally which they throw upon the life, character, or thought of their author or of her time. "I have found," says Coleridge, and the remark is as true of the novel as it is of poetry, "that where the subject is taken immediately from the author's personal sensations and experiences, the excellence of a particular poem is but an equivocal mark, and often a fallacious pledge, of genuine poetic power."[1] Woman as she was, her feelings when aroused were ever of a vehemence to overbalance her critical judgment; and in writing for the gratification of these feelings rather than from the instinct of letters she was likely, no matter at what time of life, to reproduce the emotional confusion of her earliest period. A remark that she herself makes in her memoirs concerning *le Piccinino* is significant in this connection, and justifies in closing as well as illustrates my use of the word extra-literary as a general designation for all this kind of work. *"Ce que je pense de la noblesse de race, je l'ai écrit dans le Piccinino,"* she says, *"et je n'ai peut-être fait ce roman que pour faire les trois chapitres où j'ai développé mon sentiment sur la noblesse."*[2] It is often so, too often, in fact, that the purpose of her novels early and late, as she confesses here, is to be sought and found outside of character, situation, and plot.

De Musset himself, whatever else he may or may not have stood for, was one of the few exclusively literary ascendencies to which she ever submitted. He it was who awoke her to the existence of such a thing as form and taught her all she ever learned except of herself about style. It is impossible to estimate how great was the detriment to her genius that she should have been so long under influences that, while intellectual, were

[1]*Biographia Literaria.*
[2]*Histoire de ma vie.*

in no sense literary, and should have been obliged to work her way alone out of much that was harmful to her spirit. Had her flow been less full and copious, it may well be questioned whether the stream would not have choked in the sands of sociological and metaphysical discussion with which she was surrounded, and she have ended where George Eliot began, as a mere controversialist. It is not a little singular that these two women, the greatest *littératrices* of their respective countries, should both have been for a time under the dominance of inspirations other than literary, and should have been more or less diverted from their proper paths and more or less hindered in their proper activities by philosophical speculation. Of the two George Eliot was more inclined to such thought, and never, indeed, got quite clear of the clutter of erudition, while George Sand was in reality of no great philosophical bent and never assimilated such ideas thoroughly enough to handle them with firmness.

As a result of her feeble grasp of such subjects and of the vivacity of her feelings, she was at her best when she centered her novels neither in a doctrinal *motif* nor a merely personal emotion, but in some simple episode of common life, which she had noticed and been touched by. Her masterpieces are few in number—as any one's must be—but they are perfect in their kind:—*la Mare au diable, la Petite Fadette, François le champi. Les Maîtres sonneurs,* of the same attempt as the others, errs by excessive development; it overreaches and outruns itself and in spite of much good grows wearisome by its length; while *Jeanne* and the *Meunier d'Angibault,* which are sometimes classed with these, show traces of confusion due partly to the introduction of extra-literary ideas and partly to the mixture of idyllic and social elements; so that none of these latter three can be ranked as masterpieces beside the former. Her own district of Berry, which she always loved and to which she returned more and more in later life, furnished her with the setting for these flawless gems. After the welter of passions and ideas, into which she had been cast young and in which she was long whirled, had subsided, and she could attend to the voice of her own desires; when her love of the unaffected and the natural asserted itself and she had leisure for quiet contemplation in the face of nature;—then she was quick to recognize and respond to the charm of just such characters and incidents as she met in her *Vallé noir* of the romantic name, and as she has rendered with exquisite sensibility. The simple, unpretentious life of the peasant amid his fields with his robust loves and hates, hopes and fears was a discovery in comparative humanity to French letters. The healthfulness and freshness of these idyls, full of the air of wood and lawn, the breath of morning and evening, is a revelation after the stale intrigue skulking away in the close and tainted atmosphere of city rooms. They justify to the English reader the existence of French fiction. It may be, as M. Brunetière declares, that George Sand made the French novel capable of sustaining thought;[1] it is of infinitely greater credit to her to have shown that it was possible for the French novel to carry good, clean, wholesome sentiment. No reader of modern French fiction can return to these stories without feeling that there life has been triumphantly vindicated against *naturalism,* and without feeling, too, that his heart has been purified and gladdened by contact with a great art.

Of George Sand's latest work it is hardly necessary to speak here. Its merits and demerits are essentially and, as far as we are concerned at present, those of the earlier periods. There is noticeable a constantly growing disinclination to air her personal convictions and feelings, together with a marked tendency to rationalize the action, which is quite new, and a very perceptible loss of reality, the result, perhaps, of waning enthusiasm, perhaps of overstudy of the plot, for she could do nothing well that she could not do naturally. It is better to leave her at the moment when her gift for improvisation, the heritage of the born story-teller, which I have tried to show in its strength and weakness, was at its fullest. This impression is certainly the pleasantest of her to carry away, and, what is more important, it is also the truest; for it is in respect of this quality that she holds among French novelists of the century—eminently a century of novelists—a unique position. Others may have writ-

[1]*Manuel de l'histoire de la littérature française.*

ten with equal facility; but no one has written so easily and at the same time so well. Dumas and Hugo in his prose are the only ones who have approached her in point of spontaneity and excellence. That the former is still greatly her inferior is generally recognized, so that it needs here no discussion to prove that his style is of a lower order and that his matter is such that it can impart no imperishable value to his work. With Hugo the case is rather different. He had transcendently the trick of the phrase; but catch-words do not make literature. And I think that as time goes on, whatever his fate as a poet, his fame as a novelist will lower to George Sand's, if it has not actually done so already; because he lacks, at least in prose, the sincerity that alone gives the writer's utterance weight and authority, while the trace of charlatanry in his novels, as must be the case where the thought waits upon the word, will, when they are farther removed from a fashion to which they have catered, be felt more and more to vitiate his work in this kind. I say nothing of the relative volume of these two authors' productions, just as it would be to consider such a matter in a question of their comparative spontaneity. And I say nothing of their relative importance to the historical development of fiction, nor urge that his contribution to the growth of the novel was inferior to hers, notwithstanding his place poetically in *romanticism*. I am trying merely to estimate their value for present readers and not their places in the evolution of fiction. And I am even willing to let Hugo pass as an exception to my remark; for whatever may be the weight of the two in technical literary performance, George Sand has, I think, a message for the present day to which Hugo can not pretend. To none of George Sand's serenity can Victor Hugo, or any other modern French writer that I know of, lay claim. In spite of the spiritual turmoil of her period, in spite of her personal difficulties, her work at its best is eminently serene. It possesses in a high degree the twin characters of all work that is great and sane,—simplicity and serenity. And this is assuredly the wisest lesson that can be read to our own two vices of extreme at this moment,—to our impateince and our intemperance. To the vague trouble, the haunting disquietude that disfigures almost all our work to-day, not only our literary and artistic work but our work of every kind particularly here in America, her writing offers the best contrast and correction. If there are two qualities that we lack just now, they are the qualities of patience on the one hand and of moderation on the other;—the patience to await results and to labor honestly for them, and the moderation to be satisfied with a fair day's labor and a fair wage for it. For either we neglect our task, shuffling it hastily aside to turn to something new, or else we are mastered by it and become its slaves. And the reason is, as we should find if we took the trouble to analyze ourselves, that we are not serious about high things. About low and small things we are deeply, passionately serious; we are serious about our material rewards, about the price of our work, about our popularity that people know our names and faces and the figures of the fortunes that we have made; and we allow ourselves to be diverted from the things that are really high and serious,—from the aim and purpose of our work itself, from its issue and influence. About these matters we are no more serious than was Flaubert, when he spent his leisure picking over words and shuffling the cadences of his artificial phrases. He had his reward: he founded a cult and provoked much technical discussion among the curious, and he introduced into French literature a trouble of which it is not yet rid. But he lost the hearts that George Sand won and holds at home and abroad. There is trouble in her books, to be sure, but it is the trouble of life bravely faced and nobly overcome. She does not allow her personal anxiety about her work to enter and disturb the ultimate peace of art. One feels that she, like Shakespeare, was greater than her task. It is work done without the haste of impatience or the waste of fret; and in consequence it is good and great,—done, I venture to say, in spite of our momentary aberrations, in the true spirit of English work. In its patience and moderation, in its magnificent spontaneity and naturalness, and, above all, in its serenity it is an especially opportune example to the vices of the time, to which we, no matter what our occupations, can return again and again with a sense of relief and renewal.

II.—*Notes on Certain Negative Verb Contractions in the Present*

BY LOUISE POUND

For some time lexicographers have recorded in standard dictionaries the colloquial and vulgar forms *ain't* and *hain't*, and their predecessors *an't* and *han't*. *Ain't* especially is now in so widespread usage as to deserve notation in the completer grammars also, though Dr. Sweet seems to be the only one yet to give it place. He notes, with his usual tolerance for the idiomatic or colloquial (*New Eng. Gr.*, 1900, § 1491):

"The negative forms [of *to be*] in the pres. are generally supplied by (eint) in familiar speech, which is, however, felt to be a vulgarism, and is avoided by many educated speakers, who say (aim not) instead of (ai eint), (aa ju not) instead of (eint ju)."

It was natural (1) that there should be confusion between *ain't* and *hain't*, and (2) that these forms should occur in all persons; the first because of the light quality and frequent instability of English initial *h*, the second because distinction in person and number is not usually observed in negative contractions. So the familiar *don't* for *does not, dost not*, as well as *do not*. So occasionally with *aren't*, especially in the speech of children. "Aren't I a good brother to you?" George Eliot, *Mill on the Floss*, v; "I'm no reader, I aren't," *Ib.*, iv; "I'm a pretty considerable favorite with the ladies—arn't I?" Captain Boroughcliffe, dramatization of Cooper's *Pilot* (1825), II, ii; "I'm a sort of a kind of a nonentity—arn't I?" *Ib.*

The vowel sound (ê, ei) of *ain't* and *hain't* seems, however, less expected. Obviously the normal contractions of the negative present of *to be* should be, 1st sg. *æn't* (am̹ not), 3rd sg. *in't* (is not), plur. *ân't* (are not). Similarly *hæn't* (Eng.

hân't) for *has not* and *have not.* Not (ê, ei) in any person of either verb.

In the eighteenth century, outside of contractions purely dialectal (for the numerous dialect contractions of the negative, cf. the *Eng. Dialect Dict.*, Ed. Wright), three forms are found. These are *in't, an't* (for pronunciation, see below), and *ain't, ain't* being apparently the latest form of the three, while *in't* is the first to disappear. Cf. Richardson, *Pamela* (1740), III, lxxii, "Oh, dear heart, thought I, in't it so!"; F. Burney, *Cecilia* (1782), I, viii, "However I assure you it in't true." Dr. Murray, *New English Dictionary,* quotes as his earliest example of *an't,* the plural, 1706, E. Ward, *Hud. Rev.* (1711) I, l. 24, "But if your Eyes an't quick of Motion, They'll play the rogue." His first example of *an't* =*am not* is 1737, *Hist. Reg.*, I, i, "No more I an't, sir"; and for *an't* =*is not,* 1812, H. and J. Smith, *Rejected Addresses,* 69, "No, that an't it, says he." The earliest occurrence of *ain't* noted is 1778, F. Burney, *Evelina,* I, xxi, "Those you are engaged to ain't half so near related to you as we are." Some other eighteenth century examples of *an't* (*ain't* is rarer) are: Sheridan, *Rivals* (1775), IV, i, "I suppose there an't been so merciless a beast in the world": F. Burney, *Cecilia,* I, ix, "Why, sure, madam, an't you his honour's lady?"; *Ib.*, I, iii, "It won't do; an't so soon put upon."

In the nineteenth century, *an't* and *ain't* are found side by side, with *ain't* monopolizing popularity only in the last half: Dickens, *Pickwick Papers* (1836–37), vi, "There an't a better spot"; *Ib.,* xix, "Very easy, ain't it"; *Nicholas Nickleby* (1838–39), lvi, "It an't time"; *Ib.,* iii, "Her name ain't Nickleby"; *Oliver Twist* (1837–38), xxvii, "An't yer fond of oysters"; *Ib.,* v, "Yer the new boy, ain't yer"; *Edwin Drood* (1870), v, "It ain't a spot for novelty"; Thackeray, *Vanity Fair* (1847–48), xiv, "The sneak ain't worthy of her." In America also, *an't* was widely written in the first half of last century. Sylvester Judd, *Margaret* (1855), has only *an't.* "Them an't yarbs. They won't doctor" I, v; "She an't a flower," *Ib.* Cf.

also Cooper, *The Pioneers* (1823), vii, "Ain't Marmaduke a judge?"; *Ib.*, xx, "The Squire ain't far out of the way"; W. G. Simms, *Guy Rivers* (1835), vi, "I ain't the man to deny the truth"; *Ib.*, "I ain't slow to say that"; *Ib.*, xxx, "I an't afraid."

In present American dictionaries (cf. the *Century, Standard,* etc.) ***an't*** is not entered as obsolete or obsolescent; although a fairly close survey of contemporary American colloquial and dialect literature reveals no examples, but rather, in contrast with results for the first half of the nineteenth century, the complete ascendancy of *ain't*. It would seem time to enter *an't* as dying, or dead, in America, and restricted to dialect speech in England.

In the *London Illustrated News,* April 4, 1903, occurs, in a story by "Q", the form *amn't,* "Am I captain here, or amn't I?" The story is Cornish, and the contraction probably local, hence belongs rather to dialect than to general English.

In the eighteenth century, lasting into the nineteenth, the expected *han't* was the familiar contraction of *have not, has not*. Cf. Congreve, *Way of the World*, 1700, III, iii, "Why then, belike, my aunt han't dined"; De Foe, *Colonel Jack* (1722), iii, "No, it is well if you han't"; Sheridan, *Rivals* (1775), I, i, "I doubt, Mr. Fag, you han't changed for the better"; Dickens, *Our Mutual Friend* (1864-65), xii, "Why han't you gone to Lawyer Lightwood?"; George Eliot, *Mill on the Floss* (1860), xi, "We han't got no treacle." So in America, Simms, *Guy Rivers* (1835), vi, "Han't I told you"; *Ib.*, "Here's none of us that han't something to say agin that pedler"; Judd, *Margaret* (1855), II, xii, "Marm han't said." For *hain't*, cf. Simms, *Guy Rivers,* vi, "Hain't he lied and cheated"; *Ib.*, "I, who hain't the courage"; *Ib.*, "We hain't got much law and justice in these pairts"; M. C. Graham, *Stories of the Foothills* (1895), i, "He hain't no notion o' doing that," "If I'd really had any idee . . . but I hain't"; J. Fox, Jr., *Little Shepherd of Kingdom Come* (1903), "I hain't got no daddy, an I hain't never had none."

There was confusion between *an't* and *han't,* as now between *ain't* and *hain't*. For the use of *an't* for *han't* cf. Sheridan,

Rivals (1775), IV, i, "I suppose there an't been so merciless a beast in the world"; Dickens, *Our Mutual Friend* (1864–65), xii, "Have you finished?" "No, I an't"; *Bleak House* (1852–53), viii, "No, I an't read the little book wot you left," etc. *Han't* for *an't* is found in *Pickwick Papers* (1836–37), xi, "Where's that villain Joe!" "Here I am; but I han't a willin." Examples of *ain't* for *hain't* are: Dickens, *Bleak House,* "We ain't got no watches to tell the time by"; S. O. Jewett, *A Dunnet Shepherdess* (1899), vii, "Ain't William been gone?"; *Ib., Where's Nora?* "Ain't you got the Queen's luck?" The still more illiterate *hain't* for *ain't* is heard often, but is less frequently written, "I hain't nothing but a boy," J. Fox, Jr., *Little Shepherd of Kingdom Come,* 1903; "Oh, you're a reglar tin peddler, hain't yer?"; "You ain't green," "You bet I hain't"; Julian Ralph, *Trip with a Tin Peddler,* Harper's Mag., 1903; "Dat rifle hain't neber gwine kick," *Her Freedom,* negro story by V. F. Boyle, *Century,* Feb., 1903. In the latter story was noted an example of *ain't* for *don't,* "I sho ain' want ter mairy Rias." Negro dialect has *cain't,* beside *ain't* and *hain't,* as "I cain't go in dar—no I cain't," *Ib.*

As suggested in opening, *an't* and *han't* some time ago found their way into the dictionaries, *ain't* and *hain't* more recently. Bailey enters none of the four forms; Dr. Johnson *han't* only. An edition of Webster's *Dictionary,* as late as 1855, when *ain't* had become about as widespread as *an't* (cf. examples *supra*) enters *an't, han't* only. Both Webster's *Dictionary* and the *Imperial Dictionary* (1856), based on Webster's, give the following curious etymology:

"*An't,* in our vulgar dialect, as in the phrases, I *an't,* you *an't,* he *an't,* we *an't,* etc., is undoubtedly a contraction of the Danish *er, ere,* the substantive verb, in the present tense of the indicative mode, and *not;* I *er-not,* we *ere-not,* he *er-not;* or of the Swedish *ar,* the same verb; infinitive *vara,* to be. These phrases are doubtless legitimate remains of the Gothic dialect."

With regard to the vowel sounds in the contracted forms *an't, han't*, the lexicographical evidence within reach shows not a little diversity. Was the sound the expected Eng. (â), Am. (æ, â), or was it (ê), and the difference from *ain't, hain't*, for the most part one of spelling only? Dr. Johnson, as so well known, does not mark pronunciation, nor does Richardson. Knowles, *A Pronouncing and Explanatory Dictionary of the English Language*, London, 1848, gives (â, ê) for *an't*, entering no *ain't*, and for *han't* (â), recognizing no (hênt). Webster's *Dictionary*, ed. 1855, gives (ê) only for *an't, han't*. The *Imperial Dictionary* (English), 1856, gives (ê) only, like Webster's. Dr. Murray, *N. E. D.*, 1885, gives (â) only for *an't*, reserving (ê) for *ain't*. Of present American dictionaries, the 1900 Webster gives (ê) only for both *an't* and *han't*, mentioning, however, an English *hân't*; the *Standard* gives (â) only for *an't*, but (ê) only for *han't*; and the *Century* gives (â, ê) for *an't*, with the note, "In the second pronunciation also written *ain't*", and (ê) only for *han't*.

Perhaps before affirming too much, there should be closer examination of lexicographical and other evidence that was possible from the material at hand; nevertheless it seems probable, despite the testimony, English and American, given above, that the entry of the sound (ê) for *an't* and *han't* is nothing more than a legacy from the period when *ain't* and *hain't* were not yet recorded, the older forms having to do double duty. (*v.* Knowles.) When has *a* followed by *nt* had the value of (ê)? It would be difficult to believe that *an't* and *ain't*, forms which occur side by side in so many early and middle nineteenth century texts, when orthography within individual authors is fixed, were for the most part nothing more than variant spellings for the same (ênt). If the entries are to be exact, our American dictionaries should probably adopt the *New English Dictionary* (1885) differentiation of *an't* and *ain't*, and furthermore clearly distinguish *han't* (æ, â) and *hain't*. The spelling of colloquial and vulgar forms is generally pretty phonetic. *An't* is not now written for (ênt), and in the days when it was commonly found was most probably pronounced as spelled. So with its companion form.

For the vowel sound of *ain't,* so far as I am aware, no explanation has been offered, at least no accepted explanation. The (ê, ei) could hardly be due, as I have once or twice heard suggested, to the analogy of *mayn't,* or of some such form perhaps as the dialectal *bain't.* If following analogy the contraction would more likely have fallen in with *shan't* and *can't.* Nor could it have assumed the vowel of *hain't.* The latter arose, probably, slightly later, and has itself to be explained. Rather may (ê, ei) be the result of some sort of unconscious compromise between the vowels of *am, are,* and *is.* The vowel of *are* would be moved forward and higher, and that of *am* be made higher and closer, through the influence of *is.* A compromise between (â), (æ), and (i) could hardly result otherwise. The vowel of *han't* then went the same way.

III.—*On the Variation and Functional Relation of Certain Sentence-Constants in Standard Literature*

BY R. E. MORITZ

"Surely the claim of mathematics to take a place among the liberal arts must now be admitted as fully made good. . . . It seems to me that the whole of æsthetic (so far as at present revealed) may be regarded as a scheme having four centers, viz., Epic, Music, Plastic, Mathematic. There will be a common plane to every three of these, outside of which lies the fourth; and through every two may be drawn a common axis opposite to the axis passing through the remaining two."

J. J. Sylvester, *Philosophic Magazine*, 1878 (1), p. 184.

"It therefore seems clear that mathematics can be shown to sustain a certain relation to rhetoric and may aid in determining its laws."

L. A. Sherman, *University Studies*, vol. I, p. 130.

Pythagoras taught that the essence of things is numerical relation, Aristotle that number is the means to true knowledge. In more modern times Lagrange conceived of the possibility of representing the complete history of the universe by one huge differential equation, the actual state of progress at any moment by a single time integral between the limits, minus infinity and zero. According to Herbert Spencer's definition of a law of nature, every such law can ultimately be embodied in an algebraic equation. Solvay, a Belgian scientist, recently established an equation governing the energy set free by an organism in vital phenomena for a given food supply.

Even Rhetoric bows to Number. During the closing decade of the last century it was demonstrated that all good writers lisp in numbers, that there is cadence in the essays of Bacon and Emerson and in the histories of Hume and Macaulay just as truly as in the "winged words" of a Homer, a Milton, or a Goethe. This cadence is as inaudible as the music of the spheres to ordinary ears, but it reveals its sweet measures to the patient investigator. There is a sentence-rhythm, as it has been called by its discoverer, which is the hidden mark, the cipher, the cryptogram, with which each author unconsciously endows the progeny of his pen. Measured, it yields a number, an author-constant, corresponding to the wave-length of this rhythm. And just as the wave-length of standard musical tones has been shortened in the course of time,[1] so the author-constant has gradually diminished from the beginning of English literature to the present day.

Such is a brief paraphrase of a theory set forth at some length by Professor L. A. Sherman[2] in two articles published in 1892 and 1894 respectively. In these, as in a later article by G. W. Gerwig,[3] results are announced and tentative statements are made which lead one to infer that English prose writers conform to sentence-instinct, which has all the force of a positive law in controlling their written utterances. After tabulating the sentence-lengths[4] measured in words of from 500 to 800 consecutive sentences from each of the authors De Quincey, Macaulay, Channing, Emerson, and Bartol, Professor Sherman remarks: "Now that the number of words in consecutive sentences was definitely exhibited, strange facts and features of style were indicated or suggested. The length of one sentence, it was shown, might be echoed unconsciously into the next, as notably in Macaulay's groups of seventeens. . . . But the really remarkable thing was the apparently constant sentence-average in the respective authors. Could it be possible that stylists, as eminent and practiced as these, are *subject to a rigid rhythmic law, from*

[1] Ellis, an English physicist, has shown that the concert C_4 normal tone has reduced its wave-length (in air) from 2.33 to 1.99 feet in the course of 130 years.

[2] *Some Observations upon the Sentence-Length in English Prose*, University Studies, Lincoln, Neb., vol. I, no. 2, p. 119.

On Certain Facts and Principles in the Development of Form in Literature. *Ibid*., vol. I, no. 4, p. 337.

[3] *On the Decrease of Predication and of Sentence Weight in English Prose*, *Ibid*., vol. II, no. 1, p. 17.

[4] The study of sentence-lengths was suggested by T. C. Mendenhall in his article on *The Characteristic Curves of Composition*, in Science, March 11, 1887, who in turn credits August DeMorgan, the well-known mathematician, with the priority of the thought of applying numerical analysis to the study of literary style.

which even by the widest range and variety of sentence-lengths and forms they may not escape? At once pushing the suspicion to a proof, I made, first, an extended test in Macaulay's essays: result, 23+, the number obtained before; then in Channing: average again, 25."[1] After several other tests with similar results he continues, "No evidence appearing to the contrary, *it seemed likely enough that sentence-rhythm was a universal law.*"[2] (The italics in the preceding quotations as well as in those which follow are mine.)

Mr. Gerwig occupied himself particularly with the average number of predications per sentence and the per cent of simple sentences in various authors, to the number of one hundred. His conclusions are summed up in the following words: "A very little investigation served to convince me that *the same remarkable uniformity which had been found in the average number of words used by any given author per sentence* would also hold in regard to the number of finite verbs, or predications, found in each sentence. The results obtained convinced me also that there was a uniformity in the number of simple sentences per hundred of a given author."[3] Mr. Gerwig then examined Chaucer's *Tale of Melibeus* and 2500 sentences from Macaulay, and, finding the expected uniformity, he says: "Other authors were taken in the same way until it was demonstrated that the average of 500 periods of any author who has achieved a style *was approximately the average of his whole work.*"[4] In particular he discovered that "while Chaucer and Spenser *habitually put over five main verbs in each sentence they wrote,* and less than ten simple sentences in each hundred, Macaulay and Emerson used only a little over two verbs per sentence, and left over thirty-five sentences in each hundred simple."[5]

In neither of these quotations is there any explicit statement to the effect that the principle suggested or announced is independent of the nature of the composition, but the implication

[1] *University Studies*, vol. I, no. 4, p. 348.
[2] *Ibid.*, p. 349.
[3] *University Studies*, vol. II, no. 1, p. 17.
[4] *Ibid.*, p. 18.
[5] *Ibid.*, p. 19.

certainly is that there is a single set of constants for each author. Professor Sherman speaks of "a rhythmic law from which an author may not escape," and asserts that "the determining factor (of sentence-length) in each case is the relative capacity of the author to respond to what may be called the sentence-sense in his own mind,"[1] "which, if it could have its will, would reduce all sentences to procrustean regularity."[2] Here we notice the excepting clause "if it could have its own will," indicating that the writer was conscious of some kind of limitation of the principle in question; and elsewhere we read, "to avoid complication, no consistent attempt has yet been made to determine the sentence-average in works of fiction. Here of course the matter is mainly narrative or descriptive, thus reaching the imagination of the reader more directly, also much of the language is quotation and dialogue."[3] Yet dialogue was not ruled out in comparisons. In the test case of the 40,000 sentences in Macaulay's *History of England,* it was found that "After the dialogue passages and consequent reduced averages, seemingly by a sort of reaction, full rounded periods and high averages take their place."[4] Later on, this restriction seems to have been entirely ignored, as, for instance, when Miss C. Whiting,[5] in studying the relative sentence-lengths during different periods of English literature, by an examination of 500 sentences from each of 60 authors, admits De Foe's *Robinson Crusoe,* Fielding's *Tom Jones,* Goldsmith's *Vicar of Wakefield,* Scott's *Kenilworth,* Eliot's *Middlemarch,* Howell's *Rise of Silas Lapham,* etc., alongside of Milton's *Areopagitica,* Dryden's *Discourse on Satire,* Gibbon's *Rome,* Emerson's *American Scholar,* Channing's *Self-Culture,* Spencer's *Data of Ethics,* etc.

Gerwig remarks that "the question *incidentally* arose whether a writer had the same sentence-structure in poetry as in prose,"[6]

[1] *University Studies*, vol. I, no. 2, p. 119.

[2] *Ibid.*, vol. I, no. 4, p. 353.

[3] *Ibid.*, vol. I, no. 2, pp. 129, 130.

[4] *Ibid.*, vol. I, no. 4, p. 352.

[5] *The Descent of Sentence-Length in English Prose*, master's thesis, Univ. of Neb. (unpublished).

[6] *University Studies*, vol. II, no. 1, p. 4.

and realizes that "it would be manifestly unfair to compare truncated, dramatic dialogue (such as Shakespeare's), abounding in exclamations and broken sentences, with the even flow of the *Hind and the Panther*."[1] To equalize matters, he proposes to examine only passages three or more lines in length.

But whatever limitations to the Sherman principle[2] may have been recognized by its author and Mr. Gerwig, they were all swept aside at one stroke in an article which appeared in 1897.[3] In this article, the principle of constancy of sentence-length, predication-averages, and simple-sentence-percentages is set forth as follows: "Ten years or more ago Professor Sherman, while investigating the course of stylistic evolution in English prose, made the discovery that *authors indicate their individuality by constant sentence-proportions*, personal and peculiar to themselves. This was demonstrated especially with the number of words used per sentence in large averagings. It was found that De Quincey, Channing, and Macaulay, if five hundred periods or more were taken, *evinced this average invariably*, and in the earliest as well as in the latest period of their authorship. This discovery led to the suspicion that good writers would be found constant in predication averages, in per cent of simple sentences, and other stylistic details. Acting upon a suggestion to this effect, Mr. G. W. Gerwig, then a pupil of Professor Sherman, undertook an investigation *that established the constancy of predication, as well as simple-sentence frequency, in given authors*. . . . Professor Sherman and Mr. Gerwig have thus established by the examination of a great many authors, that writers are structurally consistent with themselves; that they possess a certain sentence-sense peculiarly their own. These investigators have established that by this instinct *authors use a constant average sentence-length, and a certain number of predications per sentence, and that a given per cent of their sentences will be simple sentences*. . . . The work of these investigators cov-

[1] *Ibid.*, p. 3.

[2] In honor of its discoverer, I shall call the principle of sentence-instinct, into whatever form it may ultimately crystallize, the Sherman principle.

[3] *University Studies*, vol. II, no. 2, p. 131.

ers a large amount of material and a wide field of literature. They have examined and compared the ***works of ancient and recent authors, early and late writings of the same author,*** and ***writings of the same author of different character,*** such as ***history and dialogue, poetry and prose.*** The results thus far obtained are sufficient to show that ***it is not possible for a writer to escape from his stylistic peculiarities."***[1]

There is no uncertainty, no indefiniteness, no ambiguity in these statements. The principle as here set forth is as thoroughgoing as it is simple. A good writer can not escape from his stylistic peculiarities any more than he can change his pace or alter his voice. Whether he write history or drama, poetry or prose, these stylistic characteristics are ever present with him. They permeate all his writings. They manifest themselves in certain numbers, such as sentence-length, predication-frequency, simple-sentence-percentage, etc. Once discover these numbers and you have marks which will serve for detective purposes quite as well as a man's chirography. They are the earmarks by which to trace anonymous and disputed writings to their sources, the touchstones to disclose the spurious and the false.

That this is the thought is clear from the sequel of the article referred to. If Shakespeare's plays were written by Bacon, they will reveal the same constants as the other writings of Bacon. In the author's own words, the end sought is evidence touching the authorship of Shakespeare's plays; whether Bacon wrote Shakespeare's works, or at least whether the Baconian and Shakespearean writings were the work of one and the same person, or of different persons. Consequently, an examination is made of the prose in fifteen of the *Plays*, of Bacon's *Essays*, and the *New Atlantis*. To eliminate possible errors, arising from careless and inconsistent punctuation, all the material is repunctuated according to modern principles. Fairness is added to consistency by omitting on the Shakespearean side of the inquiry all inorganic, broken, and suspended diction. Then follow twelve pages of figures representing totals and specimen results, and finally comes the

[1]*University Studies*, vol. II, no. 2, pp. 147-148

GRAND SUMMARY

	No. of words examined	No. of sentences	Sentence-length	Predications	Simple sentences
Shakespeare	61956	5002	12.39	1.70	.39
Bacon	66524	2041	32.59	3.45	.14

Instead of adding evidence to the Bacon-Shakespeare controversy, this argument seemed to me merely a *reductio ad absurdum* of the Sherman principle itself as stated by the writer. It served to convince me of the existence of certain limitations to the principle in question, which had not been recognized, certain restrictions which had been violated. For who is not aware that dramatic prose generally, if not invariably, contains shorter sentences, consequently, other things being equal, fewer predications per sentence, and a larger per cent of simple sentences than do other forms of prose composition? The argument, based upon a difference in sentence-constants merely, could be used equally well to prove that Dryden, the dramatist, did not write the famous *Essay on Satire.* The only real information conveyed by the above summary is not the fact of variation, which is a common notion of all who are at all familiar with the writers in question, but the widely divergent results indicated by the numbers 12.39 and 32.59. Can this difference in the sentence-constants be entirely accounted for by the difference in character of the material examined? To satisfy my own mind in the matter, I was impelled to make a test. Goethe's works were at hand. I selected one of his prose dramas, *Goetz von Berlichingen,* and his *Bildhauerkunst,* a collection of essays on art. The results were:

GOETHE (*Goetz von Berlichingen*)

First five hundred periods....	8.7
Second " " "	9.2
Third " " "	8.7
Fourth " " "	7.8
Fifth " " "	8.1
Average for 2,500 periods......	8.5

GOETHE (*Bildhauerkunst*)

First hundred periods........	27.1
Second " "	32.9
Third " "	33.7
Fourth " "	27.0
Fifth " "	36.9
Average for 500 periods......	31.5

These results showed even a greater divergence than those obtained from a comparison of Shakespeare's and Bacon's prose. Possibly Goethe occupied a unique position in this respect. I continued the test with an examination of Schiller. This time I selected *Die Räuber,* a prose tragedy, and his *History of the Thirty Years War.*

SCHILLER (*Die Räuber*)		SCHILLER (*History Thirty Years War*)	
First five hundred periods	12.2	First two hundred periods	29.9
Second " " "	11.4	Second " " "	27.8
Third " " "	12.6	Third " " "	28.3
Fourth " " "	10.2	Fourth " " "	24.9
Fifth " " "	10.9	Fifth " " "	25.7
Average for 2,500 periods	**11.5**	Average for 1,000 periods	**27.3**

This would not do. Whatever sentence-rhythm these German writers manifested was certainly greatly dependent upon the particular style of composition employed. Perhaps I had made a mistake in making tests from books nearest at hand, and should have limited myself to English authors. So I took up Swift and Dryden, who, I thought, would be unobjectionable from any point of view. I examined a prose drama and an essay from each, and decided to limit my examination in each case to five hundred periods.[1]

SWIFT (*Polite Conversation*)		SWIFT (*Essay on the Four Last Years of Queen Anne*)	
First hundred periods	10.8	First hundred periods	51.6
Second " "	12.0	Second " "	49.3
Third " "	12.1	Third " "	58.2
Fourth "	13.7	Fourth " "	62.1
Fifth "	13.6	Fifth " "	53.0
Average for 500 periods	**12.4**	Average for 500 periods	**54.8**

[1] Professor Sherman expresses the opinion that three hundred periods will generally reveal the sentence-rhythm of any author who has achieved a style. *University Studies,* vol. I, no. 2, p. 130.

DRYDEN (*The Mock Astrologer*)	
First hundred periods	15.0
Second " "	16.5
Third " "	19.6
Fourth " "	17.1
Fifth " "	16.3
Average for 500 periods	**16.9**

DRYDEN (*Essay on Satire*)	
First hundred periods	45.0
Second " "	48.1
Third " "	40.1
Fourth " "	44.8
Fifth " "	33.3
Average for 500 periods	**42.3**

In these authors the divergence of constants was even greater than in the case of Goethe and Schiller. Shakespeare's 12.39 and Bacon's 32.59 seemed no longer remarkable, the ratio of these two constants being 2.6, while the corresponding ratios for Goethe and Swift are 3.7 and 4.4 respectively.

These results urged me to continue the investigation. I soon found that the sentence-constants varied not only when a comparison was made between drama and history, or essays, but in other forms of composition as well. I append a few of my results:[1]

GOLDSMITH (*She Stoops to Conquer*)	
First hundred periods	13.9
Second " "	13.1
Third " "	13.2
Fourth " "	14.9
Fifth " "	12.4
Average for 500 periods	**13.5**

GOLDSMITH (*Present State of Polite Learning in Europe*)	
First hundred periods	30.4
Second " "	24.6
Third " "	25.7
Fourth " "	23.5
Fifth " "	20.4
Average for 500 periods	**24.9**

GOLDSMITH (*The Vicar of Wakefield*)	
First hundred periods	31.2
Second " "	30.8
Third " "	27.5
Fourth " "	25.8
Fifth " "	26.5
Average for 500 periods	**28.4**

SCOTT (*Ivanhoe*)	
First hundred periods	46.2
Second " "	35.3
Third " "	33.7
Fourth " "	29.8
Fifth " "	32.2
Average for 500 periods	**35 4**

[1] In each case the count was made from the beginning of the work cited. Introductions, headings, footnotes, sentences containing long quotations, verses, and, in the case of dramas, stage directions were consistently omitted.

SCOTT

(*Auchindrane*)

First hundred periods	19.1
Second " "	21.9
Third " "	23.9
Fourth " "	19.2
Fifth " "	21.8
Average for 500 periods	**21.2**

SCOTT

(*Life of Napoleon*)

First hundred periods	46.0
Second " "	47.0
Third " "	50.0
Fourth " "	49.8
Fifth " "	46.5
Average for 500 periods	**47.9**

CARLYLE

(*Signs of the Times, etc.*)

First hundred periods	27.4
Second " "	31.8
Third " "	26.9
Fourth " "	34.5
Fifth " "	42.3
Average for 500 periods	**32.6**

CARLYLE

(*The Hero as Divinity*)

First hundred periods	25.2
Second " "	21.5
Third " "	24.3
Fourth " "	23.5
Fifth " "	23.4
Average for 500 periods	**23.6**

BAYARD TAYLOR

(*The Prophet*)

First hundred periods	15.8
Second " "	11.8
Third " "	11.3
Fourth " "	14.8
Fifth " "	16.4
Average for 500 periods	**14.0**

BAYARD TAYLOR

(*Balearic Days*)

First hundred periods	30.3
Second " "	29.3
Third " "	27.8
Fourth " "	26.6
Fifth " "	27.6
Average for 500 periods	**28.3**

LOWELL

(*Letters, vol. II*)

First hundred periods	18.8
Second " "	18.2
Third " "	22.5
Fourth " "	21.8
Fifth " "	17.2
Average for 500 periods	**19.7**

LOWELL

(*Fable for Critics*)

First fifty periods	106.5
Second " "	88.7
Third " "	87.6
Remaining 30 periods	108.3
Average 180 periods	**96.6**

HOLMES

(*Guardian Angel*)

First hundred periods	26.4
Second " "	29.0
Third " "	32.2
Fourth " "	31.1
Fifth " "	25.6
Average for 500 periods	**28.9**

HOLMES

(*The Autocrat at the Breakfast Table*)

First hundred periods	19.7
Second " "	18.8
Third " "	18.0
Fourth " "	23.8
Fifth " "	20.1
Average for 500 periods	**20.1**

LONGFELLOW (*The Spanish Student*)	
First hundred periods	12.0
Second " "	8.8
Third " "	11.1
Fourth " "	8.8
Fifth " "	10.1
Average for 500 periods	10.2

LONGFELLOW (*Hyperion*)	
First hundred periods	20.5
Second " "	25.4
Third " "	27.5
Fourth " "	24.3
Fifth " "	21.3
Average for 500 periods	23.8

The list could be indefinitely prolonged, but it was not necessary. The evidence seemed to me to show that sentence-lengths, and presumably also predication-averages and simple-sentence-percentages, are quite as much dependent upon the nature of the composition employed as upon the author's sentence instinct. In fact, I surmised that the variety of sentence-lengths which an author employs is limited only by his versatility as a writer. Acting on this surmise, I decided to make a test. Goethe was the author selected for this purpose, for he seemed to meet most nearly ideal conditions. His style is unquestioned, his punctuation is consistent and scientific. Most important of all, he is the most universal writer of the last century. He ranks high as a writer in the fields of poetry, drama, fiction, biography, art, literary criticism, travel, and science. The results[1] fully corroborated my conjecture.

[1] In securing these as well as the preceding results, the sentences were actually counted, but where the nature of the text permitted it, the number of words has been obtained by counting the lines and multiplying by the average number of words per line. In this way the drudgery of the work may be greatly shortened without impairing the accuracy of the results, since counting, like every other arithmetical operation which is not carefully checked, involves unavoidable accidental errors. To test the accuracy of my method, I determined by means of it the number of words in Chaucer's *Tale of Melibeus* and obtained 16,633 words as against 16,659 which was obtained by an actual count. In less than two hours I determined the number of words in Macaulay's *History of England* to be 979,668 as against 974,195 obtained by actual count. Assuming that this last number is absolutely correct, my estimate involves an error of ½ per cent, which may be safely neglected since it will be shown that the average sentence lengths based upon 500 sentences involve an average error of over 2 per cent.

GOETHE

(Dichtung und Wahrheit)

First hundred periods	31.5
Second " "	36.1
Third " "	30.9
Fourth " "	28.9
Fifth " "	31.1
Average for 500 periods	**31.7**

GOETHE

(Faust: Second Part)

First hundred periods	16.3
Second " "	17.2
Third " "	15.5
Fourth " "	12.3
Fifth " "	16.5
Average for 500 periods	**15.6**

GOETHE

(Die Leiden des Jungen Werthers: Book 2)

First hundred periods	20.7
Second " "	20.8
Third " "	19.1
Fourth " "	22.7
Fifth " "	18.2
Average for 500 periods	**20.3**

GOETHE

(Der Bürgergeneral)

First hundred periods	6.7
Second " "	5.1
Third " "	4.5
Fourth " "	3.9
Fifth " "	4.7
Average for 500 periods	**5.0**

GOETHE

(Die Wahlverwandtschaften)

First hundred periods	21.9
Second " "	23.6
Third " "	23.7
Fourth " "	25.1
Fifth " "	24.5
Average for 500 periods	**23.4**

GOETHE

(Faust: First Part)

First hundred periods	14.6
Second " "	15.2
Third " "	13.6
Fourth " "	13.0
Fifth " "	12.0
Average for 500 periods	**13.7**

GOETHE

(Reinecke Fuchs)

First hundred periods	18.5
Second " "	16.5
Third " "	16.9
Fourth " "	14.8
Fifth " "	16.6
Average for 500 periods	**16.7**

GOETHE

(Briefe aus der Schweiz: Part 2)

First hundred periods	23.2
Second " "	26.5
Third " "	25.6
Fourth " "	23.0
Fifth " "	28.4
Average for 500 periods	**25.3**

GOETHE

(Entwurf einer Farbenlehre: Didactic Part)

First hundred periods	28.4
Second " "	25.7
Third " "	29.3
Fourth " "	22.3
Fifth " "	26.1
Average for 500 periods	**26.4**

GOETHE

(Literatur: Recensionen)

First hundred periods	37.3
Second " "	32.9
Third " "	36.9
Fourth " "	31.7
Fifth " "	34.9
Average for 500 periods	**34.7**

GOETHE

(*Italiänische Reise: Rom.*)

First hundred periods	23.1
Second " "	23.7
Third " "	22.7
Fourth " "	21.5
Fifth " "	22.7
Average for 500 periods	**22.7**

GOETHE

(*Letters to Frau von Stein*)

First hundred periods	13.5
Second " "	12.5
Third " "	11.3
Fourth " "	11.2
Fifth " "	12.4
Average for 500 periods	**12.2**

GOETHE

(*Die Metamorphose der Pflanzen*)

First hundred periods	33.8
Second " "	34.1
Remaining 88 "	33.3
Average for 288 periods	**33.7**

If we arrange these results, including the numbers previously found for the *Goetz von Berlichingen* and *Bildhauerkunst,* in order of the sentence-length we have the following instructive table:

TABLE I

VARIATIONS IN SENTENCE-LENGTH OF GOETHE'S WRITINGS

Der Buergergeneral	5.0
Goetz von Berlichingen	8.5
Letters to Frau v. Stein	12.2
Faust: First Part	13.7
Faust: Second Part	15.6
Reinecke Fuchs	16.7
Die Leiden d. *J.* Werthers	20.3
Italiänische Reise: Rom	22.7
Die Wahlverwandtschaften	23.4
Briefe aus der Schweiz	25.3
Entwurf einer Farbenlehre	26.4
Bildhauerkunst	31.5
Dichtung u. Wahrheit	31.7
Metamorphose d. Pflanzen	33.7
Literatur: Recensionen	34.7

The above list includes romance, drama, allegory, criticism, biography, description, science, correspondence, but with the exception of *Faust* and *Reinecke Fuchs* the works are all in prose, so that the fact of variation in sentence-length appears, even if we consider prose literature alone. There can be but little doubt that an examination of all of Goethe's writings would furnish a chain of sentence-lengths varying by almost insensible gradations from five to thirty-five or forty words per sentence. Other authors may not yield such wide extremes, but it is the fact rather than the extent of variation which is essential in this discussion.

It is hardly necessary to comment further on the above results. They demonstrate absolutely the unreasonableness of applying the Sherman principle indiscriminately to various types of composition. They demand that, before we compare the sentence-constants of different authors, or the normal sentence-constants of the writers of a given period, there shall be some agreement or understanding regarding a standard type or standard types of composition.[1] Any conclusions concerning the stylistic evolution of literature, which are based upon the *principle of constancy* and a disregard of the *principle of variability* of sentence-length must be considered worthless, even if they lead to desired or plausible results. As an illustration I will only cite the investigation by Miss C. Whiting[2] on *The Descent of Sentence-length in English Prose.* Assuming the Sherman principle, it appears, from an examination of single works by each of sixty authors, ranging from Chaucer to Henry James, that there has been a decided diminution in the sentence-length. Averaging her re-

[1] Some of the specimens which I have examined, such as Schiller's *Räuber* and Goethe's *Goetz von Berlichingen*, may be objected to because of the rather abnormal nature of the compositions. They abound in exclamatory and interrogatory dialogue. I admit that they are not normal types, but it is the necessity of agreeing upon what shall constitute normal types which I desire to point out. Shall we disregard all interrogatory and exclamatory passages? If not, what proportion shall be counted out? Or shall we omit all sentences less than four, five, or six words in length? Or adopt Mr. Gerwig's rule and examine only passages three or more lines in length? The principle of variability could not be denied even though it were claimed that all of Goethe's sentence-lengths, but one, are unnatural.

[2] *Master's Thesis*, Univ. of Neb. (unpublished).

sults, I find that the normal sentence-lengths of writers before the seventeenth century, during the seventeenth century, the eighteenth century, and the nineteenth century, are represented respectively by the numbers 48, 42, 36, and 27, if the original punctuation is used, and by 42, 38, 34, and 26, respectively if the works examined are repunctuated. But in the selection of the works examined, no thought seems to have been given to uniformity in the form of composition; at any rate we find works so widely divergent in structure as Chaucer's *Tale of Melibeus,* Bacon's *Essays,* Milton's *Areopagitica,* Bunyan's *Pilgrim's Progress,* Locke's *Essay on the Human Understanding,* Taylor's *Sermons,* Hume's *History of England,* Goldsmith's *Vicar of Wakefield,* Irving's *Life of Washington,* Spencer's *Data of Ethics,* and Howell's *Rise of Silas Lapham* enjoying equal suffrage in the comparison.

Professor Sherman brought to light the remarkable fact that an author's sentence-length remains practically constant throughout a given work, an extensive test having been made on the 40,000 sentences of Macaulay's *History of England,* but here we have the incontestable fact that the sentence-length of one and the same author may vary by almost insensible degrees between limits so widely divergent as five and thirty-five. The conclusion from which there seems to be no escape is that *the sentence-length of a work depends both upon the writer's sentence-instinct and upon the particular form of composition into which his thought is cast.* That is to say, sentence-rhythm, inasmuch as it manifests itself in constant sentence-length, is a function of at least two variables x and y, where x signifies the author's sentence-sense and y the form into which he moulds his thought.[1]

---

[1] I trust that this statement will not be interpreted as contradicting or de-utilizing the Sherman principle. All that it insists upon is the necessity of so modifying the principle as to recognize the facts of variability in the sentence-constants.

The results obtained from Macaulay's *History* are in perfect harmony with these conclusions. Professor Sherman found the average sentence-length of the *History* (5 volumes containing 41,500 periods) to be 23.43. This is practically the same as the sentence-length 23.65 of *Machiavelli*, 24 for *Pitt*, and 23.00 for the *Essay on History* by the same author. Now the

I shall now endeavor to show that the principle just stated applies also to the other sentence-constants, predication-averages, and simple-sentence-percentages.

It will not be necessary to produce a chain of different predication-averages or simple-sentence-percentages corresponding to the chain of sentence-lengths which we found in Goethe. We need only show that there exists a functional relation between the various sets of constants, such that a variation in one set produces a variation in each of the other sets. Mathematically expressed, we need only show that

$$P = f(L) \qquad S = \phi(P)$$

where L=sentence-length,
P=predication-average,
S=simple-sentence-per cent,

from which it immediately follows that S itself is a function of L.

A priori we should expect no less than that the shorter sentence contains fewer predications, and that as the sentence grows shorter the percentage of simple sentences increases, the limits being respectively single predications in the one case and none but simple sentences in the other. However, inasmuch as no attempt appears to have been made to verify this prediction, the only statement that I can find bearing on it being exactly op-

History of England, particularly the second volume, contains much dialogue, which might cause us to expect a lower average than is actually the case. The explanation is, that taking the *History* as a whole Macaulay's normal style predominates to such an extent as to practically obliterate the "bearish" tendency of the dialogue passages. This can be easily demonstrated. There are in all 45 hundreds of periods whose average is less than 20 words per sentence. These we may take to represent approximately the dialogue portions of the *History*. The exact average of these 4,500 periods is 18.62, that is, 4.81 words less per sentence than the average for the entire *History*. If we replace these sentences by others of the normal length, we swell the total aggregate of words by 4,500×4.81 or 21,645 words. That is, if the portions of the *History* which contain an excessive amount of dialogue were replaced by an equal number of sentences of normal length, the five volumes of Macaulay's *History* would contain 41,500×23.43+21,645 or 993,990 words. Dividing this number by 41,500, we obtain 23.95 words per sentence, a result not essentially different from the actual average. For the data employed see *Univ. Studies*, vol. I, no. 2, p. 130; *Ibid.*, vol. I, no. 4, pp. 351, 352.

posed to it,[1] it may be worth the while to examine such data as are available, for the purpose of detecting some relation. Fortunately there is some material at hand with which to work. Among the hundreds of works examined by Mr. Gerwig[2] for predication averages and simple-sentence-percentages there are some twenty titles which also occur in the list of works subsequently examined by Miss Whiting[3] with reference to sentence-length. Both sets of data are presumably based upon the original punctuation,[4] thus making a comparison possible. While the number of works thus furnished is not as large as one would desire, it has the advantage of precluding any results which could be attributed to an unconscious bias in case the works had been selected by myself.

TABLE II

	AUTHOR AND TITLE	N	P	S	L
A	Chaucer, *Tale of Melibeus*............	400	5.19	4.5	48.0
B	More, *Life of Richard III*.............	500	3.65	15.0	36.5
C	Spenser, *View of Present State of Ireland*	500	4.68	10.6	49.7
D	Hakluyt, *Voyages of the Eng. Nation to A.*	300	4.44	8.7	56.8
E	Hooker, *Ecclesiastical Policy*..........	500	4.12	12.0	40.9
F	Sidney, *Defense of Poesie*..............	479	3.98	10.0	39.3
G	Lyly, *Euphues*.........................	500	3.50	17.0	37.1
H	Bacon, *Essays*[5].......................	1558	3.58	12.0	32.9
I	Milton, *Areopagitica*..................	500	4.87	6.0	43.7
J	Bunyan, *Holy War*[5]...................	500	3.91	10.0	37.5
K	Swift, *Tale of the Tub*.................	300	3.32	16.3	43.0
L	Hume, *History of England*............	500	3.29	12.0	38.2
M	Junius, *Letters*.......................	500	2.54	26.0	28.7
N	Channing, *Self Culture*....	500	2.57	31.0	25.9
O	De Quincey, *Confess'ns of an Opium Eater*	500	3.19	14.0	32.6
P	Macaulay, *History of England*[5]........	41500	2.30	34.0	23.3
Q	Newman, *Idea of a University*..........	100	4.65	8.0	41.7
R	Emerson, *Divinity School Address*......	100	2.14	41.0	18.0
S	Bartol, *Radical Problems*..............	300	2.33	40.3	15.9

[1] Gerwig states that "the proportion between the average number of predications and the percentage of simple sentences is approximately constant" *Thesis*, Univ. of Neb (unpublished).

[2] *University Studies*, vol. II, no. 1, pp. 31–44.

[3] *Master's Thesis*, Univ. of Neb., 1898 (unpublished).

[4] Three works, Latimer's *Sermons*, Ascham's *Schoolmaster*, and Bacon's *Essays*, presented such anomalous results that I questioned their correctness. A rough count revealed a large discrepancy between Gerwig's and Miss Whiting's figures which I attribute to a difference in the punctuation of the texts examined.

[5] Bacon's constants are taken from Hildreth's paper, and Macaulay's and Bunyan's sentence-lengths from Professor Sherman's.

In the preceding table the figures in the second and third columns are reduced from Mr. Gerwig's tables, the fourth column is made up from Miss Whiting's results:

N=the number of periods from which the averages are taken.

P=the average number of predications per sentence.

S=the per cent of simple sentences.

L=the average number of words per sentence.

If we compare the figures under P with those under L, we see that Macaulay in his history uses almost exactly ten words to each finite verb employed, and so do More, Hooker, Sidney, and Channing, while Chaucer, Spenser, Lyly, Bacon, Milton, Bunyan, De Quincey, and Newman come approximately within ten per cent of this amount. But there is a marked tendency to depart from the ratio ten when the sentence-length approaches extremes as in the case of Hakluyt, Emerson, or Bartol. Obviously, when L is less than ten, the ratio L to P must be less than ten. The data are, however, too limited to reveal much more than the mere fact of interdependence.

The fact of interrelation is perhaps more apparent from a graphical representation as shown in fig. 1. The values of L have been used as abscissas, ten times the values of P as ordinates, and the resulting points have been marked by letters corresponding to those to the left of the names of the authors in the table above. The graph shows that four points are approximately in range with the line whose equation is

$$L=10\,P, \tag{1}$$

while all the points but four lie between the dotted lines, making an angle of only six degrees with the former.

A more striking, though less obvious, relation exists between predication-averages and simple-sentence-percentages. An inspection of the table convinces one that there is some sort of reciprocal relation, the larger P going with the smaller S and vice versa, but much more than this is not warranted by the figures in our table. Nor should we expect the table to reveal any very definite relation, when we consider the uncertainty of the numbers involved. There is a large probable error in all

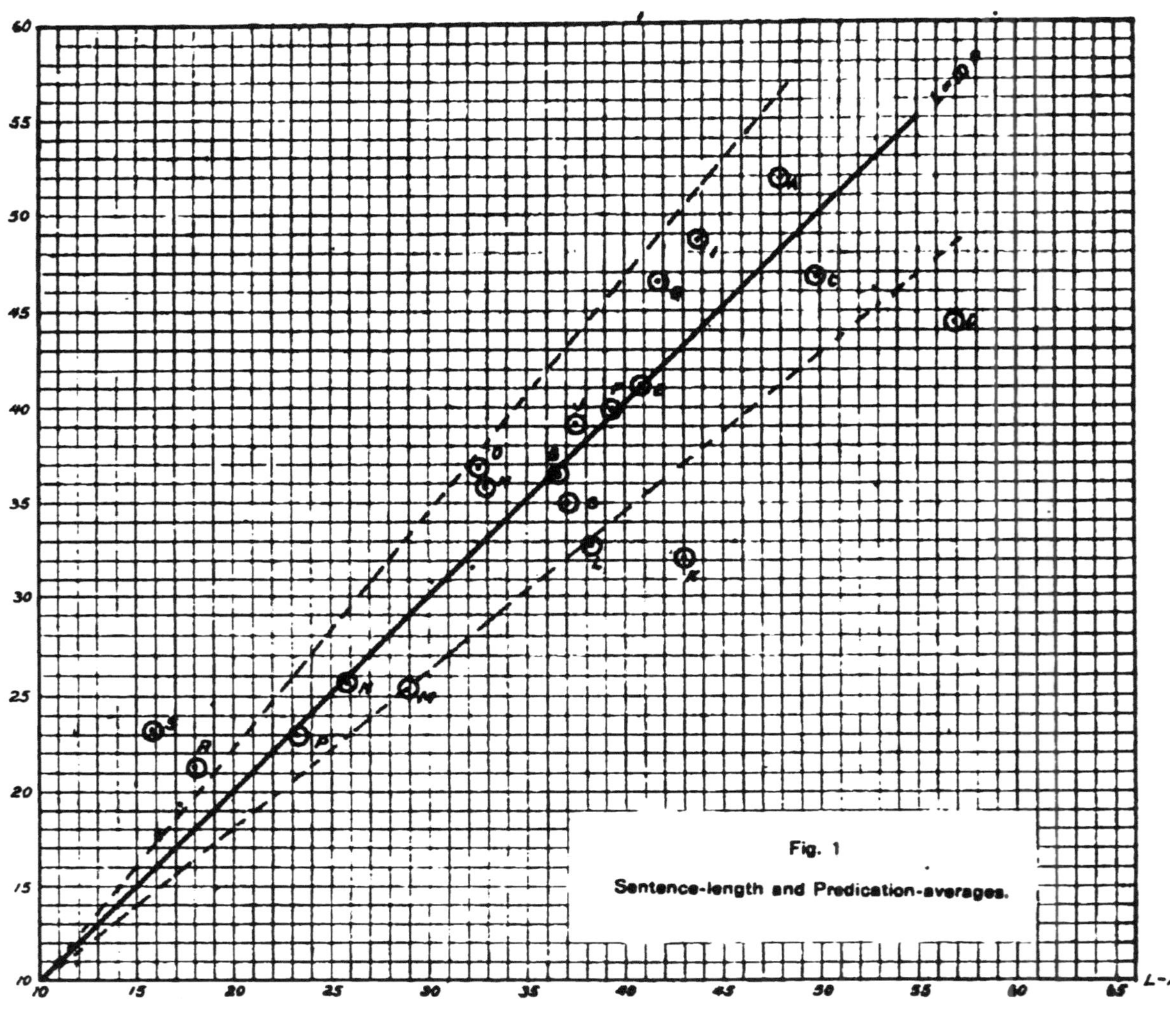

Fig. 1

Sentence-length and Predication-averages.

the figures employed, but the magnitude of the probable error in the S's can only be appreciated by a glance at the figures from which some of the averages are gotten. Thus we find Lyly's 17 to be the mean of the numbers 26, 14, 20, 15, 8,[1] these numbers representing the simple-sentence-percentages for consecutive hundreds. Similarly, De Quincey's 14 results from the numbers 10, 19, 15, 7, 21.[2] If larger averagings had been made, essentially different results would have been obtained in many cases as the following example shows:

VARIATION IN SIMPLE-SENTENCE-PERCENTAGES FOR CONSECUTIVE HUNDREDS IN

Spenser's View of the Present State of Ireland:[3]

Average	of	300	sentences	yields	12.0	per cent	simple	sentences
"	"	400	"	"	11.5	"	"	"
"	"	500	"	"	10.6	"	"	"
"	"	600	"	"	10.5	"	"	
"	"	700	"	"	9.7	"	"	
"	"	800	"	"	9.0	"	"	
"	"	900	"	"	8.9	"	"	
"	"	1000	"	"	8.0	"	"	

Even so great a stylist as Macaulay gives averages for S so widely divergent as 41 and 27 for different parts of his *History of England,* though each average is based upon 500 consecutive periods.[4]

Although disastrous for immediate progress, these facts did not lessen my confidence in a more definite law than that which was thus far apparent. The first problem was how to reduce the probable errors to manageable proportions. Of course, the obvious solution consists in making larger averagings, say, of 10,000 or more sentences from each work, but this involves an enormous amount of unattractive labor. To avoid this, I devised a seemingly crude experiment on Mr. Gerwig's figures for English prose works.[5] I struck averages of both P and S for

[1] *University Studies*, vol. II, no. 1, p. 38.

[2] *Ibid.*, p. 34.

[3] *Ibid.*, p. 42.

[4] *Ibid.*, pp. 25–26. The numbers represent the averages for the 27th to the 33d, and for the 284th to the 289th hundreds respectively.

[5] *Ibid.*, pp. 31–44. These figures comprise averages of about 60,000 periods, taken from 71 different authors.

each five hundred periods from the same work, and then grouped together all the averages for which the predication-average is between 1.50 and 2.00, similarly those for which the predication-average is from 2.00 to 2.25, from 2.25 to 2.50, etc. We thus get the following table:

TABLE III

	P=1.50-2.00	
	P	*S*
Macaulay.....	1.88	48
Symons	1.84	58
	P=2.00-2.25	
	P	*S*
Bartol	2.10	44
"	2.10	44
"	2.10	44
Blaine	2.23	39
Channing	2.14	40
"	2.04	44
Emerson	2.14	38
Macaulay......	2.22	32
"	2.22	32
"	2.07	38
"	2.07	38
"	2.18	40
"	2.17	36
"	2.17	36
"	2 17	36
"	2.17	36
"	2.17	36
Phelps	2.03	50
	P=2.25-2.50	
	P	*S*
Emerson	2.28	38
"	2.49	36
Everett	2.27	32
"	2.27	32
Forum	2.42	32
Geikie..........	2.34	32
Grant	2.31	31
James	2.45	24
Macaulay	2.31	32
"	2.31	32
"	2.41	32
"	2.26	32
"	2.26	32
"	2.:6	32
"	2.26	32
"	2 26	32
Phillips.	2.47	53
Shelley........	2.48	26
	P=2.50-2.75	
	P	*S*
Channing	2.58	31
"	2.58	31
"	2.51	31
"	2.69	30
Darwin	2.64	21
Disraeli	2.57	27
Fiske...........	2.69	20
Greeley	2.56	26
Junius.........	2.54	26
Lowell	2.54	23
"	2.74	21
"	2.74	21
Pater	2.74	26
Shaftesbury...	2.61	28
	P=2.75-3.00	
	P	*L*
Arnold	2.77	20
Browning.	2.91	25
Choate	2.88	30
George	2.92	23
Goldsmith	2.95	18
Hamerton.....	2.85	20
Higginson	2.85	21
Munger.	2.93	26
Newman	2.97	16
Shakespeare ..	2.76	31
Thoreau.......	2.86	25
	P=3.00-3 25	
	P	*S*
Bacon	3.12	19
Carlyle	3.12	18
Franklin	3.04	19
Holland	3 03	21
Johnson.......	3.23	16
Mandeville....	3.08	22
Stevenson.....	3.01	24
White	3.15	15
	P=3.25-3.50	
	P	*S*
Ascham.	3.49	19
Coleridge	3.33	19
Gladstone.....	3.43	16
Hume	3.29	12
Huxley........	3.36	16
Moore.........	3.38	11
Ruskin	3.50	18
Scott	3.36	16
	P=3.50-4.00	
	P	*S*
Addison.......	3.67	12
Barrow........	3.74	20
Bolingbroke ..	3.65	14
"	3.65	14
Bunyan	3.91	10
De Quincey ...	3.69	14
Howell........	3.74	11
Luke	3.62	10
Lyly	3.50	17
More	3.65	15
Sidney	3 98	10
Swift..........	3.69	13
Very	3.67	11
Wordsworth ..	3.87	17
	P-4 00-4.50	
	P	*S*
Hakluyt.......	4.22	12
Hooker	4.12	12
Steele	4.02	10
Chaucer	4.17	8
"	4.17	8
	P=4.50-5.00	
	P	*S*
Dryden	4.89	6
Latimer	4.75	13
Milton.........	4.87	6
	P=5.00-5.50	
	P	*S*
Chaucer.......	5.25	4
Spenser	5.44	8
"	5.44	8

Let us now average the *P*'s and *S*'s of each separate group and we obtain:

TABLE IV

INDEX	PREDICATIONS PER SENTENCE BETWEEN	AVERAGES		$P\sqrt{S}$
		P	S	
1	1.50 and 2.00	1.86	53.0	13.54
2	2.00 " 2.25	2.14	39.1	13.38
3	2.25 " 2.50	2.34	32.9	13.41
4	2.50 " 2.75	2.62	25.9	13.33
5	2.75 " 3.00	2.88	23.2	13.87
6	3.00 " 3.25	3.10	19.2	13.59
7	3.25 " 3.50	3.39	15.9	13.52
8	3.50 " 4.00	3.70	13.4	13.55
9	4.00 " 4.50	4.17	10.0	13.19
10	4.50 " 5.00	4.84	8.3	13.94
11	5.00 " 5.50	5.38	6.7	13.92

The average values of *P* and *S* which we have thus arrived at, aside from the general reciprocal relation already referred to, manifest a uniformity which can not possibly be attributed to chance. Take the square root of 53.0, the first number under *S*, and multiply it by 1.86, the corresponding number under *P*, result:

$$1.86\sqrt{53.0}=13.+$$

Next take the square root of 39.1, the second number under *S*, and multiply it by 2.14, the corresponding value for *P*, and the result is again 13.+ Proceeding in like manner with the following pairs of numbers we obtain

$$1.86\sqrt{53.0}=13.+$$
$$2.14\sqrt{39.1}=13.+$$
$$2.34\sqrt{32.9}=13.+$$
$$2.62\sqrt{25.9}=13.+$$

. . . .

$$4.84\sqrt{8.3}=13.+$$
$$5.38\sqrt{6.7}=13.+$$

The exact values are found in the last column of our table. In short, we have quite uniformly

It will be interesting to test this law on some particular work not included in our table of averages. Macaulay's *History of England* is the only work available for this purpose, for it is the only work whose constants P and S have been determined with sufficient accuracy, without tampering with the composition or punctuation of the author. Using $S=34.2$,[1] that one of the two constants P and S which is most readily determined by count, our formula gives

$$P=\sqrt{\frac{184}{S}}=\sqrt{\frac{184}{34.2}}=2.32,$$

and below are the limits within which they could vary and still satisfy our law.

Upper limit.......	56.7	42.8	35.8	28.6	23.7	20.4	17.1	14.4	11.3	8.4	6.8
S	**53.0**	**39.1**	**32.9**	**25.9**	**23.2**	**19.2**	**15.9**	**13.4**	**10.0**	**8.3**	**6.7**
Lower limit.......	48.0	36.8	30.8	24.6	20.3	17.5	14.8	12.3	9.7	7.1	5.8

Now not only are the limits comparatively narrow, but in most cases S occupies a mean position between them.

I am aware, however, that strange numerical relations may occur where there is no law. In evidence of this I take liberty to quote an example from C. S. Peirce's essay, *A Theory of Probable Inference*, published in the Johns Hopkins University *Studies of Logic*, an essay which every one who ventures upon the field of scientific induction would do well to peruse. The first five names of poets and their ages at death, taken from a certain biographical dictionary are,

Aagard	48
Abeille	76
Abulola..	84
Abunowas.............	48
Accords	45

Now, although no sane person would expect a law connecting the digits of the numbers representing the ages at deaths of poets, it is nevertheless true that for the given five numbers,

1. The difference of the two digits, divided by *three*, leaves a remainder of *one*.

2. The first digit powered by the second, the result divided by *three*, leaves a remainder *one*.

3. The sum of the prime factors (including one) of each number is divisible by *three*.

[1] Gerwig gives 34. But Gerwig gives only the integral parts of the values for S, while the values for P are given to two decimal places. To be consistent, he should have either retained the first decimal places in S, or have dropped also the second decimal places in P.

which coincides nearly with the value 2.30, determined by the laborious process of actually counting the finite verbs in over 40,000 periods.

I now offer the following as a tentative statement of the Sherman principle:

1. *Good writers manifest their individuality by the unconscious use of certain average sentence-proportions which remain nearly constant throughout a given form of composition.*

2. *These sentence-proportions vary widely, not only in various authors, but also for different forms of composition employed by the same author.*

3. *These proportions are interrelated, and in some instances at least their laws of dependence may be mathematically expressed.*

The principle as now stated opens an almost unexplored field for investigation. The present inquiry has been limited to the constants L, P, and S, but there are many other sentence-proportions, such as the percentages of various kinds of clauses and phrases, the relative abundance of adjectives, conjunctions, and other parts of speech, the ratio of Anglo-Saxon derivatives to those derived from other languages, the average length of the words employed by an author, etc. Are these proportions more or less variable than those already considered? Which of these are independent, and which are interrelated, and by what laws? Which of these are most persistent and hence most characteristic of a given author? These and many other questions press for answers.

Concerning the constants L, P, and S many questions remain open. We have seen that L depends at least upon two factors, the author's sentence-sense and the form of composition employed; but are these the only factors upon which it depends? If so, can the element of variation in L due to the form of composition be determined and the residue, due to the author's sentence-sense alone, be set free? Can the relation between sentence-length, sentence-instinct, and sentence-form be as definitely expressed as we expressed the relation between P and S?

These are problems which it seems to me need not remain unsolved.

The greatest immediate need is for more abundant data. In the data thus far obtained it has been assumed that 300 periods will furnish with comparative accuracy the constants sought, and in but few instances have more than 500 periods from single works been examined. Neither the probable nor the actual error arising from this hypothesis has been determined.

In order to get some light on this question I have computed the following table from averages of 400 periods and 500 periods each of 50 authors:[1]

TABLE V

AVERAGES FROM FIFTY AUTHORS OF	AVERAGE DEVIATION IN THE AVERAGES OF 400 AND 500 PERIODS	LIMITS OF DEVIATION		PROBABLE ERROR
		Positive	Negative	
	Per cent	Per cent	Per cent	Per cent
Sentence-length	2.35	9.5	6.3	1.6
Predication-averages	2.06	4.4	9.7	1.9
Simple-sentence-percentages	5.98	28.8	11.2	5.3

This table shows that the constants *L*, *P*, and *S* obtained from an average of 400 sentences, will differ by 2.35 per cent, 2.06 per cent, and 5.98 per cent respectively from the values of these constants based upon averages of 500 sentences, and in some cases variations amounting to 9.5 per cent, 9.7 per cent, and 28.8 per cent respectively may occur. We should, therefore, even by examining 400 sentences from each of 50 authors, obtain less than 2 per cent accuracy in the constants *L* and *P*, and about 6 per cent accuracy in *S*. The numbers in the last column show the probable deviation between 400 and 500 sentence-averages for any other work whose constants have not yet been determined. It appears, therefore, that so long as averages are based upon 500 or less periods the fractional parts in *L* and *S* may be entirely neglected, and in *P* everything after the first decimal place may be disregarded.

[1] The table is based upon Miss Whiting's results for *L* and Mr. Gerwig's tables for *P* and *S*.

IV.—*On the Errors in the Methods of Measuring the Rotary Polarization of Absorbing Substances*

BY FRED J. BATES

In the half-shade polariscope the settings are made with the two halves of the field of equal intensity. Let MO and MO, fig. 1, represent the directions of the vibrations of the wave-lengths in the three similar beams as they emerge from a transparent substance in a magnetic field. Let A be the amplitude and λ_1 to λ_n the wave-lengths, λ_n being the shorter. Let EE be the direction of vibration after passing into the analyzing nicol, and a_n to a_1 the angles which the vibrations in each half of the field make with the normal OH. We then have as the condition for equal illumination of the two halves of the field:

$$(A_1 \sin a_1)^2+(A_2 \sin a_2)^2+\cdots\cdots+(A_{n-1} \sin a_{n-1})^2+(A_n \sin a_n)^2=$$
$$(A_1 \sin a_n)^2+(A_2 \sin a_{n-1})^2+\cdots\cdots+(A_{n-1} \sin a_2)^2+(A_n \sin a_1)^2, \quad \text{(I)}$$

where each term is the intensity of the wave-length as it reaches the eye.

Consider any two wave-lengths, λ_s and λ_r, fig. 1, and let them have amplitudes A_s and A_r.
Then

$$(A_s \sin a_s)^2+(A_r \sin a_r)^2=(A_s \sin a_r)^2+(A_r \sin a_s)^2, \qquad \text{(II)}$$

Let $A_r^2=KA_s^2$ (III)
then

$$\frac{\sin^2 a_s}{K}+\sin^2 a_r=\frac{\sin^2 a_r}{K}+\sin^2 a_s, \qquad \text{(IV)}$$

since, by transposing, (IV) gives

$${}^1/_K(\sin^2 a_s - \sin^2 a_r)=(\sin^2 a_s - \sin^2 a_r), \qquad \text{(V)}$$

which can be true only when $K=1$. Consequently (V) can only be made to hold, and still maintain the condition (III), by changing the angles between the directions of vibration and the normal, that is, *OH* takes the position *OH'*.

Hence (I) for light of any amplitude becomes

$$(A_1 \sin [a_1 \pm \delta a])^2 + (A_2 \sin [a_2 \pm \delta a])^2 + \cdots + (A_{n-1} \sin [a_{n-1} \pm \delta a])^2 + (A_n \sin [a_n \pm \delta a])^2 =$$

$$(A_1 \sin [a_n \mp \delta a])^2 + (A_2 \sin [a_{n-1} \mp \delta a])^2 + \cdots + (A_{n-1} \sin [a_2 \mp \delta a])^2 + (A_n [\sin a_1 \mp \delta a])^2 \quad \text{(VI)}$$

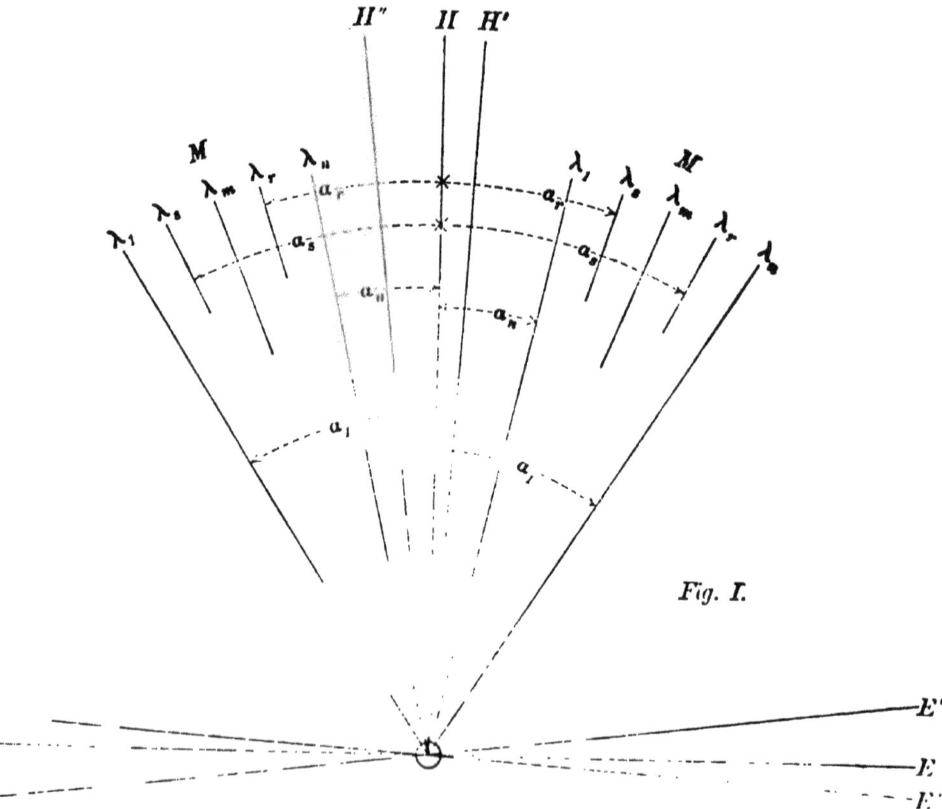

Fig. I.

Let us now consider a specific example and solve for δa. As the source of light take the wave-lengths from 566 $\mu\mu$ to 580 $\mu\mu$ cut from the solar spectrum. $(a_1 + a_n)$ (fig. 1), which is the angle between the direction of vibration in the nicols of the polarizing system, is obtained by direct measurement when there is no rotation of the light.[1] We thus find

$$(a_1+a_n)=20.4'.$$

Also for

λ in $\mu\mu$	Rotation
566	4.77°
580	4.53°

That is

$$(a_1-a_n)=14.4'$$
$$\therefore a_1=17.4'$$
$$a_n=3'.$$

Considering now the luminosity curve, *S*, of the sun[2] and the transmission curve of a fuchsin solution,[3] *F*, in fig. 2, and taking $n=5$ we have

TABLE I

λ IN $\mu\mu$	a IN MINUTES	RELATIVE INTENSITY FROM SUN	PER CENT TRANSMISSION IN FUCHSIN SOLUTION.
580	17.4	86	21.7
576	13.3	94	10.5
573	10.2	99	5
570	7.1	103	2.3
560	3.0	104.5	1

where $\lambda=573\ \mu\mu$ for example is taken as the mean intensity of the interval from $571.5\ \mu\mu$ to $574.5\ \mu\mu$.

By successive approximation (VI) is solved with sufficient accuracy, and we obtain by considering only the luminosity curve:

[1] The analyzer was set for blackness of one-half of the field and then rotated to a similar position for the other half. The angle of rotation is (a_1+a_n).

[2] Marscart, *Optique*, I, p. 104, pl. I.

[3] This curve was kindly determined for the writer by Professor B. E. Moore. The concentration, parts by weight, was 0.000024.

$$86 \sin^2 (17.4'+.4')+94 \sin^2 (13.3'+.4')+99 \sin^2 (10.2'+.4')+ 103 \sin^2 (7.1'+.4')+104.5 \sin^2 (3'+.4')=$$

$$86 \sin^2 (3'-.4')+94 \sin^2 (7.1'-.4')+99 \sin^2 (10.2'-.4')+ 103 \sin^2 (13.3-.4)+104.5 \sin^2 (17.4-.4) \qquad \text{(VII)}$$

Hence, with a transparent substance, giving a rotation of about 4.77° for 566 $\mu\mu$, the normal to the principal axis of the analyzing nicol, for the light source considered, will be rotated from the median position OH through an angle, δa=.4′ (=.006°) to a position OH'.

In a similar manner we now change the coefficients of equation (VII) to correspond to the luminosity curve resulting from passing sunlight through a fuchsin solution (see Table I) and we get approximately

$$86\times 21.7 \sin^2 (17.8'-4.1')+94\times 10.5 \sin^2 (13.7'-4.1')+ 99\times 5 \sin^2 (10.6'-4.1')+103\times 2.3 \sin^2 (7.5'-4.1')+ 104.5\times 1 \sin^2 (3.4'-4.1')=$$

$$86\times 21.7 \sin^2 (2.6'+4.1')+94\times 10.5 \sin^2 (6.7'+4.1')+ 99\times 5 \sin^2 (9.8'+4.1')+103\times 2.3 \sin^2 (12.9'+4.1')+ 104.5\times 1 \sin^2 (17'+4.1').$$

Hence the normal moves through an angle of 4.1′ from OH' to OH''. Thus it becomes evident that the change in the luminosity curve of the light reaching the eye, due to the absorption of the fuchsin solution, gives with the half-shade system an apparent rotation of the plane of polarization of approximately 4.1′ (0.065°) under the conditions assumed above. Under the same conditions the magnitude of this effect observed experimentally was 0.055°.

Errors due to the change in the luminosity curve similar to the one just discussed also enter in a proportionately greater degree into Wiedemann's[1] and allied methods of measuring the rotation of plane polarized light in and near an absorption band.

[1] G. Wiedemann, *Pogg. Ann.* 82, p. 215. 1851.

In these methods there is used as a measure of the rotation the displacement of either a black space or a clear space in a spectrum. The former is produced by a quartz plate cut at right angles to the optic axis and an analyzing nicol; the latter, which was first used by A. Schmauss,[1] by an additional plate of selenite, which gives a channelled spectra with the interference bands absent for those wave-lengths whose planes of vibration coincide with one of the principal axes. From space to space in these methods is a rotation of 180°, and the maximum sensibility seems to be obtained with about 1.5 mm. of quartz between the nicols. The working width of the space thus obtained depends upon the observer, but is assumed to be about 40 $\mu\mu$. Both methods are essentially photometrical since the position of the space is determined by the condition that the light at equal distances on each side of the central region be of the same intensity. Even if the clear space is located by having an equal number of the interference bands on each side of the central region, the process is nevertheless photometrical because the bands do not entirely disappear at any portion of the space, and the bands at which the observer ceases to count must be of equal intensity.

When either of these methods is used to measure the rotation in the edge of an absorption band, there is a shift of the space referred to, and consequently an apparent rotation of the light due to the absorbing substance. Consider, for example, the dark space method. Let the space be produced by 4.5 mm. of quartz and the direction of vibration of wave-length 589.6 $\mu\mu$ be exactly at right angles to its direction in the analyzing nicol. Under this condition S', fig. 3, is the luminosity curve of the sun as it reaches the eye; and F' the same curve modified by the insertion into the path of the light of a 1 cm. length of the alcoholic fuchsin solution 0.000024. The ordinates of S' are obtained by resolving the direction of vibration, giving it its relative intensity from curve S, fig. 2, along the direction of vibration in the analyzer. F' is obtained by giving each ordinate of S' its relative transmission from curve F, fig. 2. It will be observed from the fact that the luminosity curve of the sun is not a horizontal

[1] A. Schmauss, *Ann. d. Phys.* 2, p. 280. 1900.

straight line that, even on the curve S, the center of the space, no matter what its width, is not on wave-length 589.6 $\mu\mu$, which is the point of zero intensity, but at some distance to the left. Really this is the only wave-length that is completely absent, but for several $\mu\mu$ on each side of it the eye can detect no change in the intensity of the blackness and so is compelled to judge by means of points of equal intensity on each side of the central region. The space is thus located with an apparent center to the left of 589.6 $\mu\mu$. Assume that the boundary of the black space is defined by the wave-lengths 582 $\mu\mu$ and 594 $\mu\mu$, which are of equal intensity, thus giving it a space with a working width of 12 $\mu\mu$.

With the transparent solvent in the field it is evident that the position of the black space will be determined as though its center is at C (fig. 3). When the fuchsin solution replaces the solvent, the center, passing to curve F, is found to be at C', or a change of position of 0.7 $\mu\mu$. This gives a rotation of 0.6° when there has been no change in the position of the planes of polarization.

The conditions assumed above being very similar to those used by various investigators were easily realized experimentally. An auxiliary cell with the fuchsin solution was thrown into the path of the light and the analyzing nicol rotated until the black space was in its original position. The following data were thus obtained for this part of the spectrum:

FUCHSIN SOLUTION

OUT	IN	DIFFERENCE
16.00°	16.60°	0.60°
15.95°	16.43°	0.53°

CALCULATED ROTATION	OBSERVED ROTATION
0.62°	0.56°

The above is also just as applicable to the clear space method.

Such close agreement is of course partly accidental, since consecutive measurements can not be made by Wiedemann's method closer than 0.10°. The magnitude of the error will depend upon what width of space the observer uses.

As the shift of the space was toward the red, thus giving an apparent positive rotation, it is evident that as the space moves farther into the absorption band of the fuchsin this positive rotation would increase until it attained a maximum in the edge of the band. From there on it would gradually decrease and become zero at some point in the interior where the absorption is the same for both edges of the space. As the space approaches the other edge of the band from the interior a gradually increasing rotation would be observed, attaining a maximum in the edge of the band, and then decreasing to zero as the space leaves the band. This maximum is, however, in the opposite direction from that previously mentioned. Since the former is a positive, this is a negative rotation; that is, for solutions it would drop below the rotary dispersion curve of the solvent.

When the half-shade is used the apparent rotation due to the error increases very perceptibly with, but is not directly proportional to, an increase in the strength of the magnetic field, because a doubling of the field gives a corresponding increase in the angle ($a_1—a_n$), all other conditions remaining the same. In the black and clear space methods, however, the maximum value of the error varies only very slightly, for small rotations, with the intensity of the field, because the rotation of the substance studied constitutes only a small fraction of the total rotation. The thickness of quartz necessary to give the desired space in the spectrum is at least 1.5 mm. This gives an initial rotation of 32.5° at the *D* line, to which is added the rotation, for example, of 1 cm. length of water for a field of 8000 lines, making a total 34.2°. If the field be doubled the total becomes only 36°, which is sufficient change to affect the maximum error only slightly. In the three methods mentioned the maximum value of the spurious rotation will be nearly directly proportional to the concentration for dilute solutions, and the sharper the absorbing band the more marked the anomaly.

These apparent rotations, however, are not entirely due to the simple cutting down of the amplitudes of certain wave-lengths by the absorbing substance. Superimposed upon this is the Purkinjie effect, or change in the luminosity curve for different intensity. This effect is ordinarily negligible, but under certain conditions it may become considerable. This, we are led to suspect by a study of König's[1] luminosity curves for gas lights at different intensities. Similar curves for sunlight are apparently lacking, so that no calculation bearing upon this point could be made. With the wave-lengths from 573 $\mu\mu$ to 587 $\mu\mu$ as the source of light and a rotation of 4°, the setting of the analyzer differed by 0.02° when the light intensity was reduced 44 times by the rotating sector. The angle between the principal axes of the nicols in the polarizing system was several minutes of arc.

In fig. 2, *R* is the luminosity curve resulting from passing sunlight through a 1 cm. thickness of alcoholic fuchsin solution, concentration 0.000024. It is obtained by modifying the ordinates of the luminosity curve *S* by the transmission curve *F*. Whenever the difference in the rotation of two substances is being measured, or either the optical center of gravity of the light source is being determined for absolute measurement with the half-shade system, or the position of the observing telescope is being calibrated in wave-lengths for either the clear or the black space methods, the errors due to the use of two luminosity curves of which *R* and *S* are a specific example always enter the measurements unless proper compensation be made. These curves may be almost identical for transparent and vary greatly for absorbing substances. In the former case the errors are ordinarily negligible; in the latter they may become very large. From the theory it is evident that the angle will be a maximum at those points in the spectrum where the comparative slope of the two curves is such as to give the greatest difference in the intensity of the longest and shortest wave-lengths reaching the eye. Consider curves *S* and *R* with the half-shade system. At 600 $\mu\mu$ the error would be approximately the same as that calculated in equation (VI), (0.006°), since the conditions are about the same.

[1] *Beiträge zur Psychologie und Physiologie der Sinnesorgane.*

At about 567 $\mu\mu$ it would seem to be a maximum, because here the variation in the intensity of the extreme wave-length used will be greatest. At some point in the interior of the band it will become zero. In passing from the absorption band the error will not be nearly so marked as it was upon entering. Curves *L* and *H*, fig. 2, obtained experimentally, are given to show how the error changes in this particular instance. Their ordinates are differences in the rotations, one small division representing 0.005°. Curve *L* is for the alcoholic fuchsin solution 0.000024. *H* is the same with a concentration of 0.000012. A cell of water giving 4.5° rotation at the *D* line was between the poles of the magnet. The angle between the elements in the half-shade system was 1.10°. The intensity of the light source was 14 $\mu\mu$. A 1 cm. length of the solution placed outside the magnetic field was thrown in and out of the path of the light and the differences in the rotation of the water noted.

The effects of bleaching must be carefully guarded against. After sunlight has been passed for five minutes through solutions of such substances as fuchsin and cyanin the absorption is greatly diminished.

Much careful work has been done in determining the optical center of gravity of different light filters[1] for various parts of the spectrum, and the calibration is done with substances with no particular selective absorption. These filters let through light differing in wave-length by not less than 34 $\mu\mu$ or 40 $\mu\mu$. Hence a slight difference in the luminosity curves will give, with the half-shade system, a comparatively large error in the rotation.

When the non-absorbing substance was in the field, a rotating sector was experimented with to equalize the intensity of the light; but, since this diminishes the amplitude of all waves equally and changes the luminosity curve only in proportion to the Purkinjie effect, the spurious rotations were not eliminated. The insertion into the path of the light, outside the magnetic field, of an optical thickness of the substance equal to that in the field in such a manner as to compensate for the unabsorbed beam,

[1] H. Landolt, *Das optische Drehungsvermögen*, p. 387. Chr. Winther, *Zeitschr. Phys. Chem.*, July, 1902, p. 161.

seems to be the simplest method of keeping the luminosity curve unchanged. The elimination of these effects should in many cases bring about a more perfect agreement, especially for solution of the measurements of different investigators on the rotation of plane polarized light, both natural and magnetic.

The opportunity is here taken to thank Professor D. B. Brace for his helpful suggestions.

V.—*The Magnetic Rotary Dispersion of Solutions of Anomalous Dispersing Substances*[1]

BY FRED J. BATES

Becquerel[2] in 1880 observed the anomalous rotary dispersion in oxygen. Others have failed to verify his results. In 1898 Macaluso and Corbino[3] observed it in sodium vapor. They found that the rotation increased with great rapidity as the absorption band was approached, attaining a maximum in the edge of, and dropping to a small positive rotation in the interior of, the band. Zeeman[4] has also studied this phenomenon and finds for sodium vapor, not too dense, a negative rotation in the interior of the band.

Biot,[5] Arndsten,[6] Landolt,[7] Nasini[8] and Gennari, Cotton[9] and others found anomalous rotary dispersion in optically active liquids, and in 1896[10] Cotton observed a tendency toward what he interpreted as an anomaly in ferric chloride, copper acetate, and several other liquids, when they were placed in a magnetic field. He found as an absorption band of a certain solution was approached, going from red to violet, an increase in the rotation above what would have been obtained had the solution been transparent. In the solution of some other substance in which he could approach a band from violet to red he found the rota-

[1] Read before the American Association for the Advancement of Science at its Washington meeting, 1902–3.

[2] H. Becquerel, *Compt. rend.* 90, p. 1407. 1880.

[3] Macaluso and Corbino, *Compt. rend.* 127, p. 548. 1898.

[4] P. Zeeman, *Proceedings Royal Academy of Amsterdam.* May 31, 1902.

[5] M. Biot, *Ann. de chim. et de phys.* (3) 10, p. 5. 1844.

[6] M. A. Arndsten, *Ann. de chim. et de phys.* (3) 54, p. 403. 1858.

[7] H. Landolt, *Beibl.* 5, p. 298. 1881.

[8] R. Nasini and G. Gennari, *Zeitschr. f. physik. Chemie* 19, p. 113. 1896.

[9] Cotton, *Ann. de chim. et de phys.* (7) 8, p. 347. 1896.

[10] Cotton, *L'Eclairage electrique* 8, p. 162 and 199. 1896.

tion dropped below the normal. This he interpreted to indicate two points of inflection in the rotary dispersion curve for each band. He never succeeded, however, in establishing a curve for both sides of the same band, nor in the interior of the band. In 1900 Schmauss[1] observed an apparent anomaly in fuchsin, cyanin, naphthalene-red, eosin, and didym glass. He studied the first four in very dilute alcoholic solution, the greatest concentration by weight for each substance being 0.000024, 0.000039, 0.000005, 0.000004 (parts by weight) respectively. His rotary dispersion curves for these liquids all show four points of inflection, and a very marked rise above the curve of the solvent as the violet region is approached. The fuchsin solution 0.000024 gave a rotation of 2.81° for wave-lengths 544 $\mu\mu$ and 2.34° for 545 $\mu\mu$. The former is 0.44° greater than that of the solvent, and the latter 0.13° less. For wave-length 450 $\mu\mu$ the solution gave a rotation of 0.58° greater than that of the solvent. In subsequent papers[2] he studied didym glass, liquid oxygen, and solution of litmus, anilin blue, and three rare earths.

Schmauss's results indicate that the anomalous effects increase with increasing concentration; that the maximum rotation is independent of the field strength, and that the negative rotation in the interior of a band decreases with increasing field strength. The observations of Zeeman on moderately dense sodium vapor as well as all of Schmauss's results seem to confirm the theoretical deductions of Voigt.

The writer has been studying solution of anomalous dispersing substances and thus far has not succeeded in observing an anomaly. The substances were fuchsin, cyanin, anilin blue, and litmus. The first two were in alcoholic, the latter two in water solutions.

The method employed consisted in placing a tube of the solution between the pierced poles of an electromagnet. Light parallel to the lines of force was successively passed through a half-shade polarizing system, the tube, and an analyzing nicol. After measuring the rotation, the tube was removed and the rotation

[1] A. Schmauss, *Ann. d. Phys.* 2, p. 280. 1900.

[2] A. Schmauss, *Ibid.*, 8, p. 842, 1902, and 10, p. 853. 1903.

of a similar tube filled with the solvent noted. Since the tubes are the same length the difference between the two rotations is the rotation, for the wave-length of light used, due to the presence of the dissolved substance.

The source of light was the sun. The desired wave-lengths of sufficient homogeneity were obtained by means of a spectral system similar to that described by Doubt.[1]

The instrument used in making the measurements was a sensitive-strip spectropolariscope.[2] The polarizing system consists of two thin pieces of Iceland spar mounted in a cell of α-monobromonaphthalene and transmitting the ordinary ray. The smaller of the two pieces covers but half of the field of the larger one, and the angle between their planes of polarization can be varied at will. Both the color and polarizing systems will be treated in detail by the writer in a paper soon to be published.

The conical pole-pieces of the electromagnet used tapered from 20 cm. to 5 cm. in diameter. With 23 amperes at 500 volts and the pole-pieces separated 17 mm., a field of over 18,000 lines per sq. cm. was obtained. This intensity was used with the litmus and anilin blue solution, but was allowed to drop to 15,000 lines in studying the fuchsin and cyanin. This was sufficiently near saturation to prevent the slight fluctuations of the current interfering with the setting of the polariscope.

A star-shaped brass frame, fig. 1, rotating on pivot bearings, was rigidly supported between the poles of the magnet. The tubes rested in the V-shaped slots and were held in place by spring-back clips. A slight rotation removed one tube from the path of the light and substituted another. The tubes were 1 cm. in length and closed with cover-glasses each 0.2 mm. in thickness.

Since we compare one tube with another it is necessary that their optical paths be the same. Factors tending to make their paths unequal may be introduced, first, by the tubes not being of the same length; second, the brass frame failing to keep their axes parallel; third, the cover-glasses not being of the same thickness. These difficulties were all eliminated by surrounding the

[1] T. E. Doubt, *Phil. Mag.* Aug., 1898.

[2] D. B. Brace, *Ibid.* Jan., 1903.

frame with a stationary glass cell (100 mm. × 50 mm. × 12 mm.), fig. 1, filled with the solvent. If, now, the tube containing the solution be rotated into the path of the light, it displaces solvent equal to its own length. The tube containing the solvent serves merely as a carrier for the cover-glasses, which compensate for those closing the tube containing the solution. If the cover-glasses are not of the same thickness the only change that can occur in the optical path will be caused by replacing of the extra thickness of glass by the solvent when the next tube is thrown in. A difference of 0.02 mm. in the thickness of the cover-glasses would give an error of 0.004° in the rotation. Thus it is evident no mechanical errors can appreciably affect the measurement.

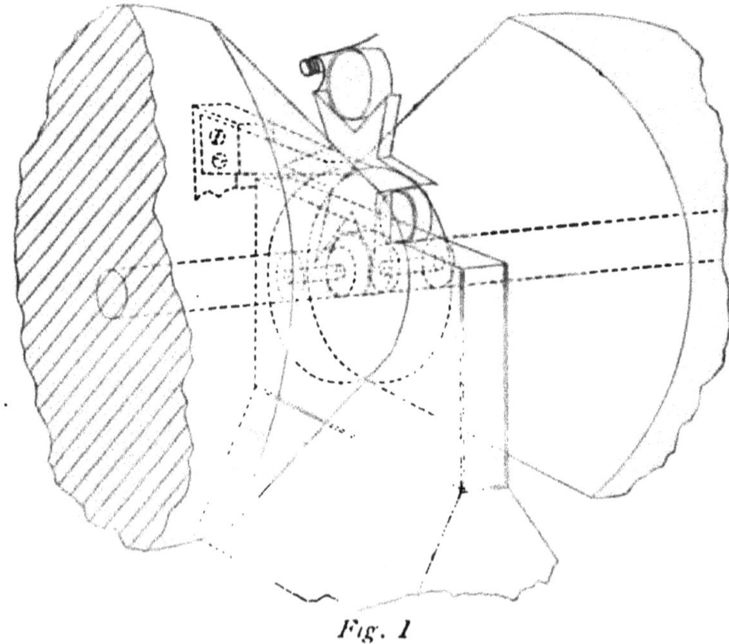

Fig. 1

In the usual methods for measuring the rotation of plane-polarized light there appear in the observing telescope, when the magnet is excited, several partially overlapping images of the field. This is due to the successive internal reflections of the beam of light within the tube. When the analyzing nicol is set for the extinction of the ray that has passed through the liquid but once, the images produced by the rays that have passed through it, three, five, etc., times, become relatively bright, since the rotation is proportional to the length of the liquid traversed. An indistinctness results. These reflected rays were eliminated from the field of the observing telescope by making the ends of the stationary cell slightly prismatic.

With such a sensitive polarizing system all glass, even when quite thin, shows more or less double refraction, and the elimination of this becomes a serious problem. By examining a great many microscope cover-glasses a sufficient number of optically good ones was obtained. The thinner the glass the freer it is likely to be from strains; but if it is less than 0.2 mm. in thickness the pressure of liquid against it will produce not only a lens effect and destroy the adjustment of the optical system, but also make the glass doubly refractive.

In constructing the tubes a cover-glass with a diameter 2 mm. greater than that of the tube was laid on a level surface and the tube placed on it. Without touching tube or cover-glass, either beeswax or a mixture of fish glue and glycerine, depending on the solvent, was flowed around them. In this way tubes were finally obtained whose end-plates showed practically no double refraction. The glass on the sides of the cell containing the tubes being 1.5 mm. thick, holes were bored and windows of cover-glass were mounted in the manner just described.

The analyzing nicol could be rapidly rotated when necessary. The setting, however, was made by means of an accurate screw carrying a drum graduated to 0.005°. A tenth of the scale divisions representing this quantity could be readily estimated by the unaided eye.

With the apparatus as described and the magnet excited, successive settings for different tubes could be made in a few seconds, and the field was neither distorted nor contained any extraneous light. Successive settings on a non-absorbing solution could be made whose extreme values differed from each other by less than 0.007°; and with the sensibility diminished to give sufficient light to read through the absorption band itself of the liquids studied, the error was less than 0.01°.

In comparing a non-absorbing with a highly absorbing solution the difference in the intensity of the light reaching the eye is very great. The eye is thus unfitted for rapid setting from the non-absorbing to the absorbing substance, and it is not desirable to allow a considerable interval of time to elapse between such observations. In order to overcome this and also to eliminate spurious rotations which are discussed in a succeeding paper, an auxiliary cell of the solution with the same absorbing power as the solution tube was placed before the collimating slit and outside the magnetic field. By means of the lever the observer threw this cell into the path of the light simultaneously with the tube of solvent. The luminosity curve, as well as the intensity of the light reaching the eye, was therefore the same for both tubes.

The tubes, one filled with solvent, the other with solution, were placed in the star-shaped frame, the lower portion of which was emersed in the cell filled with the solvent, fig. 1. A setting was now made with the solution, say, in the magnetic field; this tube was then thrown out and the tube containing solvent thrown in, and a new setting made. The next settings were from solvent to solution. The time required to exchange the tubes was a fraction of a second.

The measurements are given in Table I. The numbers opposite a wave-length are the differences in thousandths of a degree in the rotations for centimeter lengths of solvent and solution. The minus sign before a number indicates that the rotation of the solution was less than that of the solvent. The concentrations are parts by weight.

Wave length	CYANIN	FUCHSIN	FUCHSIN	Wave length	LITMUS	ANILIN BLUE
	Alcoholic solution concentration 0.(00019	Alcoholic solution concentration 0.000025	Alcoholic solution concentration 0.0)0012		Water solution concentration 0.0013	Water solution concentration 0.00008
λ=μμ	Thousandths of a degree			λ=μμ		
746				710	5 0	
720	5 —2 5 20	—2.5 —1 .5 7.5	—7.5 —10 —2.5	674	0 10	5 —5
668	2.5 5 0	2.5 —5	—12.5 —5 0	646	5 0 20 15 5	—2 0
650	0 —5 10	0 —1	0 0	622	30 30 15 20 0	—5 7.5 0
627	—12.5 5 10	12.5 —10	—10 —2.5	602	0 —15 10 0	0 10 5
606	—15 —10 2 5 2.5	5 5	12.5 15 5 2.5	593		0 10
589	—7.5 —15 —2.5 12.5 —12.5 —5 0	—10 2 5	—17.5 12.5 5	585	10 25 25 10 10	5 —10
573	—2.5 —20 0 0 —25 —5	—27.5 —7.5 25 —5 7.5	—12.5 —5	570	0 15 0 0 10 15	25 15
558	7.5 5 12.5 —7.5	10 25 —22.5 7.5	—5 —10 5	555	10 15 —5	5 5

Wave length	CYANIN Alcoholic solution concentration 0.000019	FUCHSIN Alcoholic solution concentration 0.000025	FUCHSIN Alcoholic solution concentration 0.000012	Wave length	LITMUS Water solution concentration 0.0013	ANILIN BLUE Water solution concentration 0.00008
λ=μμ	Thousandths of a degree			λ=μμ		
543	—10 15 5	—15 —15	7.5 5 0	541		0 —1
531	—7.5 —2.5 —2.5	10 15	—20 —2.5 —2.5	528		15 0 20 10
520	0 —7.5 12.5	—2.5 20 7.5	0 20 2.5	517		0 15 10 5
509	—15 —7.5 7.5	—2.5 0	12.5 12.5 7.5	510		—5 10
498	5 0 0	10 5 5	—20 —7.5 —2.5 10 0	506		—5 15 0
489	12.5 17.5 10 —20	—15 25 35 17.5		496		10 5 0
480	—15 —5 0 0	5 32.5 —10	10 —10 —20 —2.5 20	494		10 0 15 5 20 5
472	17.5	—5 2.5 —10		485		
464		—35 —20	—15 20 0 —40 35 30			

In addition to the above wave-lengths the solutions were examined for many other points in the spectrum. The method being a differential one and not dependent on successive readings of the absolute rotation of solvent and solution, the phenomena sought for, if present, could be observed directly. Hence a solution could be examined in a very short time. The differences between the solvent and solution, for wave-lengths not greatly absorbed by the latter, could have been made much smaller, had it been necessary, by increasing the sensibility of the polarizing system.

The differences noted are irregular and apparently are merely experimental errors. With a polariscope giving consecutive settings within 0.005°, considerable difficulty would naturally be expected in eliminating the various sources of error sufficiently to make this sensibility available. Even after the elimination of double refraction and mechanical errors there are the effects of bleaching and perfect compensation for the absorbing solution to be met. Unless the liquid in both the solution tube and the compensating cell is perfectly fresh a difference in rotation of perhaps 0.01° will be observed. At the time, for instance, the data for the anilin blue solution at wave-length 517 $\mu\mu$ were taken there was actually present a difference of 0.01° between solution and solvent. Upon looking for it at some subsequent time it would perhaps have vanished. The above data seem to show that there is not sufficient anomalous magnetic rotary dispersion in the solutions studied to be measured with the means available. Mr. Williams, fellow in physics, in a forthcoming paper has studied the dispersion of alcoholic fuchsin solutions by means of channeled spectra. He finds no anomaly in the dispersion curve of a fuchsin solution thirty-six times the concentration 0.000024. Hence, if we accept the various formulae, for example Becquerel's where

$$\omega = \frac{2\pi e}{\theta V_o} \lambda \frac{dn}{d\lambda}$$

which have been presented as criteria for the magnitude of the anomalous rotation, it is evident that we can not expect to observe an anomaly in dilute solutions.

The writer wishes to acknowledge his indebtedness to Professor D. B. Brace for assistance which has made this investigation possible.

Volumes I and II of UNIVERSITY STUDIES are each complete in four numbers.
Index and title-page for each volume is published separately.
A list of the papers printed in the first two volumes may be had on application.
Single numbers (excepting vol. I, no. 1, and vol. II, no. 3) may be had for $1.00 each.
A few copies of volumes I and II complete in numbers are still to be had.
All communications regarding purchase or exchange should be addressed to

THE UNIVERSITY OF NEBRASKA LIBRARY

LINCOLN, NEB., U. S. A.

JACOB NORTH & CO., PRINTERS, LINCOLN.

Vol. III. October, 1903 No. 4

University Studies

Published by the University of Nebraska

CONTENTS

I. Regeneration in Hydromedusae
George Thomas Hargitt 275

II. Some Peculiar Double Salts of Lead
John White 307

III. The Mémoires de Bailly
Fred Morrow Fling 331

IV. On the Representation of Numbers as Quotients of Sums and Differences of Perfect Squares
Robert E. Moritz 355

LINCOLN, NEBRASKA

Entered at the post-office in Lincoln, Nebraska, as second-class matter, as University Bulletin, Series 8, No. 13

UNIVERSITY STUDIES

VOL. III — *OCTOBER, 1903* — NO. 4

I.—*Regeneration in Hydromedusae*[1]

BY GEORGE THOMAS HARGITT

I. INTRODUCTION

Considerable work has been done on the regeneration and grafting in the Hydrozoa by Loeb, Bickford, Driesch, Hargitt, Peebles, Morgan and others. The results of these experiments have been, at times, somewhat antagonistic. The work among the Tubularians has been largely limited to *Tubularia crocea* and *T. mesembryanthemum*, and the histological work has been neither extensive nor conclusive. The work of which this paper is a record was therefore undertaken to review certain experiments and to throw some light on disputed or inconclusive statements, as well as also to carry out careful histological studies on the regenerated structures. The experimental work was done during the summer of 1901 at the marine laboratory at Woods Hole. The histological work was started the next winter at Syracuse University, and completed in the spring of 1903 at the University of Nebraska.

To my father, Dr. C. W. Hargitt, I am greatly indebted for supervision and helpful suggestions during the early part of the work.

[1]Studies from the Zoological Laboratory, The University of Nebraska, under the direction of Henry B. Ward, No. 57.

II. Material

Material for the experiments was found in Great Harbor, Vineyard sound, and surrounding waters. *Tubularia crocea, Eudendrium ramosum,* and *Pennaria* were found in great abundance on the piles of the U. S. Fish Commission docks. *Tubularia tenella* and *Tubularia larynx* were dredged from a shelly bottom in Vineyard sound. This latter species has a different habit from the others. While *T. crocea* and *T. tenella* grow in thick bunches or clusters of a moderate height, *T. larynx* seems to twine about older individuals of the same or possibly other species. It attains a considerable height also. *T. tenella* and *T. larynx* were operated on within a few hours, the other forms within 30 to 60 minutes.

III. Methods

The hydroids were cut with a pair of sharp scissors into the desired lengths and placed in watch glasses, petri-dishes, finger bowls, etc., in fresh sea water and covered with glass plates to exclude the dust. The water was changed at least once a day, oftener if conditions demanded. At first the water was taken from the tap in the laboratory, but as this seemed to have an injurious effect on the hydroids, it was later taken directly from the open harbor. In grafting, the cut surfaces were brought into close contact and held in place by freshly shaven bits of lead. The hydranths were always removed before grafting, because, if left, their movements would disturb the pieces and thus destroy the contact.

Specimens were killed in Gilson's fluid, and in corrosive acetic acid to which picric acid had been added. Staining was done *in toto* with borax carmine, and the resulting sections were very satisfactory for general structure and form, though detailed work on the nuclei, etc., could not be done on such slides. On the slide staining was done with Heidenhain's iron-hematoxylin, and also with Delafield's hematoxylin, followed by running up to absolute alcohol and differentiating with picric acid. This latter method was very good for differentiating the entoderm granules, etc., but in general the best results were obtained with iron-hematoxylin.

IV Experimental Work

I. Regeneration

From his work on *Antennularia antennina,* Loeb (1893, p. 42) drew the following conclusion: "In *Antennularia* gravitation not only determines the place of origin of various organs, but also the direction of their growth." He states that rhizoids grow from the lower side or end, and hydranths from the parts directed upward, regardless of the position of the stem. Driesch (1897), working on *Antennularia ramosa,* found that other factors beside gravity influenced the regeneration. Morgan (1901) found for *A. ramosa* that roots usually form at both ends. He suggests, however, that "the development or presence of roots on the basal end prevents development of roots on the apical end." Stevens (1902), working on the same form, concludes that the kind of regeneration is determined neither by the polarity of the piece nor by "orientation with respect to gravity," nor by condition existing at the other end of the piece; but that certain parts of the stalks show a tendency to form roots and other parts to produce stems.

To determine whether gravitation had any effect on the regeneration in *Tubularia crocea,* I took pieces with branches and placed the distal end of the main stalk in sand. The piece was fixed so that the ends of the branches cleared the sand and were entirely surrounded by water. In no case did stolons form, though hydranths formed quite rapidly and readily. It was further noted that hydranths formed sooner on the ends of branches, whether they extended laterally or downward, than they did on the ends of the main stems which pointed directly upward (fig. 5). This would seem to indicate quite clearly that gravity has no influence on the regeneration in this species, at least on the regeneration of the stems, and furthermore that regeneration of hydranths occurs more rapidly at the distal ends of stems.

Loeb (1893) also states that in *Margelis* and *Pennaria* the kind of regeneration and the direction of growth are determined by contact: If the end of the piece was in contact with a solid object a stolon was formed; if the end was surrounded by water a hydranth was formed. That contact has no effect on the regeneration in *Tubularia crocea, T. tenella,* and *T. larynx* is shown by the following examples: In several experiments the pieces were simply laid flat on the bottom of the dish. Regeneration was as follows: In *T. larynx* hydranths almost always formed at both ends; stolons never formed. In *T. crocea* and *T. tenella* hydranths developed at the distal end first and quite often at the other end also; stolons very rarely formed. In several pieces of *T. crocea* a stolon-like growth along the bottom of the dish was the first indication of regeneration. This growth continued for quite a distance and then turned upward away from the dish and formed a hydranth. Experiments on *Eudendrium* and *Pennaria* confirmed Loeb's results with *Margelis* and *Pennaria,* at least to a great extent. Pieces of *Pennaria,* when laid flat on the bottom of the dish, almost invariably formed stolons from the ends of the stems which were in contact with the glass. It was only from those branches which were completely surrounded by water that hydranths developed. This formation of stolons took place as readily from the distal as from the proximal end In one case regeneration in *Pennaria* was very similar to that in *Tubularia crocea,* viz., the formation of a stolon-like growth along the bottom of the dish and later the growth upward and the formation of a hydranth (fig. 8). In general, *Eudendrium* showed the same phenomenon, though the influence of contact was not so marked. Indeed, in one instance, the cut end of a branch in direct contact with the glass formed a hydranth. When *Tubularia* and *Pennaria* were placed in sand no stolons formed from the end in the sand, though this may have been due to the lack of oxygen, as Loeb (1892, p. 64) suggests.

Driesch (1897), Peebles (1900), and others state that when the end of *Tubularia* is cut obliquely the tentacle anlagen will be laid down obliquely. Peebles found this variable, however, and in a later paper (1902) she obtained the following results: When long and short pieces were grafted by oblique surfaces, the resulting anlagen were sometimes obliquely and sometimes squarely placed. Pieces of equal length grafted by oblique surfaces developed anlagen squarely placed. While I gave no particular attention to this feature, nor investigated the conditions active in bringing about this phenomenon, in general it was noted that both long and short pieces which had been cut obliquely had the anlagen squarely placed.

The results of some of the principal experiments are summarized below.

Experiment 5.—Twenty pieces of *Pennaria* were placed in a dish, the bottom of which was covered with sand, in the following way: The pieces were stuck into the sand so that the main stem was vertical. The branches did not touch the sand, but were entirely surrounded by water. Some pieces were fixed with the distal end in the sand, others with the proximal. Several pieces of stolons were also fixed upright in the same manner. Regeneration was very slow, though it was never as rapid in this species as in *Tubularia* or *Eudendrium*. After several days considerable new growth had taken place and new perisarc secreted around this growth, but then the coenosarc had withdrawn, leaving the perisarc empty. This growth occurred on all the pieces and was quite irregular, extending in all directions. After this had continued for several days, hydranths began to develop. The first one appeared on the proximal end of a stem (distal end in the sand), though very soon another one formed on the distal end of the stem (proximal end in the sand). Hydranths continued to develop for a considerable time, seeming to form as readily and at about the same rate from the proximal as from the distal ends. A hydranth also formed from the upper end of one of the stolons (fig. 7). In several specimens hydranths formed on the distal ends of branches which pointed downward and later turned directly upward (fig. 6). The first indication of a developing hydranth is the formation of a knob by the coenosarc. A thin perisarc is secreted around this, and the knob may increase in size a little or become elongated. The perisarc is ruptured, the knob of coenosarc pushes out and forms a hydranth, the tentacles first appearing as small buds or outgrowths and reaching mature size by new growth. In one specimen in which the distal end was in the sand a new stolon-like growth formed from it and pushed up through the sand.

Experiment 11.—This experiment with *Tubularia crocea* was conducted in the same manner as experiment 5. The pieces, however, were all placed with the distal ends in the sand. In 24 hours a hydranth had formed on the end of a branch (fig. 5). In 36 hours more three more hydranths had formed, all on the ends of branches, some extending laterally, others downward; 24 hours later four hydranths had developed on other branches, one extending laterally and three pointing downward. Two hydranths had regenerated at the end of the main stem, i. e., at the proximal end. After this the pieces began to degenerate and soon died. Thus, in this experiment, covering a period of about 80 hours, ten hydranths regenerated. Of these, eight developed at the distal ends of branches and two at the proximal ends of stems, which would seem to indicate a rather marked polarity. The formation of hydranths at the proximal ends of stems was always much slower in this species than the development at the distal ends.

Pennaria was found growing under entirely different conditions in different habitat, though the morphological differences were slight. The most abundant was that growing on the piles of docks in the harbor. The other form was attached to floating eel grass and matured later than the previous form (cf. Hargitt, 1901, p. 224). Experiments to determine regeneration were tried with both forms.

Experiment 13.—*Pennaria* from the piles of docks.—The pieces were simply laid flat on the bottom of the dish. Regeneration was rather slow. After some time the coenosarc emerged from the perisarc. If a hydranth was to regenerate the coenosarc enlarged into a knob from which tentacles budded and the hydranth formed. If, on emergence from the perisarc, the coenosarc came in contact with the bottom or the side of the dish it flattened out to form holdfasts (fig. 10). The behavior of these pieces resembles very closely that of *Margelis* noted by Loeb (1893). Every branch or stem which came into contact with the dish formed holdfasts, and hydranths formed only from the free ends. Moreover, the holdfasts formed with equal facility from both ends. One piece anchored itself to the bottom of the dish by the distal end and then assumed an upright position. Several hydranths formed from the branches and the proximal end of the stem (fig. 15). As already suggested, hydranths formed rather slowly. The first one was completely regenerated at the end of 42 hours, and after this they formed continually for ten days. In this time 40 to 50 developed, though not all were present at any one time. Some would form and degenerate or be withdrawn into the perisarc and others form.

EXPERIMENT 14.—*Pennaria* from eel grass.—This experiment was conducted at the same time and in the same manner as experiment 13. The behavior of the pieces and the method of regeneration were the same as in the preceding experiment, holdfasts forming at the places of contact with about equal rapidity and ease. The first hydranth was completely regenerated within 30 hours, though after the first they did not develop as rapidly nor as abundantly. This may have been due to some harmful effect of the water, or more probably to a less healthy condition of the stems.

EXPERIMENT 16.—Pieces of *Tubularia crocea* placed with the distal ends in the sand as in experiment 11.—In about 18 hours a hydranth had completely regenerated on the end of a branch pointing downward; 24 hours later three more had formed, two of them on branches extending downward. After this, hydranths formed on branches and on the ends of the main stems, i. e., at the proximal ends. Sometimes the hydranths on the ends of drooping branches hung downward; at other times they turned and grew upward. In this experiment, as in experiment 11, there seemed to be a strong tendency for hydranths to form at the distal ends of branches rather than at the free (proximal) ends of the stems, and when they formed on the end of the stem regeneration was much slower.

EXPERIMENT 23.—*Tubularia crocea* cut in pieces of about 3 mm. length and placed on the bottom of the dish.—No change was noticeable for about two days and then in several pieces there seemed to be a collection of red pigment at one end. No further change or development took place although the pieces were kept two weeks. Several other like experiments were tried, with both *T. crocea* and *T. tenella.* In no case, however, did any growth take pace nor any sign of regeneration occur, except the collection of red pigment. This latter condition was found in perhaps four or five pieces out of one hundred or more.

EXPERIMENT 24.—*Tubularia crocea* cut into pieces 12 to 20 mm. in length and placed on the bottom of a dish. Some were cut obliquely, some square across. In 24 hours the tentacle anlagen had developed in several pieces (fig. 3); 36 hours later the anlagen had been formed in twelve more pieces, and 24 hours after this ten new pieces showed the anlagen. After this the anlagen developed in the pieces gradually, so that in ten days all but five pieces out of fifty showed them. The proximal series seemed to be developed first. The general features of the development of the hydranths have been considered by other authors. First the coenosarc withdraws and the cut edges close in. Whether this closing in is a simple bending over of cut edges (Morgan, 1901), or due to amoeboid motion of the whole membrane (Morgan, 1902), or to a centripetal force acting on the cells at the cut edges (Stevens, 1902), I did not determine. After this membrane has been formed a thin perisarc is secreted over the open ends and the circulation of the enteric fluid begins, the longitudinal ridges of entoderm disappearing in the hydranth region. Then the tentacle anlagen are folded off, appearing as ridges. This folding seems to take place along the entire length of the tentacle, beginning at the distal end. The proximal anlagen are formed first. All this has taken place within the perisarc and seems to be a transformation of tissue as stated by Bickford (1894) for *T. tenella.* When the analgen have been completely formed, the thin perisarc over the end of the piece is ruptured, the coenosarc elongates, the hydranth is forced out and assumes the perfect hydranth form. The gonads were never noted as developing till later. In several pieces this process was followed only to the formation of tentacle anlagen; then the coenosarc was pushed out of the perisarc and the tentacles appeared as small knobs or buds, attaining their full size by further growth.

EXPERIMENT 25.—*Tubularia tenella* was cut into pieces of 12 to 20 mm. and arranged as in experiment 24. In about 48 hours several pieces showed the formation of one line of tentacles. The development then proceeded rather slowly, hydranths being formed gradually. The process of development of hydranths differs somewhat from that in *T. crocea.* The end closes over in about the same way, circulation of fluid begins, and red pigment is deposited in the hydranth region. Then instead of the formation of the tentacle anlagen the hydranth body is separated from the rest of the stem by a constriction (fig. 14*a*). The tentacles appear not as ridges, but as small buds (figs. 14*b*, 14*c*), and the hydranth is pushed out of the old perisarc (figs. 14*d*, 14*f*). The tentacles continue to grow until they reach their normal length.

EXPERIMENT 27.—*Tubularia larynx.*—The pieces were cut in about the same lengths and arranged in the same manner as in the previous experiments. In 24 hours only one piece showed the presence of tentacle anlagen (this was at 8:00 P.M.). The next morning nine or ten fully formed hydranths were found. The next day all that had not yet developed hydranths were found to have formed them. (However, several pieces from the extreme basal region did not regenerate at all during the progress of the experiment). Most of the pieces had hydranths at each end, and there seemed to be a strong tendency to form in this way; moreover, there was no noticeable difference in the time of development of hydranths of either end. All the hydranths were cut off. In 24 hours all showed the tentacle anlagen, and a little later hydranths were fully formed. One piece which had regenerated a hydranth cast it and formed another. This one was removed and another developed. This one was also cut off, and still a new one regenerated. Thus in just a week one piece regenerated four hydranths one after the other, the last three being formed in four days. Since I was forced to leave Woods Hole at this time, the regeneration of this piece could be followed no further.

In this species the regeneration differs from that of *T. crocea* and *T. tenella.* In every case under observation the first sign of regeneration was the emergence of the coenosarc from the old perisarc. A new perisarc was secreted about the coenosarc, and a slight annulation was always present in the new perisarc at its union with the old. After the emergence of the coenosarc the red pigment was deposited near the end, and tentacle anlagen were formed, resembling somewhat those in *T. crocea* (figs. 13*a*, 13*b*). The anlagen were much longer, however, and were usually twisted or twined spirally around the perisarc (fig. 13*a*), and the distance between the two series of tentacles was much greater. The proximal row was formed first as in *T. crocea,* but the distal tentacles seemed to form more as those of *T. tenella,* i. e., as buds. After the tentacles had been formed the hydranth emerged from the new perisarc, and took on the perfect hydranth form. The distal tentacles assumed their full size by further growth. Often gonads began to form before the hydranth emerged from the perisarc, or at any rate immediately afterwards (figs. 11–13). Another difference in the laying down of anlagen is as follows: While in *T. crocea* the distal anlagen were formed very close to the end of the coenosarc, in *T. larynx* they developed a considerable distance back of the end, thus leaving quite a mass of undifferentiated coenosarc between the distal row of tentacles and the distal end of the coenosarc (figs. 13*a*, 13*b*). After the hydranth emerged from the perisarc this seemed to be absorbed into the rather short hypostome of the hydranth (figs. 11, 12).

2. GRAFTING

Experiments in grafting were tried upon *Eudendrium ramosum, Eudendrium dispar, Pennaria tiarella,* and *Tubularia crocea.* The results confirmed those of Hargitt (1899) that union of the same species takes place equally well whether grafted orally or aborally. However, I was unable to secure a union of *Eudendrium ramosum* and *E. dispar* as he did, though the experiments along this particular line were not extensive enough to determine whether this was not due to some unfavorable condition.

The hydroids were operated upon as soon as possible after they were obtained. They were cut into the desired lengths with a pair of sharp scissors, as they left a clean surface and crushed the stem very little if at all. The cut pieces were then placed in petri-dishes, watch glasses, etc., filled with fresh sea water. The cut surfaces were carefully brought into as close contact as possible and held in place by pieces of freshly shaven lead. The hydranths were removed in all cases, as the movements of hydranths and tentacles would destroy the close contact of the pieces. As soon as the pieces had stuck together, the lead was removed and the pieces allowed to develop freely. In *Eudendrium* and *Pennaria* the pieces were very soon held firmly in place without the lead, because of the holdfasts developed from the ends of the stems and branches in contact with the glass. After the coenosarcs had united a new perisarc was secreted over the point of union. In *Tubularia crocea* union was complete in from 18 to 24 hours. At the end of this time the circulation of fluid could be observed taking place between the two pieces. In the uniting of the coenosarcs the pieces were often pushed apart slightly (fig. 4). However, a new perisarc was soon secreted over this and the wound entirely covered. The formation of hydranths took place quite rapidly after union was complete, sometimes forming at both ends. *Eudendrium* usually united in 24 hours, though hydranths were not fully formed till 24 to 48 hours later. The method of union was similar to that of *Tubularia*, and new perisarc was formed over the point of union (fig. 1). The grafting of *Eudendrium* and *Pennaria* was tried, but no union took place, the coenosarc of *Pennaria* forcing itself out into stolons (fig. 9).

In *Pennaria* the time necessary for complete union was from 24 to 48 hours. Hydranths developed slowly, not being fully formed till 24 to 48 hours later. A feature quite common in the grafting of *Pennaria* was the formation of stolons from the point of contact of the two pieces (figs. *2a*, *2b*). *Pennaria* found growing on the piles of docks was successfully grafted with *Pennaria* from floating eel grass, though, as already mentioned, the habits of the two were quite different.

V. Histology

In her work on *Tubularia tenella* Bickford (1894, p. 422) says, "In the case of this hydroid . . . the regeneration appears to be largely a direct transformation of a stem portion over into the body portion of the new hydranth." Since in her experiments no definite histological demonstration of this was undertaken, it has seemed desirable to repeat her experiments and to work out critically the histology of the processes involved. Also to determine to what extent the same processes of transformation are operative in *T. crocea* and *T. larynx*.

Stevens (1901), who has worked out the histological changes involved in the development of the tentacles in *T. mesembryanthemum*, states that the proximal tentacles form by a folding of both ectoderm and entoderm, and the distal tentacles by a rod or column of entoderm being separated from the rest of the entodermal tissue, and the ectoderm folding around it and pinching off the tentacle. In regard to the increase in surface of the ectoderm made necessary by this process of folding, she says, "The increase in surface of the ectoderm is due partly to cell division and partly to a change in the cells from a more to a less columnar form. Her drawings show mitotic figures in both ectoderm and entoderm cells. This does not seem to be the case in *Tubularia crocea, T. tenella*, and *T. larynx*. The cells do sometimes lose their typical columnar form, but not to any great extent. Therefore there must be cell division to account for this increase in surface of the ectoderm. Special attention was directed toward the staining and to the observation of mitotic figures, and yet in no case was the slightest evidence of mitotic division presented. Nuclei were found which gave evidence of amitotic division, and the various stages in this process were also presented, so that the cells probably divide directly. However, the resting nuclei very commonly contain two nucleoli, and it is hard to distinguish between this resting condition and the early stages of amitosis. So it was not possible to prove definitely the presence of amitosis in all cases.

While this is a somewhat unusual histologic phenomenon, and the wholly negative character of the evidence is insufficient to warrant definite conclusions, still it would seem rather remarkable that a critical examination of many hundreds of sections should have failed to reveal the presence of mitosis if at all prevalent.

TUBULARIA CROCEA

Distal Tentacles.—The distal tentacles seem to form about as Stevens (1901) states for *T. mesembryanthemum*. Several of the entoderm cells are squeezed away from the enteric cavity into the entodermal tissue, forming a sort of column as shown in figs. 22, 23. This causes the ectoderm to push outward slightly. These entoderm cells are gradually forced further and further away from the enteric cavity, sometimes assuming a position entirely within the ectoderm, against the outer wall of the entoderm, and with their long axes at right angles to the long axes of the other cells. This of course pushes the ectoderm outward still more, forming a ridge. The ectoderm then gradually folds around this column of cells, finally enclosing it, the edges of the ectodermal folds fusing and thus pinching off the tentacle. Fig. 24 shows a condition sometimes found; a greater number of entoderm cells are pushed outward to form the entodermal column, the enclosing by the ectoderm and the completion of the tentacle being brought about in the usual way.

Proximal Tentacles.—The proximal tentacles are formed somewhat differently, seeming to be the result of complex folding brought about in the following manner. The first evidence of developing tentacles is a slight folding of both ectoderm and entoderm (fig. 16). The folds, at first rather loose and sinuous, gradually crowd closer together till the edges touch (figs. 17, 18). By this time the folds have elongated radially and a sort of entodermal column is thus formed, though it is not as distinctly marked as in the distal tentacles. At this stage the whole region of the developing tentacles is composed of a series of narrow elongate folds (fig. 19). The entodermal column is composed of two rows of cells with their edges closely dovetailed or else flattened and the edges overlapping. When this stage is reached the edges of the ectodermal folds begin to approach each other, gradually coming nearer and nearer (fig. 20). Finally they entirely surround the entoderm cells and thus cut off the tentacle (figs. 20, 21).

TUBULARIA TENELLA

Distal Tentacles.—The distal tentacles are developed about as in *T. crocea*. The entoderm cells are forced away from the enteric cavity, folding the ectoderm, which gradually encloses the entoderm cells and thus forms the tentacle. Fig. 30 shows the cells being squeezed away from the enteric cavity. In fig. 31 later stages are shown. Quite often the first indication of a developing tentacle is a single cell (sometimes several) lying in the ectoderm close against the supporting layer and with its long axis at right angles to the long axes of the other cells. The origin of this cell can not always be definitely determined from the series of sections in which it occurs, though it is doubtless entodermal. In other series it can be definitely traced to its original position in the entoderm. In series in which the cell can not be traced from the ectoderm into the entoderm, the hydranth had probably developed beyond the initial stage and the entoderm cell (or cells) had already undergone the migration from the normal position in the original stem. This single cell in the ectoderm seems to go through repeated divisions, forming a mass of cells, after which the development proceeds in the manner already described.

Proximal Tentacles.—The formation of the proximal tentacles is quite different from that of *T. crocea*, being accomplished by evagination in the early stages and not by folding. The evagination begins in the entoderm, the ectoderm being simply pushed outward. Figs. 25 to 28 show the course of development. In fig. 25 the entoderm has begun to evaginate, pushing the ectoderm outward slightly. In fig. 26 some of the entoderm cells have been pushed away from the enteric cavity and the folding of the ectoderm continued. Fig. 27 shows the evagination nearly completed and the cells arranged in two rows, with edges interlocking. At about this stage the ectoderm begins to surround the evaginated entoderm cells, continuing until the folds of ectoderm meet, fuse, and thus separate the tentacles as shown in fig 28. In longitudinal sections the evagination is even more distinctly shown (fig. 29). The development thus far has taken place within the perisarc as in *T. crocea.* When the regenerating hydranth emerges from the old perisarc the tentacles are not, as in *T. crocea,* fully formed (cf. figs. 14*d* to 14*f*), but only attain their final length by further growth. This would seem to indicate very clearly that, although in the beginning the formation of tentacles may be by the transformation of old tissue, the completion of the process is by new growth. Bickford's conclusions, already referred to, may therefore be true for the body of the hydranth, but the tentacles are at least partially of new growth.

TUBULARIA LARYNX

DISTAL TENTACLES.—The distal tentacles form in about the same manner as in *T. tenella.* The single cells lying in the ectoderm, against the supporting layer, are found somewhat more commonly than in *T. tenella* and it is more difficult to trace their origin, though they are doubtless entodermal. However, in several series of sections this cell has been traced to a position partly in the ectoderm and partly in the entoderm (fig. 43). In a number of series, also, entodermal cells seem to have been forced away from the enteric cavity in a manner similar to that found in *T. tenella* (figs. 36 to 38). In these figures the appearance of early stages is shown. These cells being forced away from the enteric cavity more and more form a column of cells as shown in figs 39, 40. The ectoderm gradually surrounds this column and thus the tentacle is formed. Fig. 43 shows the single cells in the ectoderm. These cells seem to divide and form a column within the ectodermal layer (fig. 41), and are gradually surrounded by the ectoderm and the tentacle pinched off (fig. 42).

A comparison of figs 41 to 43 with figs. 36, 39, 40, suggests the possibility that in some cases the first cell squeezed away from the enteric cavity is forced into the ectoderm immediately and by division forms a column of cells, which is surrounded by the ectoderm to form the tentacle; while in other cases the necessary number of entoderm cells are gradually forced away from the enteric cavity into the ectoderm and without undergoing any divisions make up the core of the tentacle, which is surrounded by the ectoderm in the usual way.

Proximal Tentacles.—In the formation of the proximal tentacles, the complex folding of *T. crocea* does not take place, nor is the very evident evagination of *T. tenella* the method, but rather a combination of the two. The development begins very much the same as in *T. tenella* (figs. 34, 35) by what seems to be an evagination of the entoderm. A sort of ridge is thus formed. The ectoderm folds around the entodermal cells (beginning at the distal end), very much the same as in *T. crocea.* By a continuation of the evagination and folding, the entoderm cells are surrounded by the ectoderm and the tentacle cut off (figs. 32, 33). The tentacle when thus cut off is of normal length. Unlike *T. crocea,* the gonads begin their formation before the hydranth emerges from the perisarc (cf. figs. 13*a,* 13*b*).

Rather strange is the occurrence of nematocysts in the entoderm and in the debris of the enteric cavity. This is most common in the regenerating stems of *T. larynx,* though occasionally found in other forms. Weismann (1883) explains the presence of nettling cells in the developing male gonad of *Clava squamata* as a protection against a parasitic fungus. He performed a number of experiments with the result that the male form could not be infected, while the female gonad, which was not provided with the nettling cells, was easily infected with the fungus. Such an explanation could scarcely account satisfactorily for the presence of the nettling cells in the entoderm. This explanation would prove to be especially inadequate since in other forms, in which the nettling cells are not found in the entoderm, the open end closes no quicker than in *T. larynx.*

Among the colonies of *T. larynx* obtained during the summer, three pieces were found to be naturally regenerating hydranths. These were killed and later sectioned. The folds in the hydranth region were extremely complex, often pinnate, and entirely unlike anything observed in the artificially regenerating hydranths of *T. crocea, T. tenella,* or *T. larynx.* Owing to the very limited amount of this material no conclusions can be drawn, and no explanation offered, unless it be possible that such regeneration was the result of abnormal conditions, though it is not easy to conceive what they might have been.

EUDENDRIUM RAMOSUM

Externally the first indication of a developing hydranth is the formation of a knob like protuberance by the coenosarc. A thin layer of perisarc is secreted around this knob, though this is later ruptured when the hydranth begins to develop further and the tentacles to form. This protuberance increases in size, elongating more or less and assuming a spherical shape. Then the tentacles bud off; at first very small, they later assume their normal length by new growth. The proboscis seems to form as a swelling or evagination at the distal end of the hydranth, and rather late in the development the mouth opening breaks through.

Transverse sections of very early stages ot the regenerating hydranths, i. e., sections through the knob-like structure previously referred to, show the layers of ectoderm and entoderm to be much wider than in the normal stem regions, the cells being extremely elongated radially (fig. 44). There are more cells present than in the normal stem, so that there must have been considerable cell proliferation. Indications of mitosis were found in both ectoderm and entoderm, though not as abundant as in the entoderm of later stages. In the latter, however, mitoses were far from common. Indications of amitotic division were observed, being quite marked in some cases.

The tentacles seem to start by evagination involving both layers, and somewhat similar to the same condition in *Tubularia tenella* (figs. 45 to 47). The further development of the tentacles takes place by new growth. The entoderm cells are arranged in a single row, being superimposed one upon another (figs. 57*a*, 57*b*). These entoderm cells were sometimes found in the process of dividing mitotically, and this phenomenon was occasionally observed in the ectoderm cells surrounding the tentacles. Amitotic division was not common among them, though occasionally found. The shape of the cells, however, was changed from columnar to cuboidal. Sometimes they were flattened even more (fig. 60), so that their length radially was less than their dimensions in other directions. This flattening of the ectoderm cells would increase the surface to a considerable extent and thus allow growth of the entodermal core of the tentacle. It is extremely doubtful, however, whether the increase in surface of ectoderm necessary to permit the growth of the entodermal column to the normal length could be accounted for entirely by the flattening of the cells. There must therefore be proliferation of the ectoderm cells, but whether this is by mitotic or amitotic division or by both could not be definitely determined, though indications suggest mitotic division. If at all prevalent, however, it seems rather strange that mitotic figures were not found more abundantly, since sections were made of all stages of development and stained especially to bring out this feature. The occurrence of amitosis may be accounted for partially as an aid in cell proliferation, and perhaps also for another purpose to be considered a little later.

In the formation of the entodermal core of the tentacle in the manner shown in figs. 57-60 (i. e., a single row of cells one on top of another), the question as to the method of development of this condition presents itself. Three explanations may be suggested:

1. Several cells from the primary evagination may divide and form a double row of cells, as in *Tubularia*, and later be forced together to form a single row.

2. The single cell usually found at the apex of the evaginated fold (figs. 45, 46, 49) may divide, and the cells arising from this division continue to divide until the entodermal core is fully formed.

3. The cell at the apex of the evagination may mark the begining of cell division, but instead of all the cells resulting from this primary division continuing to divide, only the apical cell of the developing entodermal core continues to divide till the core of the tentacle is fully formed.

That the first will not explain the facts is evident from an examination of sections of different stages, particularly of early stages. The conditions shown in figs. 45, 46, 49, are brought about by the primary evagination as already stated. This condition is always found at the beginning of the development, and in sections of later stages there is no indication of this double row of cells, except at the base of the tentacle (cf. figs. 57*a*,

57*b*). If a double row of cells formed by cell division and later these were forced into a single row, the double row would undoubtedly still be present at early stages. That this is not the case is shown by a comparison of fig. 57*a*. In this tentacle the entoderm is composed of only four cells, and these are arranged in a single row. This was always the condition after the tentacle had begun to elongate. It was more difficult to determine which of the other two suggested explanations was the more probable, because of the paucity of definite mitotic figures. The following facts, however, seem to point strongly to the second as the more probable explanation. In fig. 57*b* at "*a*" are shown two cells somewhat smaller than the others of the entodermal column. These two cells have every appearance of being the products of the division of a single large cell situated in that region, i. e., they indicate a division of one of the middle cells of the entodermal column. Furthermore, in transverse sections through about the middle portions of the developing tentacles, mitotic figures were found in the entoderm, which would not be the case if only the apical cell divided. Comparisons of some of the longitudinal sections show that the entoderm cells in the lower (proximal) end of the core of the tentacle are not of the same size, as would be the case if only the apical cell divided, some are large and some are small as though the result of intercalary division. These facts and observations show quite conclusively that the entodermal core of the tentacle is the result of the division of the single apical cell of the primary evagination, and then a continued division of all of the cells resulting from the first division. The entodermal core is surrounded by the ectoderm in early stages (figs. 48, 50), and in the further development stretches the ectoderm by the division of the cells of the ectoderm (figs. 57*a*, 57*b*, 60). The increase in surface of the ectoderm is therefore partly due to the change in form of the cells from a columnar condition to a more or less flattened condition, though there is doubtless some cell division in the ectoderm also. Figs. 52–56 show the course of development of two tentacles from the early stage in which a single ectoderm cell has been forced away from the enteric cavity, to the condition in which the tentacles are fully formed. The presence of more than one cell in sections of the tentacles as shown in some of the figures is due to the tentacle being cut somewhat obliquely instead of transversely.

In the mature hydranth the tentacles are sometimes composed of two series arranged in a single whorl. Those of one series are somewhat drooping in habit, the others more erect. The tentacles of the two series alternate in position and there is also an alternate elevation and depression of their bases. This condition of the tentacles is shown in the process of development in fig. 51. The irregular mass of polygonal cells shown in this figure is due to the section being made across the mass of cells between the enteric cavity of the hydranth and the hypostome (cf. figs. 58, 60).

The hypostome develops as a hollow sphere at the extreme distal end of the hydranth, beginning as an outgrowth of the layers of the regenerating hydranth. At a very early stage, however, all connection between the enteric cavity of the hydranth and the cavity of the hypostome is blocked up by a mass of entoderm cells, probably due to a multiplication of cells in this region. The further development is for the most part an increase in size. The shape of the cells change, the ectoderm being finally composed of very much flattened cells, the result of the increase in size of the hypostome without great multiplication of cells of the ectoderm. The entoderm is made up of a large number of columnar cells very much elongated radially and very narrow. Figs. 58, 60, show the relation of the two layers and also the mass of entoderm cells blocking the passage between the enteric cavity and the hypostome. This mass of cells is at first without regular arrangement, being crowded together into a dense mass. Later in the development, these crowded cells take on a more regular arrangement as represented in fig. 58. There is already present in this stage a sort of dividing line between the cells of the two sides, and by the separation of the layers of the two sides along the dividing line, the opening between the enteric cavity and the hypostome is formed (fig. 59). At the angle formed by the junction of the hypostome with the hydranth proper, the cells remain larger (cf. fig. 59), a condition likewise found in the normal hydranths. The stage of development of the hydranth at which the opening forms seems to vary. In fig. 59 the opening is shown and the tentacles were less than half their normal length. Fig. 60 shows the hydranth almost completely regenerated, tentacles of almost natural size, and the opening has not yet been made. The mouth opening does not form till the hydranth is otherwise completely regenerated.

The enteric cavity of the regenerating hydranth is more or less filled with a mass of debris (degenerating cells, etc.), somewhat similar in appearance to the same condition in *Tubularia.* Doubtless this mass is partly composed of some of the cells blocking the opening between the hypostome and the enteric cavity at an earlier stage. The protoplasm of *Eudendrium* does not contain nearly as many granules as that of *Tubularia,* and the longitudinal entodermal ridges of the latter are not present, so that this debris is not made up of masses of granules set free by the breaking down of entodermal ridges, etc. The origin of this debris was not fully determined. Whether this debris is cast out of the hydranth, when the mouth opening is formed, as Stevens (1902) has demonstrated for *T. mesembryanthemum* and *T. crocea,* was not determined during the course of the experimental work.

PENNARIA

The early appearance of regenerating hydranths in *Pennaria* is very similar to that of *Eudendrium,* viz., an enlarged knob or bulb-like protuberance of the coenosarc, from which tentacles seem to bud. The material on hand was all killed in the early stages of regeneration, so that the complete progress of development could not be worked out. Some points, however, in the early development seem to be sufficiently important to warrant mentioning them. One very pronounced feature is the great abundance of nematocysts in the regenerating hydranths. The ectoderm of the end and sides of the enlarged knob are very plentifully supplied with nematocysts, and they are more or less abundant in the entire region of the future hydranth, as well as on the stem just below the knob. They were not found in the entoderm as in *T. larynx*.

There is the same thickening of the ectoderm and entoderm layers found in sections through very early stages, as in *Eudendrium ramosum*, and this condition is probably the result of cell proliferation. A careful examination of sections shows mitosis to be fairly abundant, mitotic figures being found in both ectoderm and entoderm (figs. 61, 62). Therefore, in this form, as well as in *Eudendrium ramosum*, indications point to mitotic division as an active process in the development of the regenerating structures. However, indications of direct division were not lacking. In fig..63 is shown a nucleus in the process of direct division, accompanied by division of the cytoplasm, a cell wall being partly formed. In other sections nuclei were found dividing directly, but cytoplasm seemed to have taken no part in the process (fig. 64). Furthermore, amitotic division can not be considered merely as a somewhat abnormal condition brought about by artificial conditions. The same thing is found in sections made through the young hydranths developing normally on colonies growing in their natural habitat. Fig. 65 shows a considerable number of nuclei in different stages of direct division, and this is in the region of the developing tentacle. Such dividing nuclei were limited to the ectoderm.

Only a few sections through developing tentacles were obtained. Fig. 66 shows the appearance of the cells in one case. Since the core of the mature tentacle is made up of a single row of entoderm cells, as in *Eudendrium*, the same process of development would be expected. Whether this is the case could not be definitely determined owing to the lack of material of the later stages.

THE OCCURRENCE OF AMITOSIS

Other writers have referred to the occurrence of mitosis in both the ectoderm and entoderm of regenerating hydroids. Therefore on starting the histologic work, which has been described above, I expected to find approximately the same conditions. When traces of mitosis were not found, after staining particularly for mitotic figures, an explanation was sought for. As referred to under methods, the hydroids were killed in Gilson's fluid or in a mixture of corrosive acetic and picric acid. The killing was not done with the particular aim of securing mitotic figures, but rather for general histologic investigation. It may be possible, therefore, that the paucity of mitotic figures in most of the species considered is due to improper killing, and that if methods were employed especially to determine this phenomenon, mitosis wound be fairly abundant.

However, the presence of amitosis is a condition which demands some explanation. In *Tubularia crocea* particularly the occurrence of amitosis was very common, and in fact in all the species studied it was more or less evident. Comparisons of some of the drawings will show nuclei in the process of direct division. In fig. 63 this division of nucleus is being followed by a division of the cytoplasm. The rather common occurrence of this phenomenon and the relatively large amount of evidence thus brought forward is very suggestive as to amitosis being an active process in the regeneration of hydroids.

Wilson (1896), in referring to amitosis says, "It is of extreme rarity, if indeed it ever occurs in embryonic cells or such as are in the course of rapid and continued multiplication. It is frequent in cells . . . which are on the way toward degeneration." It has also been suggested that amitotic division may involve the nuclei only, and the cytoplasm does not divide; this for the purpose of increasing the nuclear surface as an aid in metabolism. Wilson further says, "In lower forms of life at least (amitosis) does not necessarily mean the approach of degeneration, but is the result of special conditions."

In the hydroids studied the amitotic nuclear division does not seem always to be followed by cytoplasmic division, though this may of course be due to the too early stages of the division. This suggests the possibility that the conditions here may be for the purpose of increasing the nuclear surface to aid in metabolism. Such an explanation might account for the presence of so many amitotically dividing nuclei as are shown in fig. 65.

Wilson refers to a paper by Ziegler (1896) in which it is

stated that amitotically dividing nuclei are usually of large size and are distinguished in many cases by a "specially intense secretory . . . activity." This "secretory activity" would be necessary for a time at least in *Eudendrium* and *Pennaria,* while the perisarc was being secreted around the protruding coenosarc, and amitotic nuclear division may thus be partially explained in these two species. In *Tubularia*, however, where the hydranths regenerate within the perisarc and only a few cells at the end secrete a new perisarc, the amitotic division could not be explained in this way. Indeed, while this direct division may be partially explained in the ways already considered, it seems to be too prevalent to be accounted for by these conditions alone. Furthermore, the cells do not seem to be in a degenerate condition, and are the cells from which the hydranths or tentacles are formed. It may be the result of special conditions, which, as Wilson states, has been conclusively proven to be the case in lower forms, though as to just what the special conditions might be it is difficult to conjecture. However, this phenomenon can not be fully explained till further investigations have been made.

Conclusions

1. Gravity has no apparent influence on the regeneration of *Tubularia crocea* and *Pennaria tiarella.*

2. Contact determines in some measure the kind of regeneration and the direction of growth in *Eudendrium ramosum* and *Pennaria tiarella.*

3. In *Tubularia crocea* there is a marked polarity, hydranths appearing more abundantly and in a shorter time at the distal ends.

4. Pieces of *Tubularia crocea* and *Tubularia tenella* of about 3 to 4 mm. length failed to regenerate.

5. In *Tubularia crocea* the method of regeneration is about as found by other authors. The cut end closes over, the liquid in the enteric cavity begins to circulate, and red pigment is deposited in the hydranth region. Tentacles appear first as longitudinal folds or ridges and are pinched off from the hydranth body to assume the final form and size while still within the

perisarc. The proximal tentacles develop first, the development beginning at the distal end.

6. In *Tubularia tenella* the first part of the regeneration is similar to that in *T. crocea.* The hydranth body is separated from the stem by a constriction before the tentacles begin to develop. Tentacles appear as rather short buds, the final form and size being assumed by new growth after the hydranth emerges from the perisarc.

7. In *Tubularia larnyx* there is a tendency for hydranths to regenerate at both ends of the stem, and there is no difference in the time of their development. The first stage of regeneration is the emergence of the coenosarc from the old perisarc, and the secretion of a new perisarc around the protruding part. A few annulations are always present close to the old perisarc. Circulation of fluid and deposition of pigment are similar to that in *T. crocea.* Tentacle anlagen are much longer than in *T. crocea,* and often twined spirally around the coenosarc. Tentacles assume their final form and size while still within the perisarc. Gonads may begin to form before the hydranth emerges from the perisarc.

8. *Eudendrium ramosum, E. dispar, Pennaria tiarella,* and *Tubularia crocea* were successfully grafted, union taking place equally well, whether the distal ends or the proximal ends were joined, or whether distal end was grafted to the proximal end. No union of *Eudendrium ramosum* and *E. dispar* was secured, nor of *Eudendrium* and *Pennaria.*

9. In *Tubularia crocea* the distal tentacles form by the separation from the entoderm layer of a rod or column of entoderm cells which are surrounded by the ectoderm to form the tentacle. Proximal tentacles form by a complex folding, involving both ectoderm and entoderm, the ectoderm gradually surrounding the entodermal fold and separating the tentacle from the hydranth body.

10. Distal tentacles of *Tubularia tenella* form as in *T. crocea.* Proximal tentacles develop partly by evagination of the entoderm and later by a new growth after the hydranth emerges from the perisarc.

11. The formation of the distal tentacles of *Tubularia larnyx* is about the same as in *T. crocea* and *T. tenella*. Proximal tentacles form by a combination of evagination and folding, along the whole length, the tentacles being of mature size and form when the hydranth emerges from the perisarc. Nematocysts are found quite commonly in the entoderm and in the debris of the enteric cavity of *T. tenella* and *T. larynx*.

12. Regeneration of *Eudendrium ramosum* is first indicated by a knob-like protuberance from which the tentacles bud. The tentacles start their development as evaginations, involving both ectoderm and entoderm. The entodermal cell at the apex of the evagination divides, and all cells resulting from this division may continue to divide till the entodermal core of the tentacle is formed, the cells being arranged in a single row. Proliferation of cells in the developing hydranth is by mitosis, amitosis also occurring. Increase in the surface of the ectoderm is brought about partly by a change in the form of the cells and partly by cell division. The hypostome develops as an outgrowth of the developing hydranth, cell division also taking place. The proximal end of the hypostome is blocked by a mass of entoderm cells till a comparatively late period. The mouth opening forms when the hydranth is otherwise completely regenerated.

13. The early appearance of regenerating hydranths of *Pennaria* is similar to that of *Eudendrium*. Nematocysts are very abundant in the ectoderm of the regenerating hydranths. The layers of ectoderm and entoderm are much thickened and the cells are more abundant, being the result of cell proliferation. Mitosis is quite abundant; and amitosis is also found.

14. Amitosis is quite abundant in the tissues of the regenerating hydranths studied. When not followed by cytoplasmic division (as seems to be the case sometimes) it may be for increasing the nuclear surface as an aid in metabolism. Amitosis may be the result of special conditions. The explanation of amitosis in regenerating hydroids can not be definitely determined on the present evidence.

Bibliography

AGASSIZ, L.

1862. *Contributions to the Natural History of the United States*, vol. IV Boston.

ALLMAN, G. J.

1871. *Monograph of the Gymnoblastic Hydroids.* London.

BICKFORD, E. E.

1894. *Notes on Regeneration and Heteromorphosis of Tubularian Hydroids.* Jour. Morph., vol. IX.

DRIESCH, HANS.

1896. *Notes on Regeneration and Heteromorphosis of Tubularian Hydroids-Bickford.* Arch. f. Ent.-mech., Bd. II.

1897. *Studien über das Regulationsvermögen der Organismen.* I.—*Von der regulativen Wachthums- und Differenzirungs-fähigkeiten der Tubularia.* Arch. f. Ent.-mech., Bd. V.

1899. *Studien über,* etc. II.—*Quantatative Regulationen bei der Reperation der Tubularia.* Arch. f. Ent.-mech., Bd. IX.

1901. *Studien über,* etc. III.—*Ergänzende Beobachtungen an Tubularia.* Archiv. f. Ent.-mech., Bd. XI.

1902. *Studien über,* etc. VII.—*Zwei neue Regulationen bei Tubularia.* Arch. f. Ent.-mech., Bd. XIV.

HARGITT, C. W.

1897. *Recent Experiments on Regeneration.* Zool. Bull., vol. I.

1899. *Experimental Studies on Hydromedusae.* Biol. Bull., vol. I.

1901. *Variation Among Hydromedusae.* Biol. Bull., vol. II.

LOEB, J.

1892. *Untersuchungen zur physiologischen Morphologie der Thiere.* II.—*Organbildung und Wachsthum.* Würzburg.

1893. *On Some Facts and Principles of Physiological Morphology.* Biol. Lect., Woods Hole.

MORGAN, T. H.

1901. *Regeneration in Tubularia.* Arch. f. Ent.-mech., Bd. XI.

1901*a*. *Factors that Determine Regeneration in Antennularia.* Biol. Bull., vol. II.

1901*b*. *Regeneration.* New York.

1902. *Further Experiments on the Regeneration of Tubularia.* Arch. f. Ent.-mech., Bd. XIII.

PEEBLES, F.

1900. *Experiments in Regeneration and in Grafting of Hydrozoa.* Arch. f. Ent.-mech., Bd. X.

1902. *Further Experiments in Regeneration and Grafting of Hydroids.* Arch. f. Ent.-mech., Bd. XIV.

RAND H. W.

1899. *Regeneration and Regulation in Hydra viridis.* Arch. f. Ent.-mech., Bd. VIII.

STEVENS, N. M.

1901. *Regeneration in Tubularia mesembryanthemum.* Arch. f. Ent.-mech., Bd. XIII.

1902. *Regeneration in Tubularia mesembryanthemum*, II. Arch. f. Ent.-mech., Bd. XV.

1902*a*. *Regeneration in Antennularia ramosa.* Arch. f. Ent.-mech., Bd. XV.

WEISMANN, A.

1883. *Entstehung der Sexualzellen bei den Hydromedusae.* Jena.

WILSON, E. B.

1896. *The Cell in Development and Inheritance.* New York.

EXPLANATION OF FIGURES

Fig. 1. *Eudendrium ramosum* grafted by distal ends.

Fig. 2*a*. *Pennaria* grafted by proximal ends.

Fig. 2*b*. *Pennaria* grafted by distal ends. st.—stolons.

Fig. 3. *Tubularia crocea.* Distal end showing the tentacle anlagen.

Fig. 4. *Tubularia crocea* grafted by proximal ends.

Fig. 5. *Tubularia crocea,* distal end in sand, new hydranth having regenerated. The unshaded portion is the old stem.

Fig. 6. *Pennaria,* distal end in sand showing regeneration. Unshaded portion is the old stem.

Fig. 7. *Pennaria,* stolon in sand showing regeneration. Unshaded portion is old stem.

Fig. 8. *Pennaria* lying flat on the bottom of the dish, with regenerating hydranth. Unshaded portion is old stem.

Fig. 9. *Pennaria* and *Eudendrium* grafted. st.—stolons from *Pennaria.*

Fig. 10. *Pennaria* showing holdfasts and hydranths. This drawing is made from parts of two different stems.

Figs. 11, 12. Regenerated hydranths of *Tubularia larynx.*

Figs. 13*a*, 13*b*. Regenerating hydranths of *Tubularia larynx,* showing the tentacle anlagen. Developing gonads in black.

Figs. 14*a*, 14*b*, 14*c*. Early stages in the development of the hydranths of *Tubularia tenella.*

Figs. 14*d*, 14*e*, 14*f*. Hydranths of *Tubularia tenella* after emergence from the perisarc.

Fig. 15. *Pennaria* anchored to the bottom of the dish by the distal end, showing regeneration. Unshaded portion is old stem.

Figs. 16–24. *Tubularia crocea.*

Fig. 16. Early stages in the folding of the ectoderm and entoderm to form the proximal tentacle. ×600.

Figs. 17, 18. Later stages of folding. ×600.

Figs. 19, 20. Ectoderm folding around the entoderm to cut off the proximal tentacle. ×500.

Fig. 21. Promixal tentacles separated from the hydranth body. ×500.

Figs. 22, 23. Two stages in the development of the distal tentacles. Column of entoderm cells being formed. Ectoderm folds around this column to cut off the tentacle. ×600.

Fig. 24. Separation of a mass of entoderm cells in a column.

Ectoderm gradually surrounds the column to form the tentacle. Amitotically dividing nuclei shown. ×600.

Figs. 25–31. *Tubularia tenella.*

Figs. 25, 26. Early stages in the evagination of the entoderm to form the proximal tentacle. ×600.

Fig. 27. Evagination of the entoderm complete and the ectoderm folding around the entoderm to cut off the tentacle. ×500.

Fig. 28. One proximal tentacle cut off and another with the entoderm entirely surrounded by ectoderm but not yet separated from the hydranth body. ×500.

Fig. 29. Longitudinal section through the developing proximal tentacle, showing the evagination of the entoderm and the formation of the tentacle. ×600.

Fig. 30. Entoderm cells being forced away from the enteric cavity to form the entodermal column of the distal tentacle. ×600.

Fig. 31. Entodermal column of distal tentacle separated from the entoderm layer and the ectoderm beginning to surround it. ×600.

Figs. 32–43. *Tubularia larynx.*

Fig. 32. Proximal tentacles separated from the hydranth body, others in the process of separation. In one tentacle the entoderm is not completely surrounded by the ectoderm. ×600.

Fig. 33. Earlier stages in the development of the proximal tentacles. Entodermal column nearly complete, and the ectoderm just beginning to surround it. ×600.

Figs. 34, 35. Very early stages in the development of the proximal tentacles, only a few cells pushed away from the enteric cavity. ×600.

Figs. 36–38. Early stages in the formation of the distal tentacles, the entodermal column shown in different stages of development. ×600.

Figs. 39, 40. Entodermal column being surrounded by ectoderm to complete the tentacle. ×600.

Fig. 41. Entodermal column, entirely outside of the entoderm layer, being surrounded by ectoderm. ×600.

Fig. 42. Completion of the process shown in Fig. 41. ×600.

Fig. 43. Single entoderm cells lying almost entirely within the ectoderm. A condition sometimes found in the early stages of development of the distal tentacles. Figs. 41, 42 are really the completion of this process. ×600.

Figs. 44–60. *Eudendrium ramosum.* 44–57 ×600; 58–60 ×290.

Fig. 44. Very early stage. Layers of ectoderm and entoderm much thickened.

Figs. 45–47, 49. Different stages in the early evagination of the entoderm.

Figs. 48–50. Transverse sections through completed or partially completed tentacles.

Fig. 51. Transverse sections through tentacles in different stages of development. The two more or less distinct series of anlagen, represent the development of the two series of mature tentacles whose bases are alternately elevated and depressed. The mass of polygonal cells are entoderm cells massed together in the region of the hypostome.

Figs. 52–56. Progress of development of two tentacles, from the condition in which only a single cell has been pushed away from the enteric cavity (fig. 56), to the complete separation of the tentacles (fig. 52).

Figs. 57, 58, 60. Longitudinal sections through the tentacles in various stages of development.

Figs. 58–60 show different stages in the development of the hypostome. In Fig. 60 a considerable amount of debris is present in the enteric cavity.

In figures of *Eudendrium ramosum* there often appear to be two cells in the sections of the completed tentacle. This is due to the fact that the section is not exactly transverse but has been made somewhat obliquely. Where three nuclei seem to be present this is due to the appearance of the nuclei of the underlying cells.

Figs. 61–67. *Pennaria tiarella.*

Figs. 61, 62. Nuclei in both ectoderm and entoderm in the early stages of mitosis. ×1000.

Fig. 63. Nearly completed amitotic division of an ectodermal cell. ×1000.

Fig. 64. Nucleus of ectoderm cell in an early stage of amitotic division. The cytoplasm has not yet begun to divide. ×600.

Fig. 65. Transverse section through the hydranth region of a young budding hydranth, which was forming normally in its natural habitat. ×1000.

Figs. 66, 67. Sections through the developing tentacles. ×600.

PLATE I

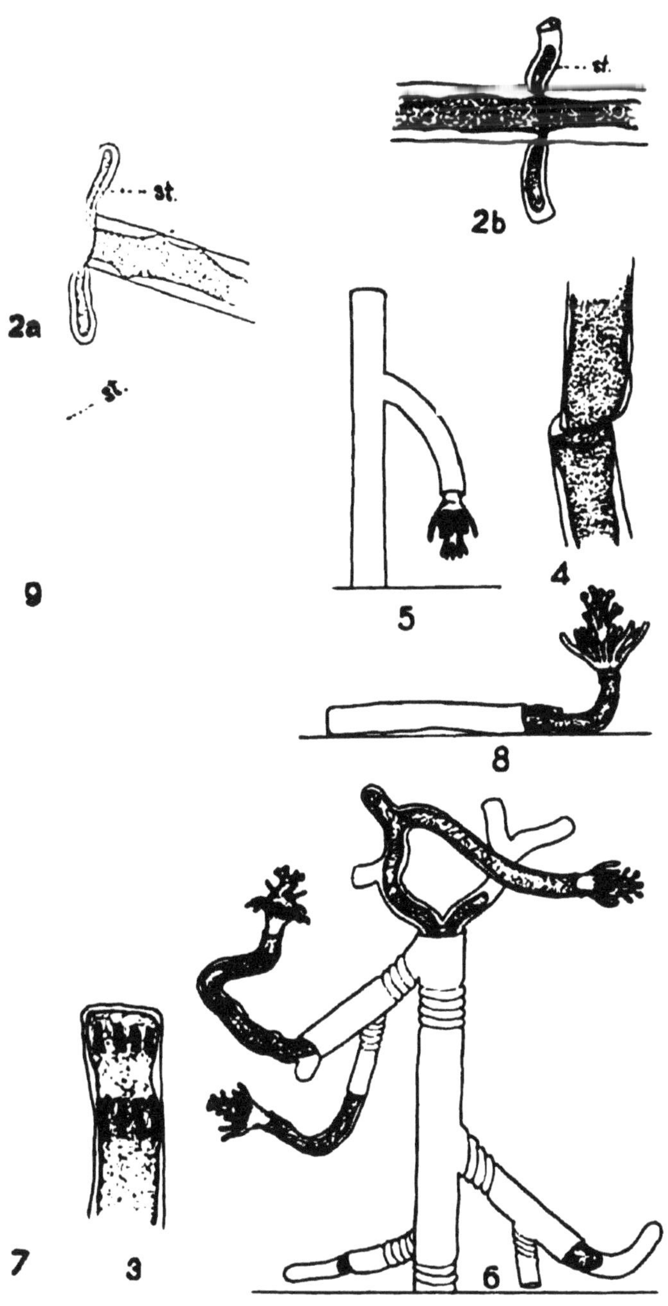

PLATE 11

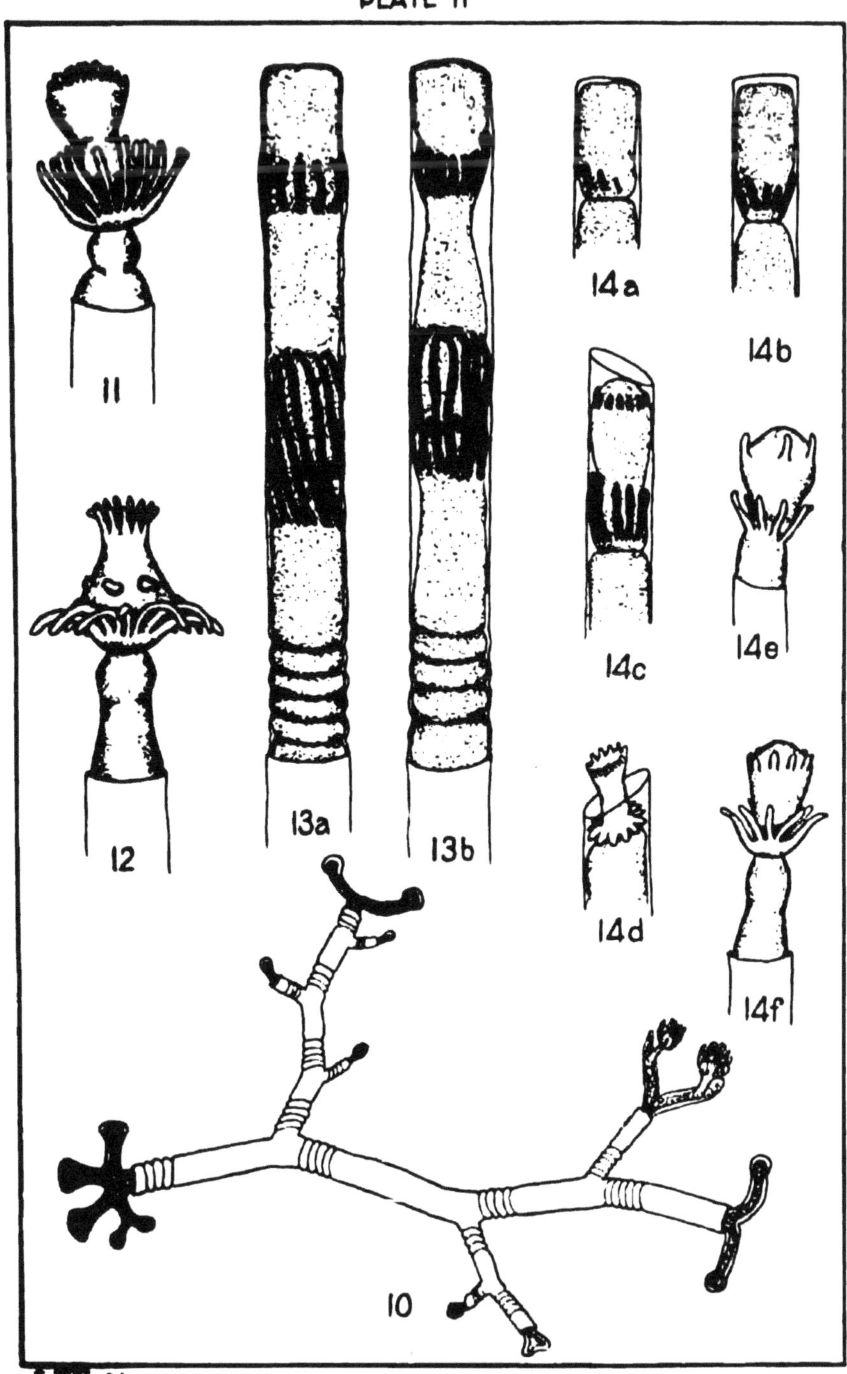

G.T.H. del.

PLATE III

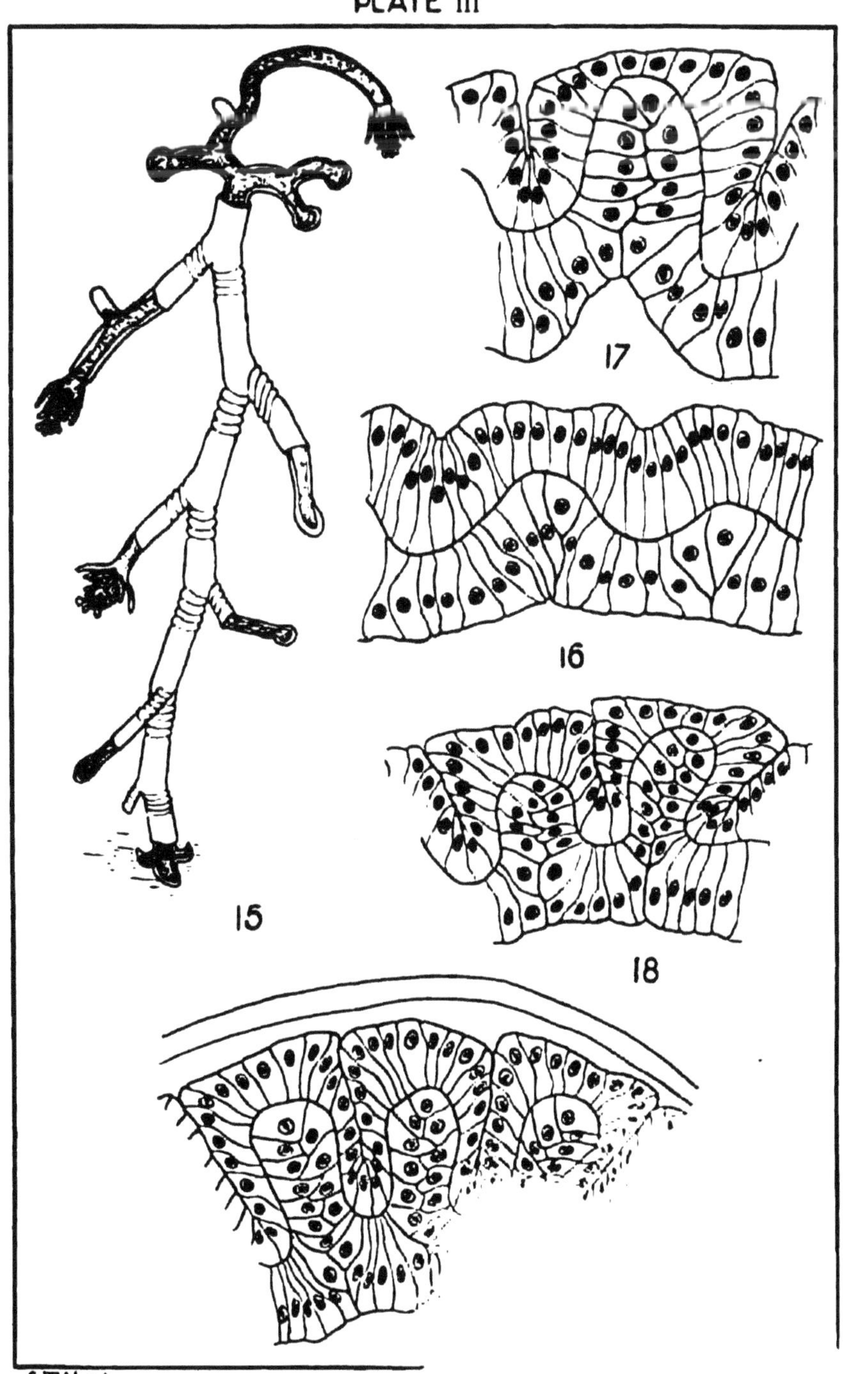

G.T.H. del.

PLATE III

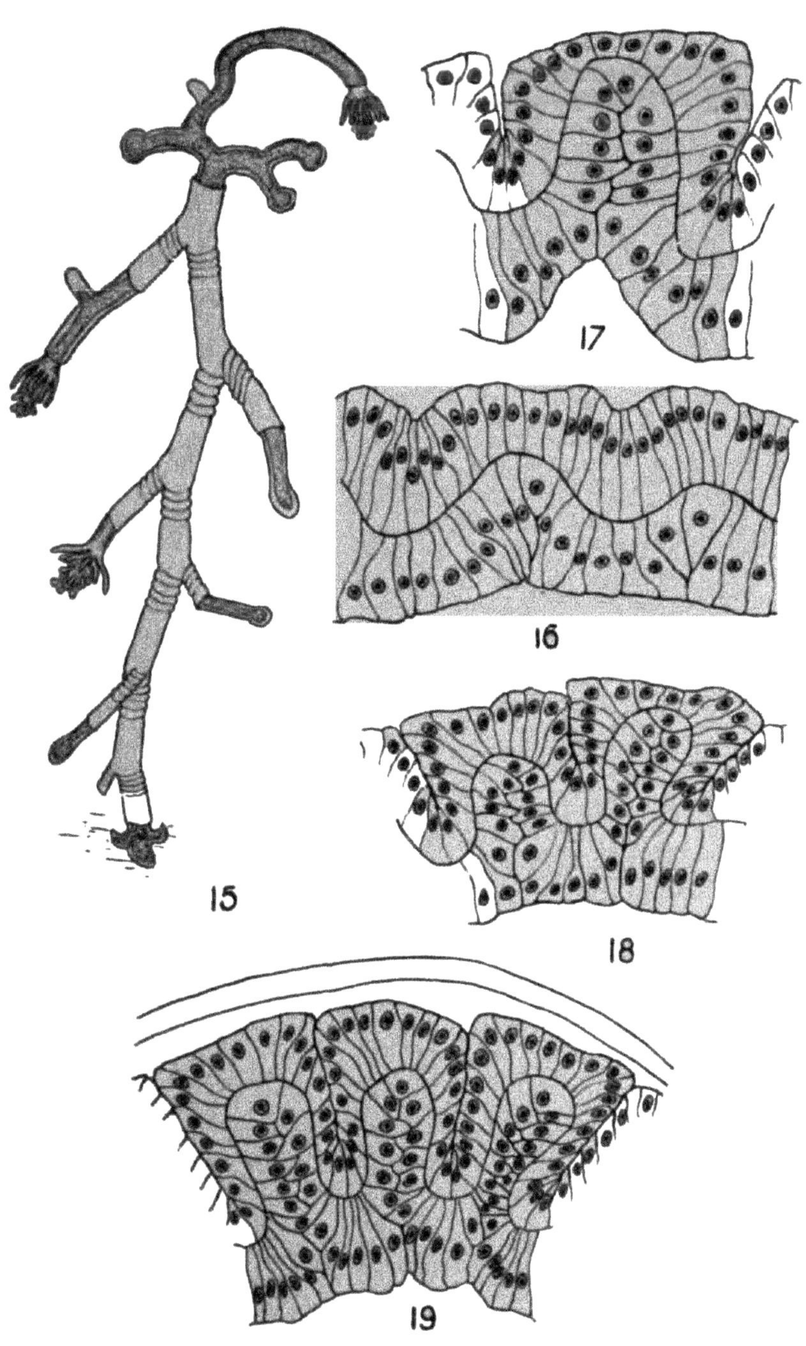

PLATE IV

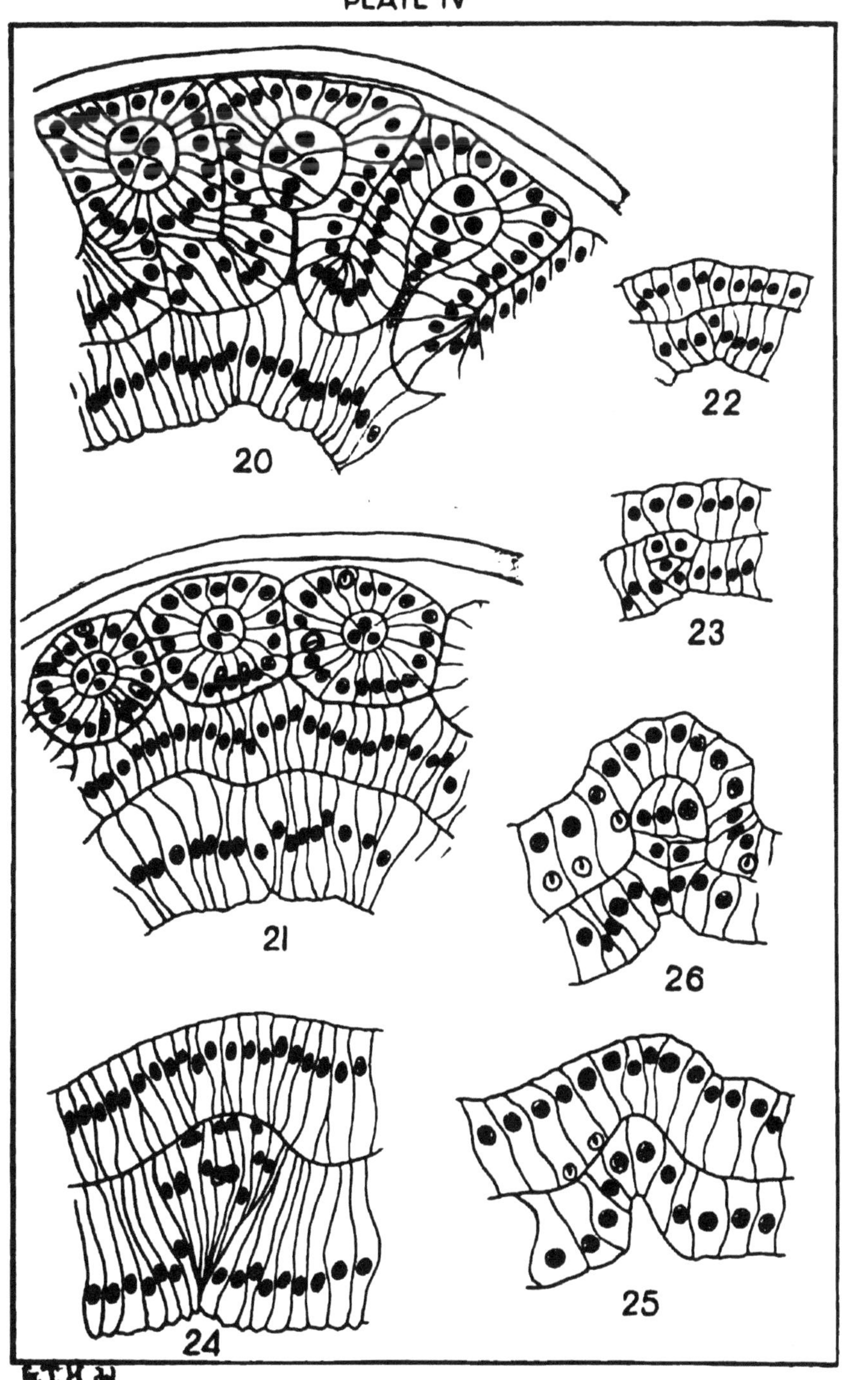

PLATE V

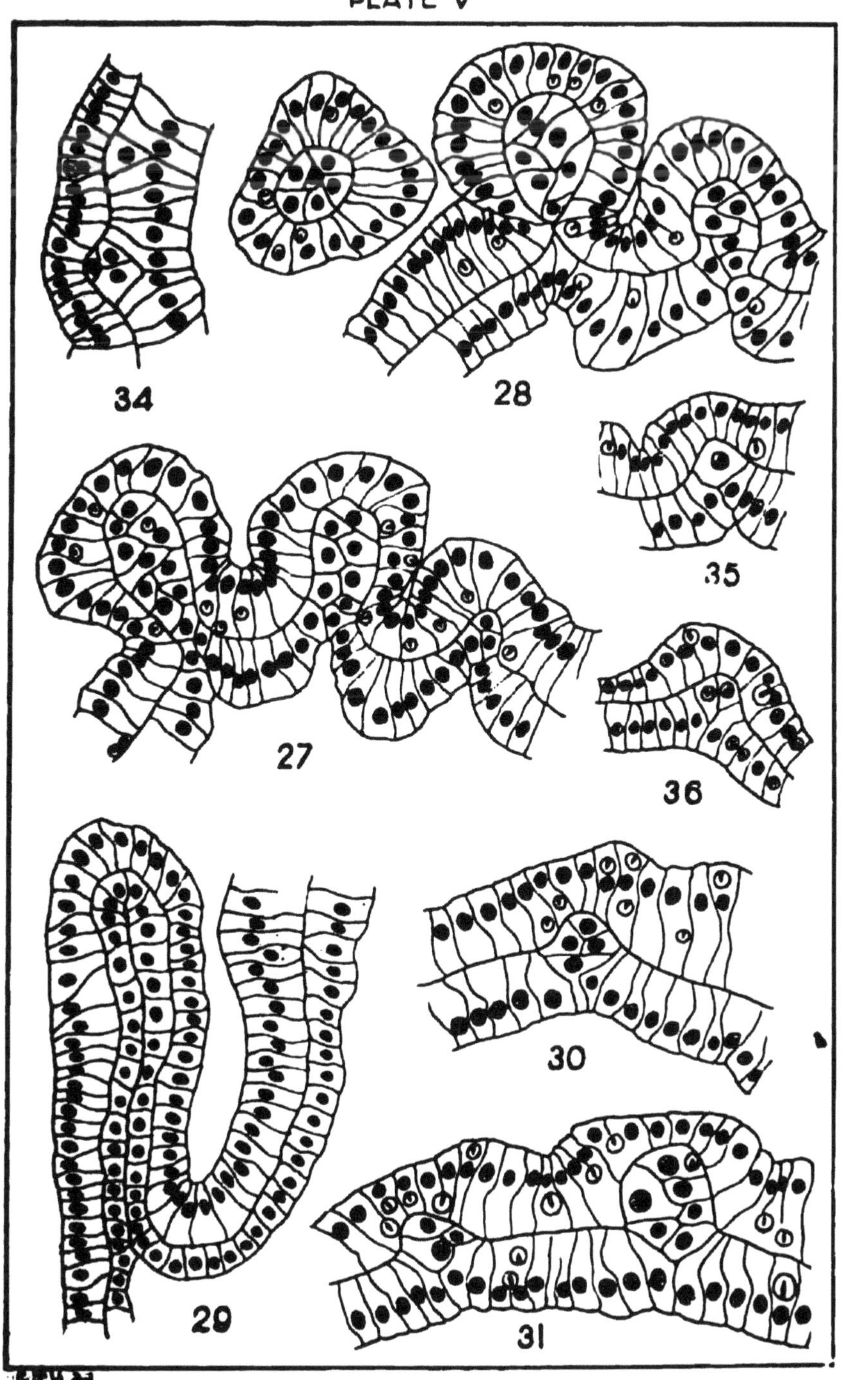

PLATE VI

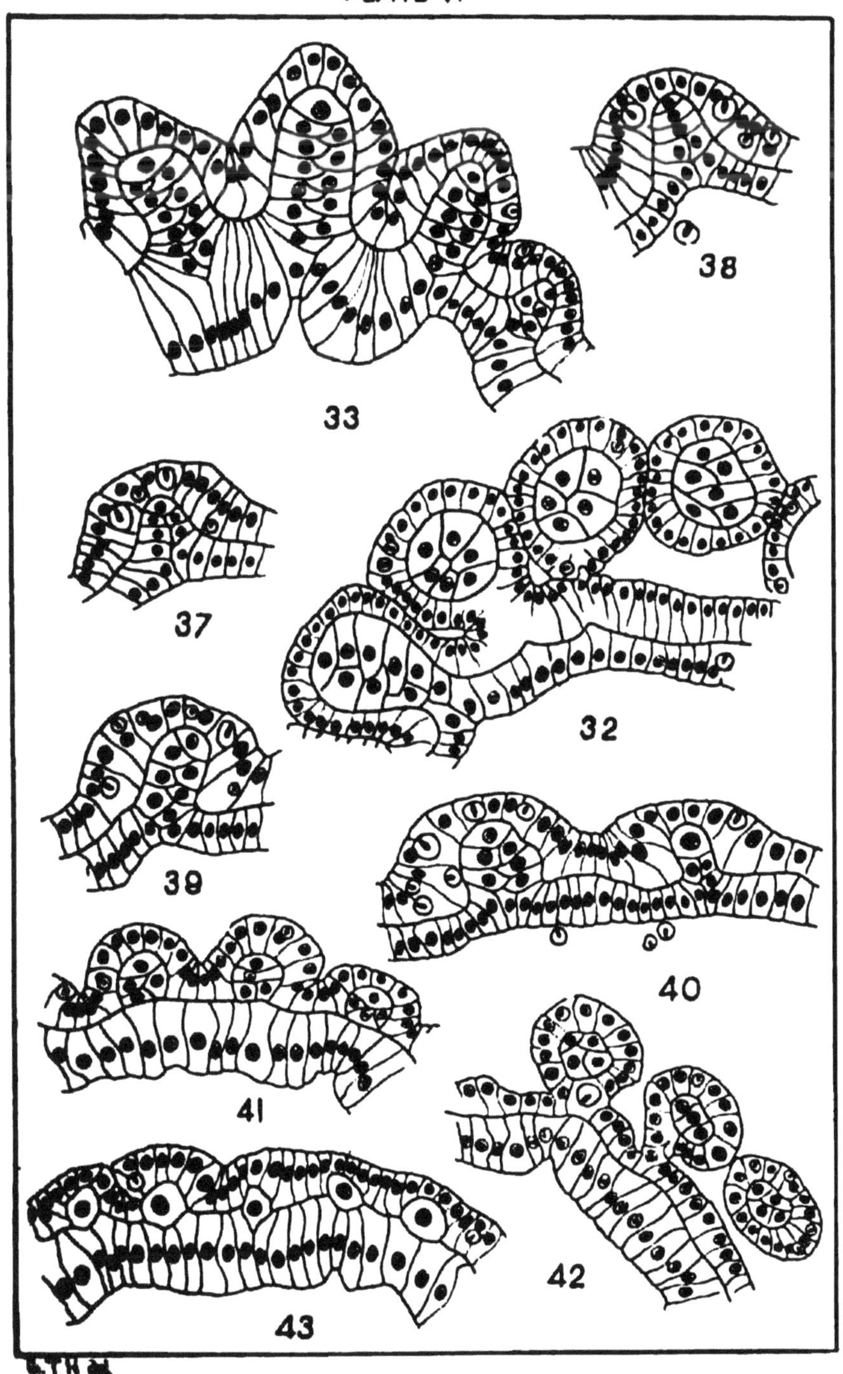

PLATE VII

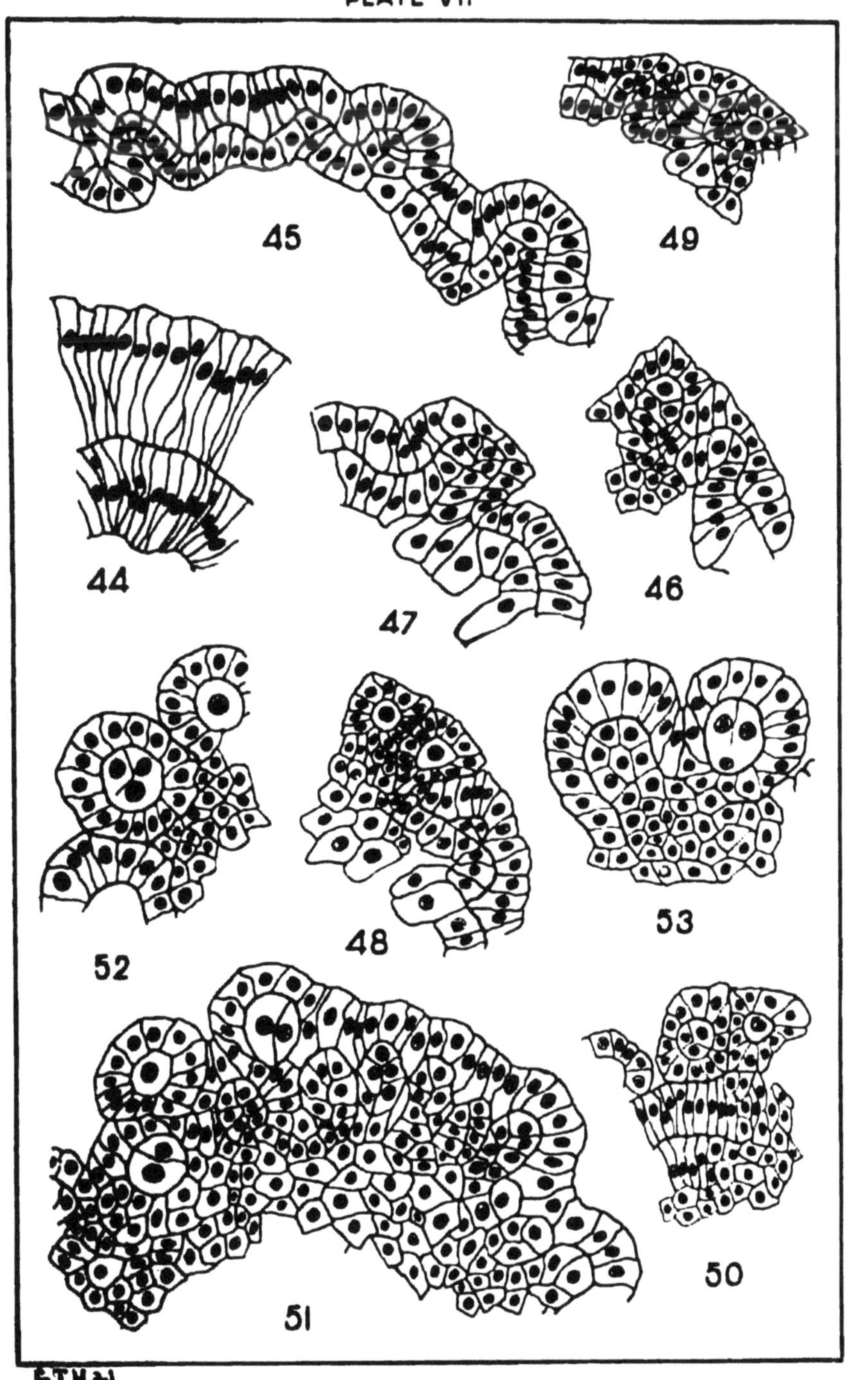

PLATE VIII

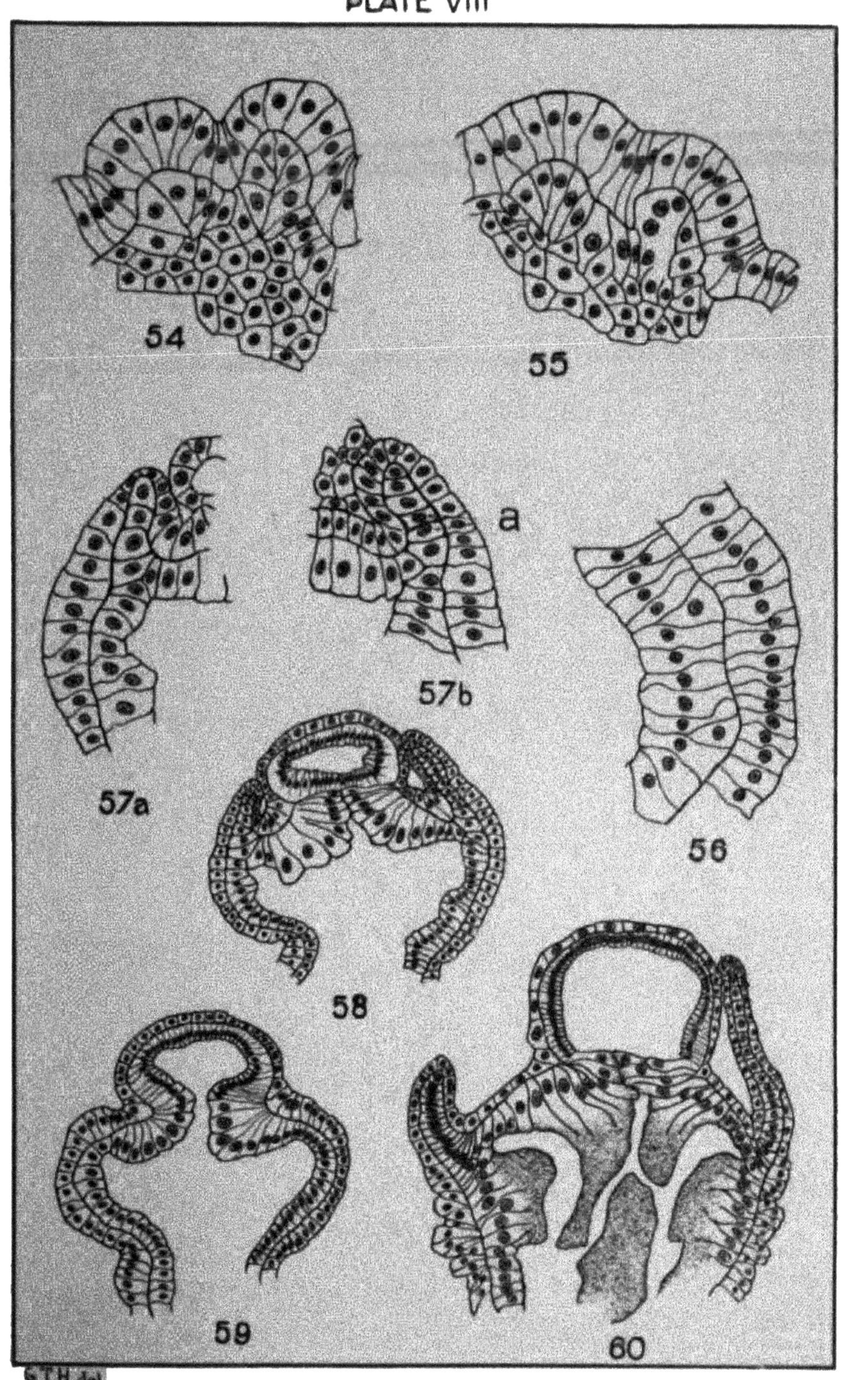

G.T.H. del.

PLATE IX

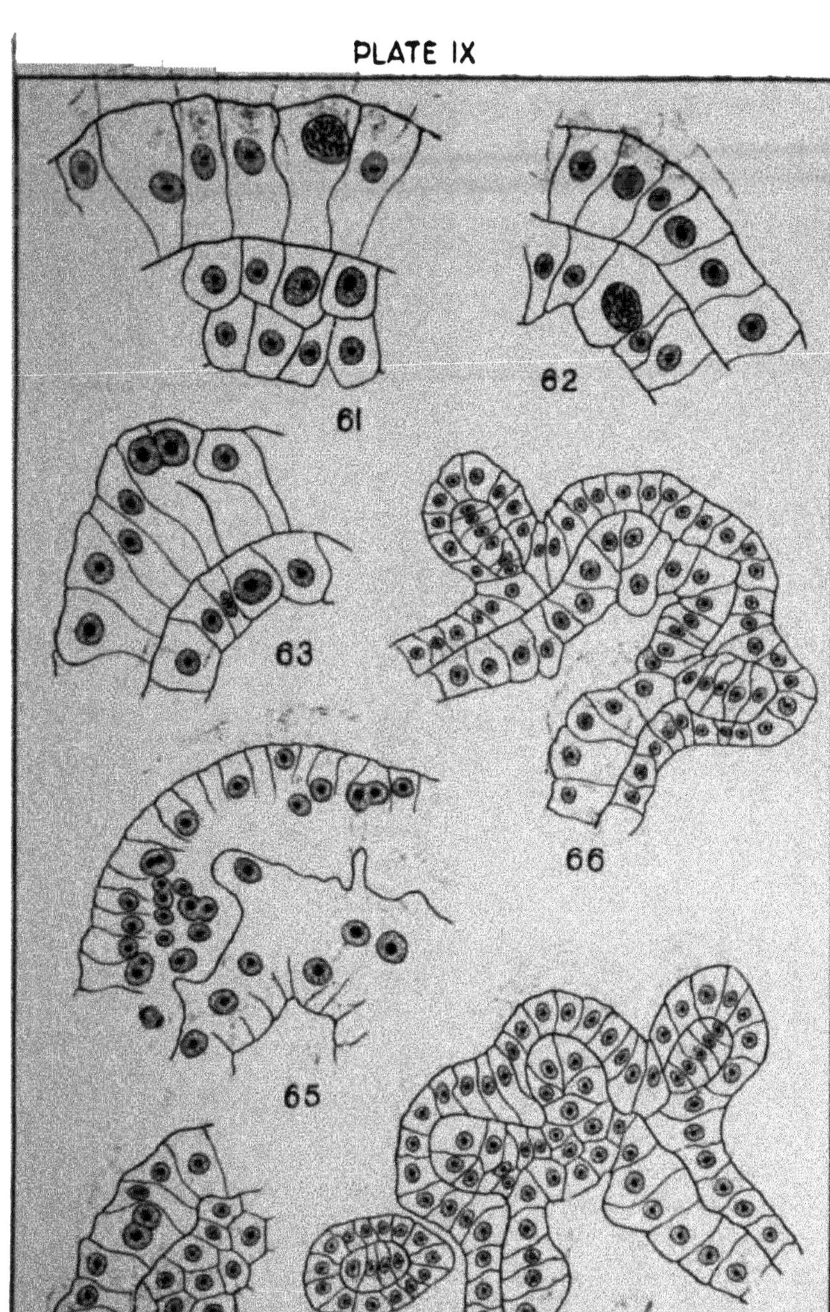

G.T.H. del.

II.—*Some Peculiar Double Salts of Lead*

BY JOHN WHITE

Although it has long been known that the sulfate of lead is markedly soluble in aqueous solutions of alkaline salts of certain organic acids, such as the acetates and tartrates of ammonium and sodium, it does not appear to have been generally remarked that this property of solubility applies also to other of the difficultly soluble lead salts.

In the course of some other work dealing with compounds of lead, it was observed that lead iodid dissolves to an appreciable extent in a concentrated aqueous solution of sodium acetate, yielding upon evaporation a white crystalline crust, which was found to contain both lead and iodin; subsequent experiments have shown that the chlorid and bromid of lead behave like the iodid under similar conditions.

Upon search, it was found that the literature of the subject is very meager, this property having been previously observed by only a few investigators, and that none of the text-books of chemistry make mention of it. The first notice of it is contained in an article by Poggiale,[1] in which he describes a complex salt as being formed by heating together lead chlorid and basic lead acetate; to this he gives the formula

$$PbCl_5.PbOC_2H_3O_2+15H_2O.$$

The published data concerning this salt is such as to lead to the conclusion that it was very impure. Later, Carius[2] prepared a somewhat similar compound by dissolving lead chlorid in a water solution of lead acetate, to which he ascribes the formula

[1]*Ann. Chem.* (Liebig), LVI, 234 (1845); *Compt. rend.* XX, 1180.
[2]*Ann. Chem.* (Liebig), CXXV, 87 (1863).

$$Pb\begin{cases} Cl \\ C_2H_3O_2 \end{cases} + Pb\begin{cases} C_2H_3O_2 \\ C_2H_3O_2 \end{cases} + 3H_2O.$$

Judging from the results obtained during the present investigation, there is every indication that Carius's results and formula are substantially correct. About ten years later, Tommasi,[1] apparently in ignorance of the work of Carius, succeeded in dissolving the iodid of lead in potassium acetate solution, and from this obtained a crystalline compound to which he gave the formula

$$2Pb\begin{cases} I \\ C_2H_3O_2 \end{cases} + KC_2H_3O_2.$$

It will be seen later in this article that, although Tommasi did not ascribe the correct formula to his compound, he in all probability had a substance of definite composition. In the same article he calls attention to the fact that the iodid is soluble in the acetates of other bases, but he was unable to isolate any of these and so to prove that definite compounds had been formed.

Observations similar in character to the above, but which led to nothing of moment, were also made by Nickles[2] and Field.[3]

The paucity of the literature upon the subject and the uncertainty of the results obtained suggested the advisability of undertaking a thorough investigation with the object of determining the general character of this and allied reactions and the nature of the compounds, if any, which are formed. Some of the results thus far obtained are recorded in the following pages.

The iodid was selected as the best of the halogen salts of lead to experiment upon in the beginning, for, although the resultant compound might prove less stable, any change of color could be more easily detected than with either the bromid or chlorid. This has proved to be the case, for in practically every case the salt obtained was white; any decomposition would give rise to lead iodid which could be easily recognized by its

---

1. *Ann. chir* [4], XXV, 168 (1872); *Bull. soc. chim.*, XVII, 3[illegible]7.
2. *Compt.* [illegible]88 (1[illegible]
3. *J. Chen* [illegible]75 (1873).

color. The present paper will describe the results obtained with the iodid.

THEORETICAL

Carius[1] describes the preparation of a class of compounds of the general formula $Pb<^{x}_{C_2H_3O_2}$, where x may be either chlorin, bromin, or iodin. To obtain these, he heated together, in a sealed tube, lead acetate and some alkyl-halid, e. g., ethyl, methyl, or methylene chlorid. On heating to a proper temperature, reaction takes place, yielding an alkyl-acetate and what he calls chloracetin of lead in a crystallized state. Schorlemmer[2] prepared in like manner lead hexylacetochlorid, thus substantiating Carius's work; the same has been done in this investigation by repeating the experiments of Carius.

In attempting an explanation of this reaction, it is possible to make use of the results of von Ende's[3] observations upon the ionic dissociation of the halogen salts of lead. In this he has shown that these salts dissociate in two stages, e. g., $PbCl_2 = PbCl^{\cdot} + Cl'$ and then $PbCl^{\cdot} = Pb^{\cdot\cdot} + Cl'$; the first stage of ionization takes place much more readily than the second. It seems reasonable to suppose that this same argument applies to the other salts of lead, in which case the formation of compounds like $Pb<^{Cl}_{C_2H_3O_2}$ can be readily understood, for:

$$Pb<^{C_2H_3O_2}_{C_2H_3O_2} = PbC_2H_3O_2^{\cdot} + C_2H_3O_2'.$$

This first dissociation would take place with comparative ease, the second less readily, being hindered furthermore by the presence of acetic acid—glacial acetic acid being always used as a solvent; the positive ion $PbC_2H_3O_2^{\cdot}$ could then combine with the more negative Cl' ion, forming the compound $Pb<^{Cl}_{C_2H_3O_2}$. The reaction is probably reversible, that is,

[1] *L. c.*

[2] *Ann. Chem.* (Liebig), CXCIX, 142 (1879).

[3] *Ztschr. Anorg. Chem.*, XXVI, 129 (1901).

lead chlorid if heated with ethyl acetate would yield the same compound, were it not for the fact that the chlorid is soluble in ethyl acetate to so slight an extent as to make it impractical.

In the endeavor to overcome this objection, a solution of sodium acetate was used instead of the ethyl acetate, with the result that a reaction took place, presumably yielding the above compound, but only as an intermediate product, for the excess of sodium acetate required to bring about a reaction was so large that a further action took place, presumably caused by the halid-acetate first formed uniting with the metallic acetate, yielding a double compound of the general type $Pb<^{x}_{C_2H_3O_2} + M.C_2H_3O_2$, where x may be chlorin, bromin, or iodin, and M any metal.

The above formula has been written as a molecular compound, and, so far as the evidence at hand up to the present shows, this seems to be the correct method of expressing it; further evidence is, however, needed before this can be stated with certainty to be the case.

EXPERIMENTAL

As previously stated, the iodin salts are the ones described in this paper. These were found to be quite unstable, being easily split up by water and many other reagents. This action of water probably furnishes an explanation for the fact that other investigators, notably Tommasi, were unable to isolate the salts, for, unless a very large excess of the metallic acetate be used in the water solution, lead iodid will either not dissolve to any extent, giving instead a basic iodid, or, after going into solution, will be again precipitated out on cooling. Further, in endeavoring to free the salt from the excess of metallic acetate, the water used in washing it would give rise to decomposition products, one of which would always be lead iodid.

This fact, which was noticed early in the investigation, suggested the necessity of using some other solvent than water. This solvent must, however, be one in which the metallic acetate used is comparatively readily soluble, else a sufficient excess of the acetate would not be present to hold the lead iodid in solution. Alcohol, either strong or, in some cases, diluted with water, was found to best meet the requirements, but even this would not serve in the case of certain of the acetates tried, which are so slightly soluble in alcohol as to necessitate the addition of too much water to effect solution, the excess of water producing decomposition. It was found further that the readiness with which the acetate dissolves in the solvent, and, therefore, to this degree, the solvent itself, plays an important part in determining the character and composition of the product obtained; the reaction is an excellent illustration of the influence of mass. This part of the problem is being separately investigated and will appear later.

THE IODID-ACETATE OF LEAD.—It was stated on page 4, that, when sodium acetate acts upon lead iodid a product,

$$Pb<^{I}_{C_2H_3O_2} + NaC_2H_3O_2$$

is obtained, and the assumption was made that the iodid-acetate of lead was formed as an intermediate product. If this assumption is correct, it should be possible to start with the iodid-acetate and cause this to unite directly with sodium acetate. Carius[1] prepared the iodid-acetate of lead by heating together in a sealed tube at 140° C. a mixture of 1 mol. ethyl iodid, little more than 1 mol. lead acetate, and 1 mol. glacial acetic acid. Numerous attempts were made to repeat Carius's work, using the utmost precaution, and also by varying the conditions, but up to the present without obtaining the product described by him. A reaction evidently takes place, and in some cases small quantities of a white, crystalline substance were obtained, but generally, long before the reaction could have completed itself, lead iodid began to form and crystallize out, and in such quantities as to preclude the possibility of securing the white, crystalline mass in a fit condition for analysis.[2] This made it

[1] *L. c.* p. 89.

[2] Some further experiments, made just as this paper was being prepared for publication, indicate the possibility of obtaining a pure product, whether iodid-acetate or not yet remains to be determined.

impossible to effect the synthesis of the double compound by causing the molecules to unite directly.

THE SODIUM SALT.—Attempts were first made to prepare a sodium double salt by heating together lead iodid and sodium acetate in sealed tubes, using as solvent various substances, such as ethyl acetate, glacial acetic acid, water, etc. A reaction took place in each case, and when sufficient sodium acetate was used the lead iodid was completely dissolved. Generally a thin, straw-colored or white liquid was obtained, but it was found impossible to remove the contents from the tubes, for, immediately on pouring the liquid out, there ensued, in nearly every case, first a crystallization and then a rapid decomposition of the crystals; this decomposition was accompanied by a marked evolution of heat.

After considerable experimentation, the method of preparation finally adopted was as follows: A rather concentrated solution of sodium acetate in (about 80 per cent) alcohol was prepared by the aid of heat; this was used as the solvent, and to the boiling solution recently precipitated lead iodid was added in small quantities at a time, until the solution became noticeably yellow. It was observed that the previous addition of a small quantity of glacial acetic acid, usually 2 to 5 cc., very materially increases both the amount of iodid which dissolves and the readiness with which it passes into solution. If the yellow solution, after filtering, be allowed to cool in the air, marked decomposition ensues, with deposition of lead iodid, but if instead the air be excluded, this does not happen. To this end, the vessel containing the solution was cooled by placing in a vacuum desiccator over concentrated sulfuric acid, the desiccator partially exhausted by means of a good pump, and set aside to crystallize. Under these conditions nearly white—often very pale sulfur yellow—crystals were obtained.

Attempts were made to determine with certainty the proper proportions to be used in preparing these crystals, but no satisfactory results could be obtained. It was found that the satisfactoriness of the preparation depended to a very considerable extent upon the conditions of handling. The double salt is apparently more soluble than its constituents, and under practically similar conditions one would sometimes get lead iodid crystallizing out first; at other times sodium acetate. The approximate quantities which led to the best results were: 40 to 50 gr. sodium acetate in enough alcohol (about 80 per cent) to make a fairly concentrated solution, 2 to 3 cc. glacial acetic acid and 10 to 12 gr. lead iodid. From this solution, on cooling in a partial vacuum, no lead iodid separated out, but several crops of sodium acetate were obtained. These were removed as fast as formed, and if decomposition set in during the handling, by heating the solution it could be again obtained clear. This was kept up until no further separation of sodium acetate took place, but upon slow evaporation in a partial vacuum, large, almost white, crystals were obtained. It is easy to judge of the relative purity of the several crops of sodium acetate by taking a portion of the salt, pressing out rapidly between drying paper, then moistening with water or with acetic acid. Instant decomposition takes place, and as lead iodid is one of the products of decomposition of the double salt, the depth of color due to this serves as a very good qualitative test.

Several preparations of the salt were made, and the crystallized product was carefully examined each time to determine if there was a tendency to the formation of a uniform substance. This was found to be true except in one instance, to be noted later. The crystals consist of large, flat orthorhombic plates, usually admixed with a small amount of thin, pearly scales, from which they could be easily separated mechanically.

The crystal form and planes usually occurring are shown in the accompanying figure. Crystals up to 3 mm. in length by 1 to 2 mm. in width were obtained. The planes were well defined, but no measurements could be made, the necessary apparatus not being at hand.

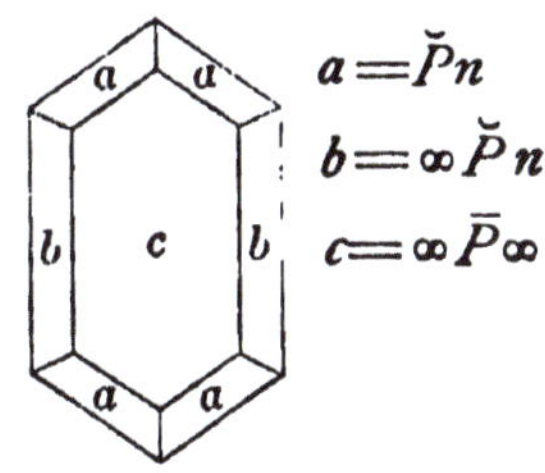

Fig. 1

To obtain the crystals in a form suitable for analysis, they were collected on a Witt filter and washed by the aid of the pump with a wash liquid prepared by mixing together in approximately equal volumes dry ethyl acetate and strong alcohol in which a small quantity of sodium acetate had been dissolved; the latter was found to be necessary to prevent (drive back) the dissociation of the double salt, which invariably happens as soon as the excess of sodium acetate has been washed out. After washing, the crystals were immediately transferred to a watch glass, placed in a vacuum desiccator, and dried in a vacuum over strong sulfuric acid, which requires only a short time. The yellowish tint can be removed by shaking them up in a stoppered flask with dry ethyl acetate, letting them stand for twenty-four hours or longer. The crystals become somewhat opaque and white, while the ethyl acetate assumes a yellowish hue. Comparative analysis made with the slightly yellow and the washed crystals shows that there was practically no difference in them; the color is probably due to a very slight superficial decomposition with formation of lead iodid.

The iodin in the salt was determined by dissolving with the aid of heat in water, slightly acidulated with nitric acid, and then precipitating as silver iodid. To estimate the lead and sodium, the solution in very weakly acidulated water was saturated with hydrogen sulfid, thus precipitating the lead. The lead sulfid was collected on a filter, dried, the paper burned, and the sulfid oxydized in a porcelain crucible by fuming nitric acid, the excess of acid evaporated off, a drop of strong sulfuric acid added, evaporated over a small flame, ignited gently, and weighed as sulfate. The filtrate from the lead sulfid was evaporated in a platinum dish on the water bath, filtered to remove the trace of lead sulfid, transferred to a weighed platinum crucible, evaporated to dryness, a drop or two of strong sulfuric acid added, and evaporated over a small flame, after addition of ammonium carbonate. The sodium was eventually weighed as the sulfate. The results proved very satisfactory. The carbon and hydrogen were determined in the usual way, by burning with lead chromate. The following results were obtained:

0.5585 gram substance gave 0.3273 gram lead sulfate and 0.0791 gram sodium sulfate.

0.4078 gram substance gave 0.2396 gram lead sulfate and 0.0590 gram sodium sulfate.

0.3957 gram of the white substance gave 0.1821 gram silver iodid.

0.3762 gram of the same gave 0.1723 gram silver iodid.

0.6849 gram of the pale yellow substance gave 0.3143 gram silver iodid.

0.2494 gram substance gave 0.1032 gram CO_2 and 0.0364 gram H_2O.

0.4424 gram substance lost on heating 4½ hours 0.0255 gram in weight.

0.2237 gram substance lost on heating 8 hours 0.0133 gram in weight.

The analyses lead to the formula

$$Pb <^{I}_{C_2H_3O_2} + NaC_2H_3O_2 . \tfrac{1}{2}C_2H_4O_2.$$

	Found		Calculated for $PbNaI(C_2H_3O_2)_2 . \tfrac{1}{2}C_2H_4O_2$
	I	II	
Lead............	40.03	40.13	40.98
Sodium..........	4.59	4.69	4.57
Iodin...........	24.86	24.76-24.79	25.12
Carbon..........	11.28	......	11.89
Hydrogen.......	1.64	......	1 60
Acetic acid......	5.76	5.95	5.95

The acetic acid of crystallization was determined by heating weighed portions of the pulverized substance in a double walled vacuum air bath, containing alcohol as the heating liquid. The heating was carried out at a temperature of 78° C. and in a vacuum reading 650–710 mm. Under these conditions the substance loses weight slowly, the weight becoming constant after 3½ to 5 hours heating. In one instance the temperature was raised, by substituting water in place of alcohol in the jacket, to 98°C., but without causing any further loss of weight. Heated in the air a further gradual loss was observed. The color changes during the drying from a creamy white to a canary yellow; when heated in air it goes finally to an orange yellow. When heated in a melting tube, the substance changes to a light sulfur yellow color at a temperature of about 90°C.; between 95° and 100°, it assumes an orange red hue; and when heated still higher, it sinters at 120° and finally melts, with partial decomposition, to a reddish, syrupy liquid at 124° to 125°C.; on cooling again it solidifies to a lemon-yellow crystalline mass.

Thus far no satisfactory solvent for the salt has been found; water decomposes it instantly, yielding first lead iodid, then a basic iodid. Acids all decompose it, giving lead iodid. It is soluble in hot alcohol containing a considerable excess of sodium acetate; otherwise neither hot nor cold alcohol appears to exert any marked solvent action, but gradually produces decomposition, as shown by the change in color. Dry ethyl acetate is without effect, while hot benzol apparently dissolves it in small amount. It is readily soluble in hot nitro-benzol, but does not crystallize out again on cooling. In dry air, as when kept in tightly stoppered flasks, it is apparently perfectly stable, but when exposed to the air under ordinary conditions it quickly turns yellow, the moisture in the air probably causing decomposition; light is without appreciable action.

It was stated above that, accompanying the large crystals, there was always found another product, crystallizing in thin, pearly white scales, and that in one instance the coarse crystals were present in the smaller amount. It was at first supposed that the thin flakes were chiefly sodium acetate, but a qualitative test having shown a high iodin content, the substance was examined more closely. Although this substance is an invariable constituent of the product obtained through the action of sodium acetate upon lead iodid, it was obtained in very small quantity in all except the one case previously noted, and the exact conditions for its preparation could not be determined. Where present in small quantity, or when the other product is obtained in large crystals, the two could be fairly well separated by picking out the coarse particles, but in general it was found to be easier to effect separation by taking advantage of the difference in the specific gravity of the two. To this end, the whole product was shaken up in a stoppered flask with dry ethyl acetate, when, upon standing a short time, the coarsely crystalline substance settled almost at once, while the lighter, fine portion remained for some time in suspension; by decanting, the separation could be effected. After several repetitions, a fairly complete separation was accomplished; the heavier portion was found, upon analysis, to contain 24.61 per cent of iodin, while the lighter contained 18.34. After making several mechanical separations of the latter, it apparently reached a constant value, the average being 18.81 per cent of iodin. This method of separation, while it leaves room for doubt as to the purity of the latter product, since one can not be sure that it may not be mixed with sodium acetate, nevertheless led to such uniform analytical results as to suggest that this was a new substance.

The following analytical results were obtained:

0.4158 gram substance gave 0.1856 gram lead sulfate.
0.3477 gram substance gave 0.1559 gram lead sulfate.
0.4727 gram substance gave 0.1514 gram sodium sulfate.
0.3477 gram substance gave 0.1109 gram sodium sulfate.
0.5179 gram substance gave 0.1803 gram silver iodid.

On drying at 78°C. for 7 hours in a vacuum of 650 mm., 0.5261 gram substance lost 0.0236 gram in weight.

The most probable formula deduced from the analysis is

$$Pb<^{I}_{C_2H_3O_2} + 3NaC_2H_3O_2 . \tfrac{1}{2}C_2H_4O_2.$$

	FOUND		CALCULATED FOR $PbNa_3I(C_2H_3O_2)_4 . \tfrac{1}{2}C_2H_4O_2$
	I	II	
Lead............	30.48	30.62	30.93
Sodium.........	10.39	10.35	10.34
Iodin............	......	18.81	18.96
Acetic acid	4.49	......	4.49

Although the above substance may be a mixture, the analytical results indicate the contrary, agreeing very well indeed with the calculated values for a definite salt. According to the theory proposed in explanation of the formation of this class of compounds, there is no reason why one molecule of the iodid-acetate should not unite with a varying number of molecules of the metallic acetate, forming compounds like the above. Some recent results obtained with another salt tend to show that this is possible.

The properties of the above salt are in general similar to those described under the first sodium salt.

THE POTASSIUM SALT.—As previously stated, Tommasi[1] obtained a potassium lead iodid-acetate to which he ascribed the formula $2Pb<^{I}_{C_2H_3O_2}+KC_2H_3O_2$. This he obtained by dissolving 1 mol. lead iodid in a hot aqueous solution of potassium acetate, containing 2 mols. This, on cooling, deposited a pale yellow, crystalline mass, which he redissolved by heating with twelve times its weight of absolute alcohol and allowed to cool in a desiccator over lime. In this way he obtained a large quantity of white crystal flakes. The formula assigned to the compound by Tommasi differs slightly from that used here, yet there is every reason to believe that the two compounds are identical, for the method used was essentially the same as that described in his paper, and the general appearance and character of the two products were alike. Tommasi gives no analytical details. The potassium salt is easier of preparation than the sodium compound, being much less soluble and more stable than the sodium salt. Alcohol of greater dilution can be used without causing decomposition; indeed it was found that unless a moderately dilute alcohol were used, very little lead iodid could be brought into solution. A strong solution of potassium acetate in 50 or 60 per cent alcohol was made, and to this, after heating to boiling, lead iodid was added in small portions at a time, until it ceased to dissolve; the addition of

---

[1] *L. c.*

acetic acid is not necessary, appearing in fact to operate against the reaction. Only very slight cooling is required for the salt to separate, which it does rapidly, coming out as glistening, pearly white scales, forming almost a solid mass in the vessel. These were redissolved by adding more alcohol and heating, when on cooling they were obtained in larger flakes. For washing, a mixture of ethyl acetate and alcohol was used, containing 2 to 3 per cent of potassium acetate, but, on account of the tendency of the thin flakes to pack together on the filter, much difficulty was experienced in washing them thoroughly, as it must be done rapidly, otherwise decomposition sets in. The analyses show that not all of the excess of the potassium acetate had been removed. The drying was accomplished in the same manner as with the sodium salt; the dry substance is fairly stable. The same method of analysis was used as with the sodium compounds.

The following results were obtained:

0.2463 gram substance gave 0.1494 gram lead sulfate and 0.0517 gram potassium sulfate.
0.2888 gram substance gave 0.1760 gram lead sulfate.
0.4028 gram substance gave 0.0811 gram potassium sulfate.
0.2795 gram substance gave 0.1355 gram silver iodid.
0.3010 gram substance gave 0.1436 gram silver iodid.

Another preparation gave:

0.2868[1] gram substance gave 0.1743 gram lead sulfate.
0.4422[1] gram substance gave 0.2686 gram lead sulfate.
0.3862 gram substance gave 0.0799 gram potassium sulfate.
0.2591 gram substance gave 0.1235 gram silver iodid.
0.4211[1] gram substance gave 0.1648 gram CO_2 and 0.0464 gram H_2O.
0.3727[1] gram substance gave 0.1463 gram CO_2 and 0.0409 gram H_2O.

---

[1] I am indebted to Miss Mary L. Fossler, of the University of Nebraska, for these results.

The formula deduced from these is $Pb<^{I}_{C_2H_3O_2}+KC_2H_3O_2$.

	FOUND				CALCULATED FOR $PbKI(C_2H_3O_2)_2$
	PREPARATION I		PREPARATION II		
	I	II	I	II	
Lead	41 42	41.62	41.50	41.48	42.14
Potassium	9.48	9.50	9.29	.....	7.97
Iodin	25.81	25.78	25.75	.....	25.84
Carbon	.....	.....	10.67	10.71	9.77
Hydrogen	.....	.....	1.21	1.23	1.23

The high value for potassium is probably due to incomplete washing, leaving a slight excess of potassium acetate adhering to the crystals. When heated in a vacuum oven the salt suffered no appreciable loss in weight and exhibited only a very slight change in color, even after long-continued heating. When heated in a melting tube, it showed no change up to 205°C., when there were signs of sintering and it turned very slightly yellow. At 208–208.5°C. it melted fairly sharp to a pale straw-colored liquid; this solidified to a white crystalline mass on cooling, which again gave the same melting point, showing that no decomposition had taken place.

From the mother liquor, after removal of the potassium salt, upon evaporaton, there was obtained long needle-like crystals which were white, but readily turn slightly yellow on standing. This salt has the appearance and properties of the double potassium lead iodid described by Remsen and Herty[1] and by Wells.[2] No particular attempt was made to purify these; an iodin estimation gave 54.30 per cent iodin, while $KPbI_3.2H_2O$ requires 57.46 per cent iodin.

THE AMMONIUM SALT.—This was prepared in the same way as the sodium salt, except that the reaction was found to work better if no acetic acid is added. The ammonium compound is so insoluble that it generally precipitates from the boiling

[1] *Amer. Chem. Jour.*, XI, 298; Ibid, XIV, 107.

[2] *Amer. Jour. Sci.*, XLV, February No. (1893); *Studies from the Chemical Laboratory of Sheffield Scientific School*, Yale University, vol. I, 250.

solution, and can not again be dissolved, as the sodium and potassium compounds are, by boiling with the mother liquor. The conditions of preparation may be varied between wider limits in the case of the ammonium salt than with the others, the same product being always obtained. A number of preparations were made, a white, highly crystalline product being obtained in each case. The method of washing and purification was the same as used with the sodium and potassium salts; the salt is quite unstable, undergoing decomposition very readily on exposure to air and is also acted upon by light.

In a few cases rather large, 1 to 2 mm., crystals were obtained. These upon examination under the microscope were apparently orthorhombic prisms; the more common type and the planes upon it are represented in the accompanying figure.

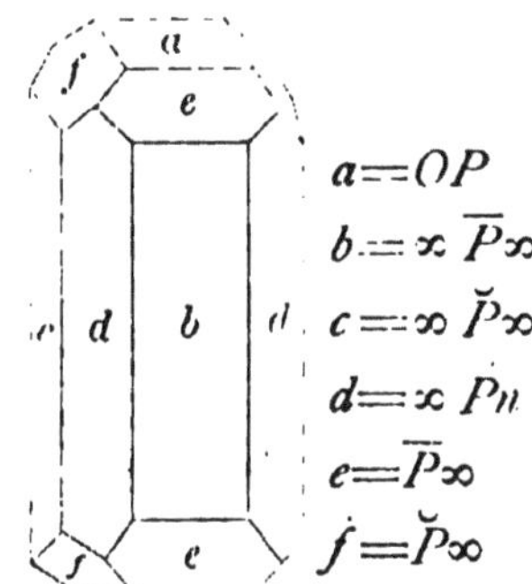

Fig. 2

The following analytical results were obtained with three separate preparations, designated as A, B, and C, respectively:

A. 0.4639 gram substance gave 0.2327 gram silver iodid.
B. 0.2873 gram substance gave 0.1430 gram silver iodid.
0.2492 gram substance gave 0.1257 gram silver iodid.
0.3501 gram substance gave 0.1748 gram silver iodid.
0.3535 gram substance gave 0.1761 gram silver iodid.
0.2774 gram substance gave 0.1788 gram lead sulfate.
C. 0.2996 gram substance gave 0.1506 gram silver iodid.
0.1948 gram substance gave 0.1254 gram lead sulfate.
0.2583 gram substance gave 0.1669 gram lead sulfate.

The lead in these was determined by direct evaporation after treatment with concentrated sulfuric acid.

The formula deduced from the analysis is

$$Pb<^{I}_{C_2H_3O_2}+NH_4C_2H_3O_2.$$

	FOUND			CALCULATED FOR
	A	B	C	$PbINH_4(C_2H_3O_2)_2$
Lead	......	44.02	43.96–44.12	44.03
Iodin..........	27.10	26.89–27.25–26.98–26.91	27.16	27.00

When heated in a melting tube the salt rapidly turns to a deep lemon color, sinters at about 157°C., and melts to an amber-colored liquid at 166° to 167°C., which does not again solidify unless cooled to a very low temperature.

There were obtained from the mother liquor, by evaporation, long slender, hair-like needles, radiating in tufts from a common center. These are white, easily soluble in the mother liquor on heating, and turn yellow very quickly when freed from it. They correspond in appearance and properties with the ammonium lead iodid described by Wells.[1] The compound in both the dry and moist condition is very sensitive to light, turning first yellow, then brownish red, and finally grayish brown. The following analytical results were obtained:

0.1430 gram substance gave 0.0678 gram lead sulfate.
0.1611 gram substance gave 0.0766 gram lead sulfate.
0.1402 gram substance gave 0.1512 gram silver iodid.
0.1374 gram substance gave 0.1476 gram silver iodid.

	CALCULATED FOR $NH_4PbI_3.2H_2O$	FOUND
Lead	32.24	32.39–32.48
Iodin..........	59.36	58.27–58.04

[1] *Amer. Jour. Sci.*, XLVI, July No. (1893); *Studies from the Chemical Laboratory of the Sheffield Scientific School*, vol. I, 283.

The appearance of this salt, as well as that of a similar compound in the waste liquor from the potassium double salt, as by-products, may be regarded as indicating the general nature of the reaction, showing that a partial interchange takes place between the lead iodid and the alkaline acetate, thus furnishing additional evidence in favor of the theory which has been proposed in explanation of the double salt formation.

THE LEAD SALT.—In his paper dealing with the salts of the type $Pb<^{Cl}_{C_2H_3O_2}$ Carius[1] states that water decomposes these, throwing out a white powder, which, however, soon goes completely into solution. He further says that the halogen salts of lead are soluble in lead acetate, and that the compounds obtained by the two methods are identical to this. Based upon a chlorin estimation, he ascribes the formula

$$Pb<^{Cl}_{C_2H_3O_2}+Pb(C_2H_3O_2)_2\ 3H_2O.$$

Although Carius mentions the fact that lead iodid acts like the chlorid, he does not give any description of the iodin compound. A repetition of his work reveals the probable cause of this omission, for when lead iodid is dissolved in a water solution of lead acetate, the *basic iodid* of lead is invariably formed; the same is indeed the case when an alcoholic solution is used, unless a relatively large amount of acetic acid be added.

The iodid is not so readily soluble in lead acetate solution as in solutions of the alkaline acetates. The following proportions used in several instances will give an idea of the solubility: 50 gr. lead acetate were dissolved in 100 cc. 93 per cent alcohol and 30 cc. glacial acetic acid; in this, at the boiling temperature, there were dissolved about 4 gr. lead iodid. The solution does not have to be cooled in a vacuum, as no decomposition takes place during the precipitation. The crystals obtained were white, well defined, and consisted of simple monoclinic forms.

---

[1] *L. c.*

Those most frequently obtained are shown in the accompanying figure. In some instances well defined interpenetration twins were obtained, the individuals being arranged in the form of a Latin cross.

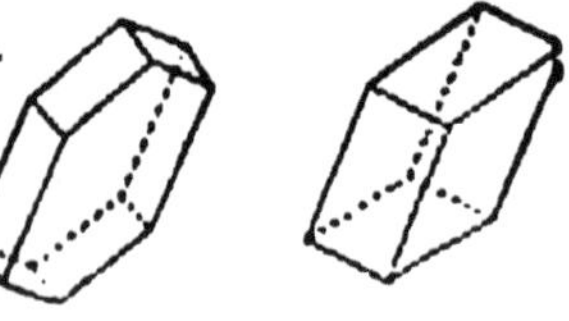

Fig. 3

The washing of the crystals is most readily accomplished by decantation, after shaking or mixing with absolute alcohol (96 to 98 per cent), containing a trace of lead acetate, the final washing being carried out on a Witt filter by the aid of the pump; the crystals were afterwards dried in a vacuum over sulfuric acid. It, like the other salts previously described, is stable only in dry air. The analyses were made in the same manner as with the ammonium compound. The following results have been obtained:

A. 0.1447 gram substance gave 0.1201 gram lead sulfate.
0.1279 gram substance gave 0.1055 gram lead sulfate.
0.2620 gram substance gave 0.0792 gram silver iodid.

After a further washing, the same substance gave:

0.3134 gram substance gave 0.0960 gram lead sulfate.
0.1523 gram substance gave 0.1263 gram silver iodid.
B. 0.1096 gram substance gave 0.0907 gram lead sulfate.
0.1074 gram substance gave 0.0333 gram silver iodid.
0.7587[1] gram substance gave 0.2866 gram CO_2 and 0.1024 gram H_2O.
0.3256[1] gram substance gave 0.1259 gram CO_2 and 0.0424 gram H_2O.

The formula deduced from these results is

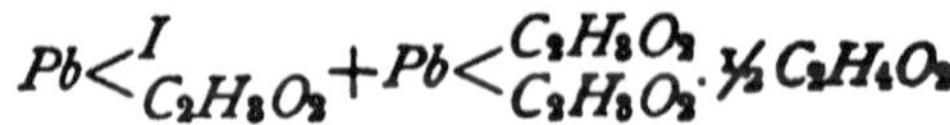

$$Pb<{I \atop C_2H_3O_2} + Pb<{C_2H_3O_2 \atop C_2H_3O_2} \cdot \tfrac{1}{2}\, C_2H_4O_2$$

[1] These analyses were kindly made for me by Miss Fossler.

	Found		Calculated for
	A	B	$Pb_2I(C_2H_3O_2)_3.\frac{1}{2}C_2H_4O_2$
Lead	56.67–56.63–56.51	56.33	55.34
Iodin..........	16.33–16.55	16.75	16.96
Carbon	....	10.31–10.54	11.23
Hydrogen	....	1.51– 1.46	1.48
Acetic acid....	3.75+	....	4.01

On heating in a vacuum bath at 77°C. for 7 hours, 0.8712 gram substance lost 0.0327 gram in weight, which corresponds closely with the required loss for the above compound; it was impossible, however, to bring it to constant weight, for a very slow, almost imperceptible diminution of weight follows from this point. This was found to be true of all compounds containing lead acetate, and is probably caused by a gradual breaking down of the lead acetate; in air this takes place rapidly. When heated in a melting tube, the salt assumes a slight yellowish tint with slight signs of sintering at 180°C.; at 192° it becomes pasty, and melts slowly to a clear, amber colored, very viscous liquid at 202–205°C. No visible decomposition takes place during the heating.

The lead salt behaves in general like the other salts described, being, however, somewhat more stable than these.

The Action of Solvents.—It has been stated that most of the substances commonly employed as solvents exert a decomposing action upon these double iodid-acetates. This is particularly true of water, the alcohols, and the acids; the ethereal salts, when thoroughly dry, seem to be without action of any sort; dry hydrocarbons likewise. It is unfortunate that no good solvent has as yet been found, since it has not been possible to settle positively the question of the structure of the salts. But the action of water and alcohol is of interest, nevertheless, since the decompositions resulting furnish some evidence bearing upon the structure. The sodium, potassium, and ammonium double salts were each pulverized and shaken up in stoppered flasks with water and with absolute (96 to 98 per cent) alcohol respectively, the time ranging from 24 hours to six days.

In each instance, the action of water was to produce an almost instantaneous decomposition, yielding lead iodid, shown by the orange-yellow color characteristic of this substance when deposited in the amorphous state; in a very short time, however, the color begins to change, turning lighter in hue. The time required for completion of this change varied in length with the size of the particles and, to some extent, with the salt, the potassium salt usually reacting more slowly than the others. The color at the end was a light sulfur-yellow. Analysis proved this to consist of basic iodid of lead. In the table the sodium salt is designated as A, the potassium as B, and the ammonium as C.

	CALCULATED FOR $PbI(OH)$	FOUND		
		A	B	C
Lead.........	58.98	58.58	......	58.97
Iodin..........	36.16	36.44	36.13–36.21	36.16

The action of absolute alcohol upon the three double salts was in general of a character similar to that of water, but here the nature of the individual salt seemed to play an important part, for it was found that the potassium salt, which is the easiest of the three to prepare, was scarcely acted upon at all by the alcohol, there being but very slight change in color, even after digesting for several days. An iodin estimation in the filtered and dried residue, after prolonged extraction, showed 25.74 per cent iodin, the theory requiring 25.84 per cent, thus proving that there had been no action. The ammonium salt, which stands next to the potassium salt in readiness of preparation, showed, after the same time of digestion, a slight decomposition; the residue gave 30.73 per cent iodin, while the original salt contains 27.00 per cent. The decomposition in the case of the sodium salt was almost complete, giving 33.64 per cent iodin, while the original salt contained 25.12 per cent and the basic iodid 36.16 per cent. Thus it appears that the degree of solubility of the salt in absolute alcohol is indicative of the degree of decomposition produced by it. It appears, however, as if a condition of equilibrium were reached, for it was found after several extractions, varying in length, using the same salt, that all yielded practically the same amount of iodin. The other products of the decomposition, the acetates of lead and of the alkali metal, could be readily detected qualitatively in the filtrate.

A study of the decomposition reactions just noted suggests that the first action of the reagent is to split up the salt into the two components, e. g.:

$$1. \quad \left\{ Pb\!<^{I}_{C_2H_3O_2} + M.C_2H_3O_2 \right\} = Pb\!<^{I}_{C_2H_3O_2} + M.C_2H_3O_2$$

and that subsequently the iodid-acetate decomposes into lead iodid and lead acetate, e. g.,

$$2. \quad 2Pb\!<^{I}_{C_2H_3O_2} = PbI_2 + Pb(C_2H_3O_2)_2.$$

It is, however, not at first sight easy to explain the formation of the basic iodid, for experiment shows that lead iodid is not acted upon to any marked extent by either hot or cold water with formation of a basic salt. Further, should it be conceded that the iodid is converted directly by water into basic iodid, $PbI_2 + H_2O = PbI(OH) + HI$, the presence of the hydriodic acid—or of its salts, since in this case it would naturally react with the alkaline acetate formed in reaction 1, yielding an alkaline iodid—could be readily detected.. Examination of the filtrates reveals traces only of iodid, such as might be expected from the lead iodid in solution. If, on the other hand, it be conceived that a part at least of the iodid-acetate is acted upon by water and thus transformed into the basic iodide,

$$3. \quad Pb\!<^{I}_{C_2H_3O_2} + H_2O = Pb\!<^{I}_{OH} + H.C_2H_3O_2,$$

then the addition of a relatively small quantity of acetic acid should serve to prevent this last reaction from taking place. It was found, however, that, in alcoholic solution, if added in small quantities at a time, a relatively large amount of glacial acetic acid could be used without affecting the result to any appreciable extent; naturally, when added in very large quantity, lead iodid was always formed. The most reasonable explanation for the conversion of lead iodid into the basic iodid, then, is to suppose reaction 2, above, to be reversible; the first stage, i. e.,

$$2Pb<^{I}_{C_2H_3O_2} \rightleftarrows Pb<^{I}_{I} + Pb<^{C_2H_3O_2}_{C_2H_3O_2}$$

would naturally take place very readily on account of the tendency to form the difficultly soluble iodid; later, however, under the gradual action of water, with the formation of a still less soluble basic iodid, the reaction reverses itself, i. e., goes in the sense

$$Pb(C_2H_3O_2)_2 + PbI_2 \rightleftarrows 2Pb<^{I}_{C_2H_3O_2}$$

and then water acting upon the $Pb<^{I}_{C_2H_3O_2}$ would convert it into the basic iodid as shown in reaction 3 above. The conversion of the iodid into the basic iodid would of course be slow, owing in part to the slight solubility of lead iodid, but more to the fact that the active mass of lead acetate is too small to force the original reaction to reverse itself. That this is the case was demonstrated by the addition of lead acetate, when the formation of the basic iodid took place quite rapidly. Indeed it was found that freshly precipitated lead iodid, which is but slightly affected by cold water, can be readily and completely transformed into the basic iodid by simply adding lead acetate and shaking. Other acetates produce a like effect, but not quite so readily as the lead acetate. The fact that acetic acid does not, in small quantities, prevent the reaction is probably because it is so slightly ionized in comparison with its salts as to produce little or no effect. It is hoped to test reaction 3 above in the near future. The impossibility of preparing the iodid-acetate has so far prevented this.

From the study of the reactions involved in the formation of these double salts, as well as those just considered in their decomposition, the conclusion may be drawn that in these we must regard the iodin as united directly to the lead, and that in all probability the double salt is of the type commonly designated as molecular. The investigation is still in progress, and it is hoped soon to furnish additional evidence bearing upon this point.

III.—*The Mémoires de Bailly*

BY FRED MORROW FLING

One of the striking characteristics of the scientific literature of modern history, sharply contrasting in this respect with the literature of ancient and medieval history, is the extreme paucity of critical source-studies. Careful, exhaustive criticisms of the sources of Greek, Roman, and medieval history abound, while some of the most frequently used sources of modern history have never formed the subject of a critical monograph. The explanation of this lamentable condition is to be found, to some extent, in the lack of training and tradition among the workers in modern history, to some extent in the thoughtlessness of the investigator, who does not realize the wisdom of formulating for those who shall come after him the results of his critical studies; but chiefly it is due to the belief prevalent among the writers of modern history that the sources that they employ do not need to be criticised. This belief may be attributed to the fact that much of the critical work in connection with the sources of modern history is so easy that it is largely unconscious; but, easy or difficult, the critical process should always be gone through with and that, too, consciously.

Perhaps no period of modern history has suffered more from this dilettanteism, as it might properly be called, than the French revolution. There is a real need to-day of a critical monograph literature upon the *mémoires* and contemporary histories of this important period of European history. Both the past and the present attitude of historians toward this material has been largely unscientific. The tendency, until a few years since, was to accept it as valuable *en bloc*, without criticism; the tendency to-day seems to be to reject it *en bloc*, equally without criticism. The last tendency is, to be sure, the less harmful, but it is not strictly scientific. *Mémoires* and contemporary histories can not be rejected as a whole on the ground that they were written a longer or shorter time after the events described, and, as the writer relied upon his memory and as his memory after such a lapse of time would certainly play him false, little that he wrote could be trustworthy. It is true that these conditions are not favorable to the making of the best kind of a historical record, but what effect the conditions have had upon the making of a particular record can not be told until it has been carefully criticised. The results of the criticism may be negative, but at least we shall know definitely why the record is worthless.

Of all the contemporary accounts of the first months of the revolution no one has been more frequently used by historians than the *Mémoires de Bailly,* and yet no detailed criticism of this work has ever been published. In its most common form, it is in three volumes, only the first two, however, having been written by Bailly. The third volume, bearing the title *Extrait des notes inédites de feu M. . . . membre de l'Assemblée constituante,* is anonymous, but was attributed by M. Tourneux to Camus. I have shown elsewhere[1] that the volume is a forgery, being a composite of extracts taken from the *Courrier de Provence,* the *Point du jour,* and *Les révolutions de Paris.*

The first two volumes were undoubtedly written by Bailly. Some years ago, M. Brette advanced the theory that the work as we have it now was not entirely written by Bailly. "It can not be doubted," he said, "that Bailly left notes, perhaps a journal; disrespectful editors have reduced what was a reliable record to the form of a historical romance. The ***Mémoires de Bailly,*** in our opinion, do not constitute a foundation for serious investigations; we shall cite them only to show, by incontestable testimony, the errors that they contain."[2] The unsoundness of this criticism was pointed out by M. Flammermont, who called attention to the fact that "the printed text conforms to the original manuscript that still exists."[3] The manuscript is in the library

---

[1] La révolution française, Nov. 14, 1902, *Une pièce fabriquée: le troisième volume des Mémoires de Bailly.*

[2] *Ibid.*, XX, p. 13, *La séance royale du 22 juin 1789.*

[3] Flammermont, J. *La journée du 14 juillet 1789*, p. CLX.

of the Chamber of Deputies at Paris. It is difficult to see how M. Brette could have overlooked so well known a fact, the editors of the edition of the *Mémoires* published in 1821 having referred to it in their introduction.[1]

No serious attempt has ever been made to determine definitely when Bailly composed his *Mémoires*.[2] It was believed by M. Naigeon that the work was really a journal kept from day to day. He noted that the style was not as correct as one had a right to expect from "an author, member of three academies," but he excused Bailly on the ground that "he wrote his journal only when he returned home at night, and in moments when his mind was more or less agitated, saddened, affected as well by what had passed during the day as by the dangers to which he would find himself exposed on the morrow."[3] M. Naigeon was misled by the title of the work, its form, and phraseology. The full title is, *Mémoires d'un témoin de la révolution ou journal des faits qui se sont passés sous ses yeux, et qui ont préparé et fixé la constitution française;* the text is arranged in the form of a journal, events being grouped under the day of the week and month when they occurred; and, finally, the verisimilitude is increased by the employment of such expressions as *"hier," "aujourd'hui," "ce matin," "demain," "ce soir."* It was not the intention of Bailly, however, to deceive anybody.[4]

The opening paragraph of the first volume makes clear that we have to do with *Mémoires* and not with a journal: "When

[1] *Mémoires de Bailly*, 3 vols., Paris, 1821: "Le manuscrit original déposé à la bibliothèque de la chambre des députés," p. II. This is the edition that I have made use of. The first edition appeared in 1805.

[2] Flammermont noted that the *Mémoires* were written "au commencement de l' année 1792," citing in support of his statement the parenthetical expression (I, p. 358), "Aujourd'hui 23 fevrier 1792." Flammermont, *La journée du 14 juillet 1789*, p. CLIX.

[3] *Mémoires de Bailly*, II, p. 411. The *Jugement de M. Naigeon sur les Mémoires de Bailly* (pp. 409-418), has little scientific value.

[4] These expressions were evidently introduced on account of the form into which Bailly threw his narrative. He wished to reproduce the events of the revolution day by day, and making much use of the historical present, he naturally used such expressions as "ce matin," "ce soir," "hier," etc. His frequent references to later events and his expressions of regret because of his inability to recall what happened on a particular day furnish sufficient proof of his honesty.

I was called to the elections," wrote Bailly, "I could not have suspected the part that would be successively given to me in the public administration nor the influence that I would have upon affairs: this part increased, this influence extended always in an unlooked for manner. I have regretted much that I did not have constantly with me a secretary to collect the facts, the anecdotes, the characteristics, the thoughts that would have merited preservation, that I might paint with more fidelity and animate by their aid the grand scenes of which I have been a witness. Reduced to my memory to retrace them at this moment in my mind (I shall show presently that the burden placed upon Bailly's memory was not excessive), and to commit them to this journal, I protest that my memory will be faithful."[1] Fortunately, the value of the journal does not depend wholly upon the correctness of this naïve protestation. It would seem reasonable to infer from this sentence and from the one immediately following that the work was composed after Bailly had closed his official career. If it is objected that this paragraph may have been prefixed to the *Mémoires* some time after they were written—a thing that does not seem probable—it is easy to show that a retrospective strain runs through the two volumes. Such expressions as *"depuis," "alors," "ce même jour," "déja," "encore"*[2] are scattered throughout the work. Other expressions like *"je ne me rapelle pas," "je m'en souviens," "je crois,"*[3] indicating that he had difficulty in recalling events on account of their remoteness, are also well distributed through the two volumes.

---

[1] *Mémoires de Bailly*, I, p. 1.

[2] *Mémoires de Bailly*, I, pp. 4, 8, 9, 11, 12, 17, 25, 30, 44, 51 [illegible] 58, 71, 107, 136, 148, 150, 155, 165, 170, 175, 176, 183, 215, 216, 217, 2[illegible] [illegible] 248, 253, 277, 291, 301, 316, 317, 320, 321, 323, 324, 325, 330 [illegible] 358, 360, 372, 373; II, pp. 10, 19, 33, 36, 55, 60, 61, 77, 78, 80, [illegible] 109, 116, 133, 139, 140, 141, 154, 164, 169, 189, 195, 206, 208, [illegible] 227, 229, 237, 241, 247, 249, 250, 252, 253, 271, 273, 282, 283, [illegible] 299, 303, 308, 310, 314, 317, 321, 322, 330, 331, 339, 346, [illegible] 372, 375, 378, 382, 389, 392, 395, 397, 403.

[3] *Ibid.*, I, pp. 1, 2, 35, 87, 94, 95, 105, 116, [illegible] 119, 246, 253, 258, 263, 317, 343, 363, 393; II, pp. [illegible] 170, 180, 230, 237, 239, 248, 257, 260, 268, [illegible] 341, 354, 367, 377, 380.

While these references should make clear that the *Mémoires* were not written from day to day, but some time after the events described, and were evidently composed at one time, they still do not fix definitely when the work was composed, between what limits the actual writing took place. It was plainly written after Bailly had retired from the office of mayor of Paris. Early in the first volume, under the date of April 29, 1789, he refers to the work as *"ces mémoires, ou journal de ma vie de trente-un mois."*[1] From April, 1789, when Bailly became an elector of the city of Paris, to November 18, 1791, when he ceased to be mayor of the city, was thirty-one months. He could not have begun his narrative before the latter part of November, 1791. This inference is supported by frequent references, in the text devoted to the year 1789, to events happening in 1791,[2] and, among others, to the *"seconde legislature"* October, 1791.[3] The *Mémoires* were not begun in November, for it was not until his retirement to the neighborhood of Nantes, probably in January, 1792, that Bailly had leisure for writing.[4] All of the first volume,

[1] *Mémoires de Bailly*, I, p. 32.

[2] *Ibid.*, I, p. 37, "La constitution proposée par les électeurs de Paris renferme presque toutes les bases qui ont é'é décrétées par l'Assemblée constituante"; p. 52, "J'ai toujours pensé, et je pense encore, qu'un peu plus de cet esprit philosophique n'aurait pas nui à l'Assemblée constituante"; p. 53, "Il en a resulté une constitution qui, malgré ses défauts, est un superbe ouvrage"; p. 68, "Il est imprimé dans le second des deux volumes que j'ai publiés au commencement de 1791"; p. 144, "Parmi ces ecclésiastiques étaient l'abbé Grégoire, devenu célèbre dans l'Assemblée nationale constituante"; p. 170, "Les principes sages que l'Assemblée nationale a eus depuis dans ses plus beaux momens"; p. 248, "Il faudra que le corps legislatif y vienne un jour"; p. 154, "Voilà ce qu'elles ont fait seules; voilà ce qui fut la base de la constitution française. Tout est sorti de là"; p. 280, "Elle (the Breton Club) a été l'origine et la source des jacobins"; II, p. 71, "Je n'en ai jamais fait les fonctions hors ma réception. Les marguillers étaient presque sans activite lorsque je suis sorti de place"; p. 144, "Pendant ma gestion" (as mayor); p. 157, "Le seul festin de ville donné sous ma mairie"; p. 164, "L'Assemblée nationale, qui a été accusée de s'être écartée de ses cahiers et de ses pouvoirs, a pourtant décrété toutes ces bases"; p. 205, "Je conserve précieusement ce vaisseau comme un titre de ce que j'ai été et pour eux et pour la ville de paris"; p. 298, "Le bon, c'est que je n'ai connu ce règlement que bien longtemps après être sorti de place"; p. 306, "Voyez son *Discours du 4 février 1791*"; p. 314, "Il me semble que le résultat de la constitution est une démocratie royale ou une monarchie démocratique."

[3] *Ibid.*, I, p. 13.

[4] In the document entitled, *J. S. Bailly à ses concitoyens*, Bailly wrote: "En décembre 1791, j'étais au Havre." The document is reprinted in the *Mémoires*, I, pp. 396–412; the sentence quoted is found on page 409.

from page 231 on, and all of the second were evidently written after the first week in February, 1792.[1] The proof is found on page 231. Here Bailly quotes from the pamphlet of Camille Desmoulins entitled, *J.-P. Brissot désmasqué par Camille Desmoulins.*[1] The pamphlet bears upon the first page the date of writing: *Paris, ce 1 févr. l'an 3e, et non 4e. de notre ère, en dépit du decret Ramond.* It is not likely that this pamphlet reached Bailly at Nantes before the second week in February. As we know that page 358 was written February 23, 1792,[2] it follows that the pages intervening between 231 and 358 must have been written, at the outside, in the two weeks preceding that date. It is probable, then, that the first two hundred and thirty pages of volume one were written during the latter part of January and the first part of February, 1792.

The part of the *Mémoires* between volume I, page 358, and volume II, page 303, was written between February 23 and June 14, 1792.[3] Pages 303 to 362 of the second volume must have been written in the same month of June, judging from an allusion on page 362,[4] and it is highly probable that the breaking off of the *Mémoires* abruptly at page 409 of volume II was due to the news of the invasion of the Tuileries on June 20,

[1]As the reference to the pamphlet of Desmoulins is found in a foot-note, it is possible, without doubt, that it may have been appended later than February, 1792, and not at the time that the text was written, but it seems highly probable that Bailly had the pamphlet before him when he wrote the text. The reasons for my belief are (1) the pamphlet could easily have reached him from Paris the second week in February; (2) it produced a sensation and it is highly probable that he received a copy at once; (3) the matter referred to entered more naturally into a foot-note than into the text; (4) he could have written without difficulty the pages between 231 and 358 after receiving the pamphlet and before February 23.

[2]"Si M. Barrère eut été écouté, bien des objets que le temps et les événemens ont amenés n'auraient pas été décrétés, la révolution aurait été moins complète; mis il nous aurait sauvé de l'anarchie qui a exposé et qui expose encore la constitution. (*Aujourd'hui 23 fevrier 1792*)." *Mémoires de Bailly*, I, p 358.

[3]"Ce vœu, jusqu'au moment où j'ecris, a été rempli (14 juin 1792)." *Ibid.*, I, p. 303.

[4]"L'Assemblée nationale fit, le 9 octobre, un décret provisoire, en 28 articles; elle institua les notables; elle régla que les procédures anciennes faites jusqu'alors subsisteraient, mais que toutes celles qui seraient faites après le décret, le seraient suivant les nouvelles formes. Il fallut élire des notables, il fallut que les juges apprissent un nouveau métier: pendant ce

1792. The writing of the *Mémoires* was interrupted for some time in the early part of June by a tour that Bailly made through western France.[1] This accounts, in part, for the fact that while the first volume was written in about a month, nearly four months more passed before the suspension of the writing in the second volume.

The evidence would seem to indicate, then, that Bailly began the writing of his *Mémoires* the latter part of January, 1792. He worked industriously through January and February, completing the first volume. After that, the work went more slowly, was interrupted, and finally ceased, probably the last week in June, 1792.

The writing was evidently done in a country house, near Nantes, where Bailly had fixed his residence after retiring from office.[2]

Mémoires, dealing with events happening in Paris and Versailles in 1789, written near Nantes in the winter and spring of 1792, even if composed by an eye-witness and one who declares that his "memory will be faithful," would not, as a rule, be looked upon as the most reliable of sources for historical work. It is true that Bailly, with a few exceptions, describes nothing but what he has seen; his *Mémoires* deal with (1) the elections of the Third Estate in Paris up to May 24,[3] when he went as a

temps, c'est-à-dire pendant deux ou trois mois, nous fûmes sans justice, les prisons se remplirent. . . . Aujourd'hui que les jurés commencent à travailler, nous nous sentons encore, plus de deux ans et demi après de cet encombrement des prisons et de cette impunité apparente des crimes." *Ibid.*, II, p. 362.

The decree was passed October 9, 1789; two years and a half from that date would give April 9, 1792; but as it was two or three months before the new courts were working, that time should evidently be added, giving June, 1792, and as these pages must have been written in the latter part of June, at the earliest, that date would also fit a reasonable interpretation of the passage.

[1] "Ce fut dans les premiers jours du mois de juin (1792) que Bailly, ancien maire de Paris, qui parcourait les départements, se rendit dans celui de la Vendée. Il voyageait avec sa femme qui prenait une part tres active aux affaires publiques." *Mémoires de Mercier du Rocher* in *Souvenirs et Mémoires*, April 15, 1899, p. 336. See also page 337. Mercier may have been mistaken about the date, and it may have been in the last days of June that Bailly appeared at Fontenay.

[2] *Mémoires de Bailly*, I, p. XXV.

[3] *Mémoires de Bailly*, I, pp. 1–71.

representative of the Third Estate to Versailles, with (2) the meetings of the Third Estate and of the National Assembly at Versailles until July 15 when he was chosen mayor of Paris,[1] and, finally, with (3) the affairs of the city government of Paris until October 2, 1789, when his record ends. Bailly was secretary of the electors in Paris, president of the assembly in Versailles, and at the head of the Paris government all through the period dealt with in the second volume of the *Mémoires*. He certainly had as excellent opportunities to see what was taking place as any man connected with the events described, and the part that he took in the events should have impressed the more important details upon his memory.

Bailly not only had unusual opportunities to see and to hear, but he is what might be called a good witness. He was in sympathy with the revolution, but with a moderate revolution, the revolution that substituted a constitutional monarchy for the arbitrary rule of irresponsible ministers.[2] He was an astronomer whose mind had been abruptly turned from astral to terrestrial affairs. His reputation was made, however, long before the meeting of the estates in 1789.[3] He was born in 1736, at the Louvre where his father held the place of "guard of the pictures of the king," a position that seems to have been hereditary in the family. The father wished his son to succeed him, and young Bailly was given lessons in drawing. His real inclinations showed themselves, however, when he received instruction in mathematics from a M. de Moncarville and later in astronomy from Claircault and the Abbé Lacaille. At the age of sixteen, he wrote two tragedies and later competed for prizes offered by different academies, his *éloge* of Leibnitz being crowned by the Academy of Berlin in 1769. His reputation was not due, to any large degree, to these literary exercises.

[1] *Ibid.*, I, pp. 71–395; II, pp. 1–11.

[2] "Dans un moment où le peuple s'était soulevé tout entier, non pas contre le roi, mais contre l'autorité arbitraire,' *Mémoires de Bailly*, II, p. 284. Also I, pp. 170, 191.

[3] The data for this notice of the life of Bailly, I have taken from the *Notice sur la vie de Bailly*, prefixed to volume one of the *Mémoires*, and from the notes supplied to the editors by the brother of Bailly and prefixed to volume three.

Under the guidance of the Abbé Lacaille he had acquired a considerable knowledge of astronomy, and in 1763 presented to the Academy of Sciences a paper entitled *Observations lunaires*, in which were collected numerous observations calculated under the direction of his teacher. On the death of Lacaille, he was elected to his place in the academy at the age of twenty-seven. In 1764 appeared his work upon *Les étoiles zodiacles*, in 1766 his *Essai sur les satellites de Jupiter*, and in 1771 an important *mémoire* upon the light of these satellites. His training as a writer and as a scientist fitted him admirably for the work that he now undertook and that occupied the years between 1771 and 1779, a *Histoire de l'astronomie ancienne et moderne*, in two volumes. This work was supplemented in 1787 by a *Histoire de l'Astronomie indienne et orientale*, in three volumes. In his history of ancient astronomy, he had disagreed with Voltaire concerning the origin of the sciences and had dedicated his volume to him. This led to a discussion in which Bailly published, in 1777, his *Lettres sur l'origine des sciences* and two years later the *Lettres sur l'Atlantide de Platon*. Both were dedicated to Voltaire. Although the hypotheses of Bailly were "more ingenious than solid," the form in which they were presented was attractive, and the "letters" had a great success.

Although destined to become a prominent figure in the revolution, Bailly does not seem to have contributed directly to the preparation of it. Invited to cooperate in the making of the encyclopedia, he refused on the ground that the government was opposed to it. It is said that at this time he was receiving a pension from the king. He certainly did receive one later.[1] "What appears certain is that the benevolence of the government opened to the historian of astronomy the doors of the Academy of Inscriptions and later (1784) those of the French Academy." In the same year in which he entered the academy, he was made

[1] "Ce jour (May 10, 1789), je fus instruit que le lendemain, au moments des nominations, on devait faire une motion tendante à exclure ceux qui tenaient directement ou indirectement au gouvernement, ceux qui avaient des pensions; cette motion m'écartait le premier, *Mémoires de Bailly*, I, p. 46; "La conduite de M. Bailly est d'autant plus remarquable, que sa fortune tout entière dépend du gouvernement." *Ibid.*, I, p. 50 See also I, p. 49.

a member of a commission to report upon the claims of Mesmer concerning animal magnetism. "In spite of the influence unfavorable to the German professor, his report was a *chef-d'oeuvre* of independence and impartiality. This report, in the opinion of the, public, did him the greatest honor." Bailly was appointed in 1786 one of the members of a commission to examine a plan for a new Hotel-Dieu for Paris. He was named reporter of the commission and his report received the approbation of the government. The outbreak of the revolution prevented the execution of his plan.

It is clear that in 1789 Jean-Sylvain Bailly was one of the most distinguished of the conservative burgesses of Paris. It may well be believed that when the king learned of his election as the first deputy of the Third Estate from Paris he remarked, "J'en suis bien aise, c'est un honnete homme."[1]

Such seems to have been the opinion of the contemporaries of Bailly, and it is the impression that one receives from the reading of his *Mémoires*. The long thin face and somber eyes of the portraits that have been preserved are not those of an intriguing politician. Bailly took life seriously. He was called "Bailly le modeste"[2] in an anecdote of the days before the revolution, and it is as a modest man above all things that he appears in the pages of his *Mémoires*. All his offices sought him, he asserts, and added, "I am certainly an example that proves that one may attain to everything and to the first honors without intrigue."[3] There is, however, reason to believe that he was not as lacking in ambition as he would have us think. It was believed that he coveted the place of secretary to the Academy of Sciences to which Condorcet was elected;[4] it is reported by one of the biographers of Bailly that when he learned that Buffon had pronounced in favor of the Abbé Maury for the French

[1] *Mémoires de Bailly*, I, p. 71.

[2] *Ibid.*, III, p. IV.

[3] *Ibid.*, I, p. 9: "Nul homme à Paris ne peut dire que je lui aie demandé ou fait demander son suffrage, pas même que j'ai témoigné aucun désir des place où je suis parvenu."

[4] *Ibid.*, I, p. IX.

Academy when Bailly was supporting another candidate, he exclaimed to the reporter of the anecdote, "M. le Comte de Buffon is my master in the art of writing, but I shall never see him again."[1] It is said that he kept his word. His ambition was, however, well concealed and free from intrigue. He realized that "men punish sometimes, by the refusal of a thing, the desire that you have shown to receive it."[2] He repeatedly announced that he did not expect to hold office although he was as repeatedly told by those about him that he would be elected to the assembly.[3] His speech in the assembly of the electors, announcing that "the larger part of his fortune was due to the favors and pensions of the government" was doubtless a thoroughly honest act, made, probably, without any eye to its possible favorable effect, but it was more helpful to him than any intrigue. "I do not think," he continued, "that I am thought of for the deputation, but I feel obliged to give this information that will forever deprive me of becoming a member of it," and he concluded that if his colleagues did him the honor to elect him, it would be his duty to refuse.[4] His election was practically assured under the circumstances

According to tradition, the city of Paris had always played the leading rôle in the Third Estate at the meeting of the States General.[5] On the third of June, Bailly was elected dean of the still unorganized commons.[6] He was one of the best known

---

[1] *Ibid.*, III, p. IV.

[2] *Mémoires de Bailly*, I, p. 56. Bailly made this observation in connection with his election as the first deputy from Paris and the failure of Target, who had distinguished himself by his pamphlets on the interests of the Third Estate and who had been president of the electoral assembly, to obtain the coveted honor. He attributed the failure of Target to the fact that he was a member of two electoral assemblies, Paris *within* and Paris *without the walls*.

[3] *Ibid.*, I, pp. 7, 8, 9, 20, 46, 56.

[4] *Ibid.*, I, p. 49. This speech was made in connection with the motion to exclude from the list of eligibles those who held office or received pensions from the government. Bailly remarked elsewhere (I, p. 56), "Ce qui me servit, c'est la motion même d'exclusion faite la veille; j'aurais, je crois, été nommé sans elle, mais elle me fait premier député."

[5] "On se souvenait que le prévôt des marchands de Paris etait membre né des états-généraux, et presque toujours le président du tiers." *Ibid.*, I, p. 30.

[6] *Memoires de Bailly*, I, p. 89.

and one of the least offensive members of the Paris delegation. Tronchet, as representative of the generality of Paris in the bureau, was out of the question. According to Bailly, Tronchet would not have participated in the election had he not urged him to do so and insisted that no occasion should be lost to maintain the influence of Paris in the assembly. Bailly presided with dignity as dean and later as president of the organized assembly. Although not aggressive, he was firm in maintaining the rights of the assembly even against the ministers.[1] His manner was conciliatory and he evidently made few if any enemies.[2] His views were those of the conservative reformers; he was friendly to Necker and devoted to the king.[3] He bore himself in the most creditable manner through the trying scenes of the early revolution, playing a prominent part in the days of June 16, 20, 22, and 23.[4] He kept himself well in the foreground of the narrative, realizing both at Versailles and at the Hotel de Ville in Paris that he was a prominent actor and that all that

[1] See his account of his conversation with Barentin, June 5, concerning the formal distinctions to be made between the reception of the Third Estate and of the other orders by the king: "J'abrégeai sur-le-champ la recherche, en protestant au ministre que, quelque légère que fût la difference, les communes ne la souffriraient pas,' *Ibid.*, I, p. 105.

[2] Note the matter of voting in the assembly (*Ibid.*, I, pp. 101–103), when he "gagna l'amitié de la presque totalité de mes collègues. On fut content de moi, et en effet je montrai à la fois et fermeté et sagesse." See further, I, p. 103, the audience with the king at the time of the death of the dauphin; I, p. 239, his attitude toward the nobility; I, p. 242, his remarks on the deputation from the Palais Royale; I, p. 194, his treatment of the dissenting deputy Martin d'Auch; I, p. 203, his treatment of the clergy and nobility after June 27; I, p. 136, where he asserts that he was esteemed by the whole assembly.

[3] *Mémoires de Bailly*, I, p. 6, "Mais si ces droits ont été recouvrés, il ne faut pas oublier qu'on le doit et à M. Necker et au roi, au ministre qui l'a proposé, et au roi, qui y a consenti: l'un et l'autre ont donné les moyens de la régénération de l'empire." "J'estimais M. Necker, et je craignais sa retraite," *Ibid.*, I, p. 225. "Le génie et les principes de M. Necker," *Ibid.*, I, p. 5. "Le despotisme n'entra point dans le caractère du roi; il n'a jamais désiré que le bonheur du peuple. . . . Puisque nous parlons des causes de la régéneration, disons que la première est dans le caractère de Louis XVI.," *Ibid.*, I, p. 6.

[4] *Mémoires de Bailly*, I, pp. 149–156, 180–194, 198–223. The conservative character of Bailly as presiding officer during these momentous days undoubtedly moderated the movement of the assembly and gave dignity to its acts.

concerned him was worthy of record.[1] In justification it should be said that in recording "the facts that passed under his eyes" he could not well have avoided speaking much of himself, and he certainly did not do so in an offensive manner. Conservative in his views, never favorable to extreme measures or methods,[2] he was considerate of those who disagreed with him. Not in harmony with the ideas of Mirabeau or of Abbé Maury, he was not blind to their ability, and praised that in them that was worthy of praise.[3] He was just toward Lafayette and acknowledged his popularity, although in a sense his rival.[4] Irritated by the disregard of his authority by the municipal assembly of Paris, he often criticised that body severely, but his point of view is not unreasonable, and the facts as found in the *Procès-verbal* might have justified a more vigorous attitude on his part.[5]

Bailly was not a great man, although he held important places. One even wonders, at times, how he went so far in the midst of men of so much greater knowledge of affairs and of so much greater political ability. In the early days of the revolution, perhaps these very things, by arousing jealousy, hampered their possessor more than they helped him. Bailly's point of view is that of the educated, middle-class reformer, who would have established a constitution that recognized the class to which he belonged and placed the ministry under the control of the assembly. He even believed that this could be done in cooperation with the king and that his pension would not suffer from

[1] He describes with evident satisfaction (I, p. 122), the applause that greeted the announcement, on June 8, of his election as president of the assembly. *Ibid.*, I, p. 256, concerning the enthusiastic treatment he received at the hands of the people of Chaillot; I, p. 278, the expression of appreciation of Bailly's services voted by the assembly. These are but a few illustrations of his naïve expressions of satisfaction over his popular successes. See II, p. 150, the letter of Marmontel, also I, p. 26.

[2] The two volumes are proof of this. His acts during 1789 and his comments upon these acts show him to be a man fundamentally opposed to extreme measures.

[3] *Mémoires de Bailly*, I, p. 8, for Maury, and I, p. 303, for Mirabeau.

[4] In spite of the *Procès-verbal* of the electors of Paris he is quite sure that he was elected mayor before Lafayette was elected commandant of the city militia. *Mémoires de Bailly*, II, p. 25; also, II, pp. 47, 135, 143, 145.

[5] *Ibid.*, II, pp. 72, 73, 92, 143, 147, 195, 261.

his acts.[1] S[illegible] a man would know little of the intrigues of the court and of the assembly, of the great underground currents of the revolution, of the serious dangers of the future. Whatever came [illegible] in [illegible] is limited vision, he described well and honestly, [illegible] rous of making a reliable record of what he had [illegible] l, even to the extent of repeatedly acknowledging [illegible] could not recall exactly what he had said or done on a [illegible] important occasion.[2]

But however well informed and honest the writer of *Mémoires* may be, no guarantee can be [illegible] infallibility of his memory. Bailly knew [illegible] lay in his record and he knew how to stren[illegible] rt, and the smaller part, of his *Mémoires* rests upon [illegible] memory; for the larger part of his work he refreshed [illegible] y by turning to the most valuable sources attainable. [illegible] sources can be had today, in writing the history o[illegible] 1789, than Bailly made use of in writing his *Mémoi*[illegible] examination of the two volumes shows that he has [illegible] following sources: the *Procès-verbal* of the electors of [illegible] is, the *Récit des séances des députés des communes*[4] (May 5–June 12, 1789), the *Procès-verbal*[5] of the National Assembly, the *Procès-verbal*[6] of the city

[1] *Ibid.*, I, p. 50: In reply to Mirabeau's remark that Bailly's conduct was so much the more remarkable as his entire fortune was at the mercy of the government, Bailly replied, "Je n'ai pas peur, le roi est trop juste pour me punir jamais d'avoir fait mon devoir."

[2] These very frank avowals of ignorance are frequent. The major part have been given in a foot-note above.

[3] *Procès-verbal des séances et délibérations de l'assemblée générale des électeurs de Paris, réunis à l'hôtel de ville le 14 juillet 1789*, 3 vols. Paris, 1790.

[4] Bailly refers to this record as follows: "Les adjoints au bureau, ou les députés des gouvernements, tenaient des notes, et c'est sur ces notes qu'a été dressé le récit des seances jusqu'au 12 juin, et qui a été imprimé. Il me sert de guide." (I, p. 95). See, also, I, p. 122.

[5] *Procès-verbal de l'Assemblée nationale*, 75 vols., Paris, 1789–1791.

[6] *Procès-verbal des séances de l'assemblée des représentants de la commune de Paris.* Printed in Paris in 1789. This record of the first assembly of the representatives of the commune, July 25–Sept. 18, forms the first volume of the *Actes de la commune de Paris* (Paris, 1894), edited by M. Sigismond Lacroix. The *Procès-verbal* of the second assembly, beginning September 19, 1789, is found in volume two of the *Actes.*

government of Paris, the *Procès-verbal*[1] of the conferences at Versailles for the verification of credentials, the *Courrier de Provence*, the *Journal de Versailles*, the *Journal de Paris*, the *Gazette de Versailles*, the *Patriote français*, the *Révolutions de Paris*, the *Chronique de Paris*, and the *Point du jour*. The *procès-verbaux* served as the foundation of the narrative, supplemented by the newspapers and by Bailly's recollections and observations. The *Mémoires* contain but little, as has been stated, that did not come directly under the observation of Bailly, but, as he did not trust to his memory for his facts, the relation of his *Mémoires* to the sources that he used is a question of the first importance.

The first natural division of the *Mémoires* includes a brief introduction and an account of the elections in Paris (I, pp. 1–68). The first six pages deal with the events leading to the convocation of the States General. It contains little that is of value in the way of statement of fact and is chiefly interesting from the point of view of what Bailly thought of it all in 1792. With page 7, he reaches the preliminaries of the elections, and introduces on that page and page 8 some recollections of personal conversations touching the possibility of his election as a deputy and what the outcome of the estates would be. The account of the district assembly follows on pages 9–13. It was probably composed without any aid to the memory. I infer this from the character of the account, devoid of details difficult to remember, and the fact that he could not remember the names of all the seven commissioners selected to draw up the *cahier*. He was a member of the commission, but he could not recall the name of the seventh man and left a blank that was never filled in.[2] It is evident that he did not have the record before him at the time of writing. If the minutes of the meeting of the district—that

[1] *Procès-verbal des conférences sur la vérification des pouvoirs.* Paris 1789. Bailly cites this work twice, I, pp. 96, 97, and uses it at times when he does not cite it.

[2] "Ces commissaires, au nombre de sept, furent MM. Marmontel, Bigot, Cholet, Moreau frères, . . . et moi, à qui l'on fit l'honneur de l'admettre," I, p. 12.

of the Feuillants—have not been preserved, Bailly's account of the meeting is valuable.

The general assembly of the representatives of the districts for the election of the representatives of the Third Estate for Paris is dealt with in the pages 13–68. Pages 13–17, inclusive, contain a brief account of the meeting of the electors of the Third Estate at the Hotel de Ville for the purpose of ascertaining if the districts had all elected representatives and for the additional purpose of depositing the records of the elections, and an account of the meeting of the three estates in the great hall of the archbishopric for the opening ceremonies. These pages are evidently written from memory and contain several interesting episodes that might well be useful to the historian of the elections of Paris. The account of the proceedings of the electors of the Third Estate is full of detail, contains much quoted matter, and clearly could not have been written from memory. A hint of the source is found in the footnote on page 23 referring to the *Procès-verbal des électeurs.* A comparison of Bailly's narrative with the original record shows that his fifty pages are a condensation of the first eighty-four pages of the *Procès-verbal,* together with extracts from the *cahier*[1] of the Third Estate, one incident drawn from the *Journal de Paris* of May 20, and some important personal recollections. The work is very carefully done, and the account that Bailly gives, although briefer, is thoroughly reliable. The dependence of his text upon the *Procès-verbal* is much closer than would be imagined from the reading of the *Mémoires;* he frequently uses the exact language of the original without indicating it in any way.[2] Perhaps he felt justified in doing so as he was the secretary of the assembly and was only utilizing the record that he himself had made. It would be a mistake to suppose, however, that the existence of the original *Procès-verbal* renders the use

---

[1] The cahier was easily accessible to him, having been printed among the documents in the third volume of the *Procès-verbal d s électeurs.*

[2] Quotation marks, as a rule, are used when a speech or document is introduced into the text. The only way to appreciate the dependence of

of the *Mémoires* superfluous for the study of the electoral assembly; the *Mémoires* correct the *Procès-verbal*, or supplement it, on several important points. The *Procès-verbal* states that the decree of the assembly concerning the freedom of the press—apropos of the suppression of Mirabeau's newspaper by the government—was carried unanimously. Bailly claims in his *Mémoires* that this is not true, that when the nays were called for Marmontel had the courage to stand up alone. He attributes his failure to be elected to the States General to the discontent of the assembly at this act.[1] The discussion of the question of the eligibility of persons receiving pensions and of nobles and clergymen as representatives of the Third Estate can not be understood from the *Procès-verbal* alone. The election of the Abbé Siéyès as a representative was made possible by the failure of Bailly as secretary to record the vote of the assembly excluding ecclesiastics from the list of possible candidates.[2] The incident is an important one and the record of it is found in the *Mémoires* alone.

On the 24th of May, Bailly, with the other representatives of the Third Estate, went to Versailles. The remaining pages of volume I (68–395) and the first eleven pages of the second volume deal with the incidents at Versailles, up to the time of Bailly's acclamation as mayor of Paris. The first five pages treating of this period (I, pp. 68–72) are drawn from memory, and while containing nothing of first importance—the dress of the deputies, the reception by the king, meeting with the Abbé

Bailly upon the *Procès-verbal* is to collate the two works line by line. One illustration will be sufficient:

Procès-verbal, I, p. 3.	*Mémoires de Bailly*, I, p. 20.
La matierè a été mise en déliberation; et la très-grande pluralité ayant été d'avis que l'Assemblée ne pouvait avoir d'autres officiers que ceux qu'elle aurait élus librement. M. le procureur du roi a requis la retraite de MM. les officiers du Châtelet, lesquels se sont tous effectivement retirés, ayant M. le lieutenant civil à leur tête.	La matière mise en délibération, la très-grande pluralité fut d'avis que l'assemblée ne pouvait avoir d'autres officiers que ceux qu'elle aurait élus librement. M. le procureur du roi requit la retraite des officiers du Châtelet, qui se sont en effect retirés ayant M. le lieutenant civil à leur tête.

[1] *Mémoires de Bailly*, I, pp. 42, 43.

[2] *Ibid.*, I, pp. 48–51, 59–64.

ury, rec ion in the assembly—are not to be despised as umulative e lence in the treatment of the early history of the asser ly. On page 72, the history of the States General begins id lows here the *Récit* and the *Procès-verbal des conféren* conscientiously as he had followed the *Procès-verbal* o e electors of Paris and uses these records in much e same , condensing or employing the very language of source. For the period up to June 12, when the *Procès-verbal de l'assemblée des communes* began (I, p. 137), besides the two sources already m y made use of the *Cour-er de Provence* twice,[2] t o *ersailles* twice,[3] and the *ournal de Paris* three times.[4] n additions are of some alue, consisting, bes s nd paragraphs scattered here and there, of an ction as dean on June 3 (I. pp. 88, 89); de pt to call upon the king the same and following 91, 92, 93, 94, 102, 103, 4, 105, 106, 140, 143); made by Bailly to intro-ice order into the deliberat assembly (I, pp. 100, 101), and of the services at Notre Dam on June 11 (I, pp. 136, 137).

On June 12, the *Procès-verbal* begins and Bailly follows it closely,[5] making use, also, in the period up to July 15, of the *Courrier de Provence* seven times,[6] the *Journal de Versailles*

---

[1] The account of the session of May 25 fills nearly two pages in the *Récit;* Bailly gives the substance of it in less than half a page. The *Procès-verbal* devotes twenty-three pages to the account of the first sessions of the commissioners selected to consider the matter of how the credentials should be verified; Bailly disposes of it in three pages. Bailly's account of the session of May 26 is a skilful condensation and combination of the *Procès-verbal* of the conference of the 25th of May and the *Récit* of the 26th. The account of the session of the 27th in the *Mémoires* is pieced together from extracts taken literally from the *Récit.*

[2] *Mémoires de Bailly,* I, pp. 79, 110.

[3] *Ibid.,* I, pp. 84, 135.

[4] *Ibid.,* I, pp. 43, 87, 98.

[5] The account of the session of the 12th of June is taken from the *Récit* for the first part of the day and from the *Procès-verbal*—that began on this day—for the last part. A comparison of the *Procès-verbal* for June 22 (I, No. 1), with the *Mémoires* (I, pp. 199–202) will furnish a good example of Bailly's dependence upon his source.

[6] *Mémoires de Bailly,* I, pp. 147, 212, 278, 284, 307, 315, 333.

four times,[1] the *Journal de Paris* once,[2] the *Point du jour* seven times.[3] His own additions are numerous and often important. Mentioning only those of at least a paragraph in length, they are: an account of the critical session of the evening of June 16 (I, pp. 150–156); a correction of the *Procès-verbal* (I, p. 167); observation on the use of the word *décret* by the assembly (I, p. 171); observation on the expression *classes privilégiées* employed in the king's letter of June 17 (I, p. 175); remark on the vote *par tête* made June 18 (I, p. 175); incident of Madame de Tessé (I, p. 175); disorder in the assembly (I, p. 176); Bailly present in the street when the clergy voted to join the commons (I, pp. 178, 179); members of the assembly threatened with violence because of their opinions (I, pp. 179, 180); the events of June 20, in which Bailly was so prominent a figure (I, pp. 180–194, the personal recollections of Bailly being combined with the account of the *Procès-verbal*); the incident of the return of Martin d'Auch to the assembly (I, pp. 192, 193, 194); letter from the king, June 21 (I, pp. 196, 197); the union of the clergy with the commons, June 22 (I, pp. 198, 203); arrangements for the Royal Session, call upon the guard of the seals, and midnight interview of Bailly with members of the liberal nobility (I, pp. 204–206); account of the Royal Session, in which Bailly supplements from memory the accounts of the *Procès-verbal* and of the *Courrier de Provence* (I, pp. 206–223); admittance of the public to the hall of the commons (I, pp. 225, 226); attempt of the people to force an entrance into the hall (I, p. 233); attempt on the part of Bailly to suppress applause in the assembly (I, p. 247); valuable personal recollections on the conduct of the majority of the nobility and the minority of the clergy (I, pp. 247–264); the affair of the French Guards (I, pp. 266–268); personal notes on Villedeuil and Breteuil (I, pp. 307–310); recollections of the 14th of July, running through the accounts taken from the sources (I, pp. 359–395); the king in the assembly (I, pp. 7–11).

With the 15th of July, the last period of the *Mémoires* opens.

---

[1] *Ibid.*, I, pp 246, 304, 312, 317.

[2] *Ibid.*, I, p. 343.

[3] *Ibid.*, I, pp. 147, 192, 247, 253, 263, 275, 278.

It deals with the incidents of the mayoralty of Bailly up to October 1, 1789. The chief sources are the *Procès-verbaux* of (1) the electors of Paris[1] (to July 25), of (2) the first assembly of the representatives[2] (July 25–September 18), of (3) the second assembly of the representatives of the commune (September 18–October 1). The most of the material is drawn from these records, all of which were accessible in printed form when Bailly wrote. He constructs his journal by selecting from these records what appears to him important enough to be emphasized and especially the incidents with which he was connected. When his journal contains little under a given date, it is not due to the failure of his memory, but to the lack of interesting material in the sources that he is consulting.[3] For the most part he selects and condenses, the condensation being much greater in this second volume than in the first; at times he follows the text closely, putting the narrative into the first person, but without quotation marks.[4] In the second volume, he refers only incidentally to what is taking place in the assembly at Versailles, taking his material from the *Procès-verbal* or the newspapers.[5] For the events in Paris, he refers frequently to the papers, quoting the *Gazette de Versailles*,[6] the *Patriote français*,[7] the *Révolutions de*

[1] The *Procès-verbal des électeurs* was largely composed *après coup*. An account of its composition is given in detail in the third volume (pp. 1-53) The substance of this account is found in Flammermont, *La journée du 14 juillet 1789*, pp. X–XV. Bailly was secretary from April 26 to May 21, Duveyrier from May 22 to July 30, 1789.

[2] The exact titles of these *Procès-verbaux* of these two assemblies have been given above. Bailly claimed (II, p. 147) that the *Procès-verbal* of the first assembly was not as reliable as that of the electors, "comme ces *procès-verbaux* n'ont été rediges que longtemps après, les rédacteurs ont mis ce qu'ils ont voulu."

[3] "Les jours présents ne me fournissent rien pour mon compte; je n'ai pas grand'chose non plus à dire de l'assemblée des représentants, ses procès-verbaux montrent le vide de ses séances" (II, p. 220).

[4] Compare especially pages 476 and 477 of volume one of the *Procès-verbal des électeurs* with pages thirty-two and thirty-three of the second volume of the *Mémoires*.

[5] These passages are not numerous, deal only with such important questions as the adoption of parts of the constitution, the discussion of the August decrees, or the declaration of rights, and are treated very briefly.

[6] *Mémoires de Bailly*, II, p. 197.

[7] *Ibid.*, II, pp. 201, 229, 243, 260, 277.

Paris,[1] the *Chronique,*[2] the *Point de jour,*[3] the *Journal de Versailles,*[4] the *Courrier de Provence,*[5] and the *Journal de Paris.*[6] He does not, however, recognize his full indebtedness to the papers, using their material and often their language without any acknowledgment.[7] He undoubtedly had before him when he wrote his correspondence with Necker during August and September, 1789,[8] and possibly some of the papers or records of the *comité des subsistances* of Paris.[9] His personal recollections do not form as large a part of the second volume as they did of the first; they relate chiefly to the work of the committee engaged in procuring food for Paris and to the friction between Bailly and the assemblies of the commune. His running comment upon the acts of the assemblies, and his interpretations of events are valuable, and the historian can not afford to overlook them. They are so interwoven with the narrative that it does not seem advisable to indicate them in detail. In the first two volumes of the *Actes de la commune de Paris* are found the

[1] *Ibid.*, II, pp. 80, 81, 219, 244, 319, 325, 368, 379, 385 392, 404.

[2] *Ibid.*, II, pp. 304, 315, 325, 330, 368, 372, 379, 392, 400.

[3] *Ibid.*, II, pp. 36, 256.

[4] *Ibid.*, II, pp. 69, 169, 253.

[5] *Ibid.*, II, p. 185.

[6] *Ibid.*, II, pp. 320, 332.

[7] In the second volume, for example, Bailly makes use, without citing his source, of the *Courrier de Provence*, the material on pages 37 and 38 being taken from volume one, page 445 of the *Courrier;* of the *Point du jour*, the passage, "M. l'abbé de Montesquiou a reconnu que les membres de la minorité du clergé s'étaient trompés, et qu'ils en faisaient l'aveu à la nation avec plaisir," (I. p. 218 of the paper) is literally reproduced on page 39 of volume two of the *Mémoires;* of the *Point du jour* (I, p. 232), on page 41; of the *Révolutions de Paris* (I, No. 2, p. 12), on page 77; of the same paper (I, No. 5, p. 36) on page 265; of the *Point du jour* (III, p. 46) on page 375. These are a few examples of the necessity of collating every passage in Bailly with the sources from which it might be drawn before crediting it to Bailly as independent evidence.

[8] Bailly refers to this correspondence on pages 235, 288, 356, and 371 of volume two. The letters have been published in volume four, pages 172 to 195 of the *Histoire parlementaire de la révolution française* by Buchez et Roux. A comparison of the contents of the letters with the text in the *Mémoires* will make clear that Bailly had the letters before him.

[9] There is only a possibility that Bailly had before him when he w ote the records of the *comité des subsistances.* He said of these records (II, p. 358), "Des registres, il y en avait peu, et ils n'étaient point parfaitement en ordre."

procès-verbaux used by Bailly, and in his excellent *éclaircissements,* the editor, M. Sigismond Lacroix, has indicated most of the passages in Bailly that add anything to the records.

What is the conclusion of the whole matter? Or, in other words, what is the value to the historian of the *Mémoires de Bailly?* It is a record made by a competent eye-witness, but by a witness who followed carefully the best sources in constructing his journal, supplementing these sources by his personal recollections. It is a compilation and yet not a compilation. Bailly could not recall the events and the days when they occurred without the use of the *procès-verbaux* and the newspapers, but when he had recalled them he seldom trusted to his memory for the order of the facts and often employed in his journal the very language found in the sources that he had consulted. By so doing, he gave his approval to the account, practically saying so it occurred and not otherwise. This approval certainly has some value, just as his corrections of the record have a value, but it is not the value that we attribute to the account of an independent witness. The *Mémoires* may be safely used by those to whom the sources from which Bailly drew are not accessible; the historian will use the work only when it corrects or supplements the sources upon which it is based or when he wishes to show the point of view of Bailly himself in 1792. Before any passage of the *Mémoires* is attributed to Bailly it must be collated with all the *procès-verbaux* and newspapers that may have served him as sources of information; if found in none of these, it may safely be treated as an independent bit of information. Failure to do this in the past has led historians to quote Bailly when he is not a source[1] and to charge him with drawing from

[1] Louis Blanc makes use of Bailly's *Mémoires* in volume three of his *Histoire de la révolution francaise,* but it seems to be a mere accident when he cites matter found in Bailly alone. An examination of the passages of the *Mémoires* cited by Blanc in his footnotes will show that many of the passages had been taken by Bailly from the *procès-verbaux* or from the newspapers. In these cases, naturally, the *Mémoires* should not have been cited. Louis Blanc's critical work was not of the highest order.

works that did not exist at the time when he wrote.[1] There are *mémoires* and *mémoires.* The most of those upon the French revolution were composed as Bailly composed his journal, although, as a rule, the writers did not show the same good judgment in selecting and in using their sources; some drew largely upon the memory—Grace Dalrymple Elliott, for example, in her *Journal of My Life during the French Revolution*[2]—and are extremely unreliable, at times, absolutely worthless. Just what the value of each *mémoire* is can be determined only after a careful critical study.

[1] The editors of the edition of the *Mémoires* published at Paris in 1821, state in a footnote to page 311, that the details of a certain incident related by Bailly on pages 310, 311 were drawn from the *Moniteur* of September 15, 1789. As the *Moniteur* of that date did not exist until after the death of Bailly, it was not possible for him to make use of it. The history of the origin of the numbers of the *Moniteur* that precede November 24, 1789, is a sealed book to many of those who make use of them.

[2] *Journal of My Life during the French Revolution*, by Grace Dalrymple Elliot, London, 1859. It would be somewhat difficult to find a work written by an eye witness that contains as many errors to the page as are contained in the first chapter of this work.

IV.—*On the Representation of Numbers as Quotients of Sums and Differences of Perfect Squares*

BY ROBERT E. MORITZ

Let the simple continuant[1] whose elements $a_r, a_{r+1}, \ldots a_{t-1}, a_t$, $r<t$, are positive integers, be denoted by

$$p(a_r, a_{r+1}, \ldots, a_t) \text{ or } p_{r,t}.$$

The fundamental property of this function, which may be established by induction, but most readily by expanding in terms of the minors of the first $s-r+1$ rows or columns its equivalent determinant, is expressed by

$$p_{r,t}=p_{r,s}\,p_{s+1,t}+p_{r,s-1}\,p_{s+2,t},\quad r<t. \qquad [1]$$

Limited by the definition of a continuant as ordinarily given, s being necessarily not less than r nor greater than t, $p_{a,b}$ is obviously meaningless when $a<r$, or $b>t$, that is, when it involves elements which do not exist in the continuant under consideration, and thus no attempt seems to have been made to extend formula [1] to values of $s<r$ or $>t$. Yet it frequently occurs that general expressions based upon [1] are to be specialized in a way which necessitates s to assume values outside the limits r and t, as when we expand a continuant with reference to the kth element, and then desire to let $k=1$ or 2. Thus constituents like $p_{1,0}$ or $p_{2,-1}$ may occur, which vitiate the results and require that values of s which give rise to such expressions be treated as separate cases.

This is quite unnecessary. Since $p_{a,b}$ is meaningless as a continuant for $a<r$ or $b>t$, it may be given any meaning we please. We shall therefore define it in a purely formal way by the fundamental equation [1].

[1]For definition and fundamental theorems see Chrystal, *Algebra*, Part II, p. 466.

Put in [1] for s, $r-1$ and we have

$$p_{r,t}=p_{r,r-1}\,p_{r,t}+p_{r,r-2}\,p_{r+1,t}.$$

This equation is obviously satisfied for $p_{r,r-1}=1$, $p_{r,r-2}=0$.

For $s=r-2$, we have

$$p_{r,t}=p_{r,r-2}\,p_{r-1,t}+p_{r,r-3}\,p_{r,t}$$

hence $p_{r,r-3}$ must equal 1.

For $s=r-3$,

$$p_{r,t}=p_{r,r-3}\,p_{r-2,t}+p_{r,r-4}\,p_{r-1,t}$$

which is satisfied for $p_{r,r-4}=0$, $p_{r-2,t}=p_{r,t}$.

Proceeding similarly we find successively

$$p_{r,r-1}=1,\ p_{r,r-2}=0,\ \ldots,\ p_{r,r-2k+1}=1,\ p_{r,r-2k}=0,$$

$$p_{r-2,t}=p_{r-4,t}=p_{r-6,t}=\ldots p_{r,t}.$$

Similarly by putting $s=t$, $t+1$, etc., we obtain

$$p_{t+1,t}=1,\ p_{t+2,t}=0,\ \ldots,\ p_{t+2k-1,t}=1,\ p_{t+2k,t}=0,$$

$$p_{r,t+2}=p_{r,t+4}=p_{r,t+6}=\ldots=p_{r,t}$$

so that if the continuant under consideration is $p_{r,t}$, $r<t$, we have

$$\left.\begin{array}{l} p_{r,r-2k+1}=1,\ p_{r,r-2k}=0,\ p_{r-2k,t}=p_{r,t} \\ p_{t+2k-1,t}=1,\ p_{t+2k,t}=0,\ p_{r,t+2k}=p_{r,t} \end{array}\right\}\ k \text{ positive.} \qquad [2]$$

By definition $p_{r,r}$*$=p(a_r)=a_r$, hence for $s=r$ [1] becomes

$$p_{r,t}=a_r\,p_{r+1,t}+p_{r+2,t}, \qquad [3]$$

and similarly for $s=t-1$ we have

$$p_{r,t}=a_t\,p_{r,t-1}+p_{r,t-2}. \qquad [4]$$

The relation between continuants and continued fractions, which has been deduced in various ways,[1] is most readily established by repeated application of [3] and [4]. For [3] gives us successively

---

*We could with equal reason write $p_{r,r}$ instead of p_r, but I wish to reserve the symbol $p_{r,r}$ for a later use.

[1] S. Günther, *Näherungswerthe v. Kettenbrüchen*, S. 31, Habilitationsschrift, Erlangen 1872.

$$p_{r,t}/p_{r+1,t} = a_r + \frac{1}{p_{r+1,t}/p_{r+2,t}},$$

$$p_{r+1,t}/p_{r+2,t} = a_{r+1} + \frac{1}{p_{r+2,t}/p_{r+3,t}}$$

$$\cdot \quad \cdot \quad \cdot \quad \cdot \quad \cdot \quad \cdot \quad \cdot$$

$$\cdot \quad \cdot \quad \cdot \quad \cdot \quad \cdot \quad \cdot \quad \cdot$$

$$p_{t-1,t}/p_t = a_{t-1} + \frac{1}{p_t \;/\; p_{t+1,t}} \cdot$$

Now $p_t = a_t$ and $p_{t+1,t} = 1$ by [2], consequently

$$\frac{p_{r,t}}{p_{r+1,t}} = a_r + \cfrac{1}{a_{r+1} + \cfrac{1}{a_{r+2} + \cdots + \cfrac{1}{a_{t-1} + \cfrac{1}{a_t}}}} \qquad [5]$$

Dirichlet[1] has introduced the symbol $(a_r, a_{r+1}, \ldots, a_t)$ to represent the fraction [5]; we may therefore write

$$\frac{p_{r,t}}{p_{r+1,t}} = (a_r, a_{r+1}, \ldots, a_t), \qquad r<t. \qquad [6]$$

Similarly we obtain from [4]

$$\frac{p_{r,t}}{p_{r,t-1}} = (a_t, a_{t-1}, \ldots, a_r). \qquad r<t. \qquad [7]$$

Finally, since by the definition of a continuant

$$p_{r,t} = p_{t,r} \qquad [8]$$

we have also

[1] Dirichlet, Werke, Bd. 2, S. 141.

$$\frac{p_{t,r}}{p_{t,r+1}} = (a_r, a_{r+1}, \ldots ; a_t), \quad r<t, \qquad [9]$$

and

$$\frac{p_{t,r}}{p_{t-1,r}} = (a_t, a_{t-1}, \ldots , a_r), \quad r<t. \qquad [10]$$

The formulae just deduced lend themselves admirably to the study of continued fractions, in fact many known results have thus been repeated by Sylvester, Günther, Muir, and others. I wish to add one application, which opens the way to some interesting results in the theory of numbers.

Let

$$x = (a_0, a_1, a_2, \ldots , a_n, y) \qquad [11]$$

where the a's are positive integers, and y arbitrary. Expressed as a quotient of continuants, by [6] we have

$$x = \frac{p(a_0, a_1, a_2, \ldots , a_n, y)}{p(a_1, a_2, \ldots , a_r, y)}$$

and then by [4]

$$x = \frac{y p_{0,n} + p_{0,n-1}}{y p_{1,n} + p_{1,n-1}}. \qquad [12]$$

If x is a pure recurring continued fraction, y is equal to x, and [12] gives us the quadratic in x,

$$x^2 p_{1,n} + x(p_{1,n-1} - p_{0,n}) - p_{0,n-1} = 0$$

from which

$$x = -\frac{p_{1,n-1} - p_{0,n}}{2p_{1,n}} \pm \frac{\sqrt{(p_{1,n-1} - p_{0,n})^2 + 4p_{1,n} p_{0,n-1}}}{2p_{1,n}}, \qquad [13]$$

that is,

Every pure recurring continued fraction is equal to some quadratic surd number.

Similarly, for the mixed recurring continued fractions

$$y = (b_1, b_2, \ldots , b^k, x)$$

where x is a pure recurring continued fraction we have

$$y = \frac{p(b_1, b_2, \ldots , b_k, x)}{p(b_2, \ldots , b_k, x)} = \frac{x p(b_1, b_2, \ldots b_k) + p(b_1, b_2, \ldots b_{k-1})}{x p(b_2, \ldots b_k) + p(b_2, \ldots b_{k-1})},$$

and therefore, since x is a quadratic surd number, we have

Every recurring continued fraction, whether pure or mixed, is equal to some quadratic surd number.

We return to equation [13]. By [6]

$$p_{0,n}/p_{1,n} = (a_0, a_1, \ldots, a_n) > 1$$

since a_0 and a_n are positive integers. Consequently

$$p_{0,n} > p_{1,n-1}, \quad \text{or} \quad p_{1,n-1} - p_{0,n} \neq 0,$$

and hence,

A pure recurring continued fraction can not be equal to a pure quadratic surd number.

If $a_0 = 0$, [3] gives us

$$p_{0,n} = p_{2,n}, \qquad p_{0,n-1} = p_{2,n-1},$$

and [13] goes over into

$$x = -\frac{p_{1,n-1} - p_{2,n}}{2p_{1,n}} \pm \frac{\sqrt{(p_{1,n-1} - p_{2,n})^2 + 4p_{1,n}p_{2,n-1}}}{2p_{1,n}}. \quad [14]$$

Now by [3] and [4]

$$p_{1,n-1} - p_{2,n} = a_1 p_{2,n-1} + p_{3,n-1} - (a_n p_{n-1,2} + p_{n-2,2})$$

hence in order that the rational constituent in [14] may vanish we must have

$$a_1 p_{2,n-1} + p_{3,n-1} - a_n p_{n-1,2} - p_{n-2,2} = 0,$$

or

$$a_1 = a_n + \frac{p_{2,n-2} - p_{3,n-1}}{p_{2,n-1}}. \quad [15]$$

Furthermore $p_{2,n-1}/p_{2,n-2} = (a_{n-1}, \ldots, a_2)$ and $p_{2,n-1}/p_{3,n-1} = (a_2, \ldots, a_{n-1})$ hence $p_{2,n-1}$ is greater than either $p_{2,n-2}$ or $p_{3,n-1}$ or their difference $p_{2,n-2} - p_{3,n-1}$. $(p_{2,n-2} - p_{3,n-1})/p_{2,n-1}$ is a proper fraction, and [15] is satisfied only when simultaneously

$$a_1 = a_n \quad \text{and} \quad p_{2,n-2} - p_{3,n-1} = 0.$$

We now treat $p_{2,n-2} - p_{3,n-1}$ precisely as we treated $p_{1,n-1} - p_{2,n}$ and find that $p_{2,n-2} - p_{3,n-1}$ can vanish only, if

$$a_2 = a_{n-1} \quad \text{and} \quad p_{3,n-3} - p_{4,n-2} = 0,$$

and this in turn leads to the general condition $a_{k+1} = a_{n-k}$.

Moreover, if these conditions prevail, the rational part of [14] vanishes, for in that case, $p_{1,n-1}=p_{n-1,1}=p_{2,n}$. We have therefore:

The necessary and sufficient conditions, that [13] may represent a pure quadratic surd, are

$$a_0=0,\ a_1=a_n,\ a_2=a_{n-1},\ a_3=a_{n-2},\ \ldots,\ a_{k+1}=a_{n-k}.$$

With these conditions, the recurring continued fraction becomes $x=(0,a_1,a_2,\ldots,a_2,a_1,x)$. The last few terms of this fraction are

$$+\cfrac{1}{a_2+\cfrac{1}{a_1+\cfrac{1}{x+}}} = +\cfrac{1}{a_2+\cfrac{1}{a_1+\cfrac{1}{0+\cfrac{1}{a_1+\cfrac{1}{a_2+}}}}} = +\cfrac{1}{a_2+\cfrac{1}{2a_1+\cfrac{1}{a_2+}}}$$

With this equation [14] may be written

$$x=(0,\dot{a}_1,a_2,\ldots,a_2,2\dot{a}_1)=\sqrt{\frac{p_{2,2}}{p_{1,1}}}, \qquad [16]$$

or its reciprocal

$$y=(\dot{a}_1,a_2,\ldots,a_2,2\dot{a}_1)=\sqrt{\frac{p_{1,1}}{p_{2,2}}} \qquad [17]$$

where the stars denote the beginning and end of the recurring elements, and for the sake of convenience $p_{1,1}$ and $p_{2,2}$ have been written for $p(a_1,a_2,\ldots,a_2,a_1)$ and $p(a_2,a_3,\ldots,a_3,a_2)$ respectively.

We now need only to take the result just obtained, namely, that every recurring continued fraction which represents a pure quadratic surd takes the form [17], in conjunction with the theorem that every pure quadratic surd can be expressed as a recurring continued fraction, to arrive at the equations.

$$\sqrt{\frac{L}{M}}=y=\sqrt{\frac{p_{1,1}}{p_{2,2}}},$$

or

$$\frac{L}{M}=\frac{p_{1,1}}{p_{2,2}},\quad L>M, \qquad [18]$$

for every non-quadratic rational number $N=L/M$, that is:

Every non-quadratic rational number can be expressed as a quotient of two simple reciprocal[1] *continuants, whose elements are positive integers, and of which one is formed by omitting the initial and final elements of the other.*

Moreover, since every quadratic surd can be expressed as a continued fraction, with unit numerators and positive integral denominators, in one way only, the representation of non-quadratic rational numbers by means of quotients of continuants is unique, there is a one-to-one correspondence between the domain of non-quadratic rational numbers and the totality of simple reciprocal continuants with a single cycle of positive integral elements. To each number N of this domain corresponds one definite continuant which we may call *its* continuant, and each number of the domain belongs to one or the other of two classes according as its continuant is of even or odd order. We may attach the order of its continuant to the number itself, and speak of numbers as having odd or even order, which of course has nothing to do with the odd- or evenness of the numbers themselves; 6, 19, 28, 51 are of an odd order, 10, 13, 58, 97 of even order.

If we admit continuants of more than a single cycle of elements, the one-to-one correspondence ceases to exist. For any pure quadratic surd is represented equally well by each of the forms

$$\sqrt{\frac{L}{M}}, (L>M), = (a_1, \underset{\cdot}{a_2}, \ldots, a_2, \underset{\cdot}{2a_1}),$$
$$= (a_1, \underset{\cdot}{a_2}, \ldots, 2a_1, \ldots, a_2, \underset{\cdot}{2a_1}),$$
$$= (a_1, \underset{\cdot}{a_2}, \ldots, 2a_1, \ldots, 2a_1, \ldots, a_2, \underset{\cdot}{2a_1}),$$
$$= (a_1, \underset{\cdot}{a_2}, \ldots, 2a_1, \ldots, 2a_1, \ldots, 2a_1, \ldots, a_2, \underset{\cdot}{2a_1}),$$
etc.,

consequently to the same number L/M corresponds each of the continuants

$$\left.\begin{array}{c} p(a_1, a_2, \ldots, a_1, a_2), \\ p(a_1, a_2, \ldots, 2a_1, \ldots, a_2, a_1), \\ p(a_1, a_2, \ldots, 2a_1, \ldots, 2a_1, \ldots, a_2, a_1), \\ p(a_1, a_2, \ldots, 2a_1, \ldots, 2a_1, \ldots, 2a_1, \ldots, a_2, a_1), \\ \text{etc.,} \end{array}\right\} \quad [19]$$

[1] A reciprocal continuant is one in which $a_k = a_{n-k}$ for every value of k.

containing respectively one, two, three, four, etc., cycles of elements.

We will use the shorter symbols $p(a_1, \ldots, a_1), p_1(a_1, \ldots, a_1)$, $p_2(a_1, \ldots, a_1), p_3(a_1, \ldots, a_1)$, etc., to represent the above continuants, and generally write $p_n(a_\lambda, \ldots, a_\mu)$ for the continuant in which the element $2a$, recurs n times between the initial and final elements a_λ and a_μ. We may then write

$$\left.\begin{aligned} \frac{L}{M} &= \frac{p(a_1, \ldots, a_1)}{p(a_2, \ldots, a_2)} = \frac{p_1(a_1, \ldots, a_1)}{p_1(a_2, \ldots, a_2)} \\ &= \frac{p_2(a_1, \ldots, a_1)}{p_2(a_2, \ldots, a_2)} = \ldots = \frac{p_n(a_1, \ldots, a_1)}{p_n(a_2, \ldots, a_2)} \end{aligned}\right\} \quad [20]$$

Let us consider the first of the continuant quotients [20] for a number whose order is even; in that case its continuant has two equal middle elements, a_k, and by means of [1] we obtain

$$\begin{aligned} p(a_1, \ldots, a_1) &= p(a_1, \ldots, a_k)p(a_k, \ldots, a_1) \\ &\qquad + p(a_1, \ldots, a_{k-1})\, p(a_{k-1}, \ldots, a_1) \\ &= p^2(a_1, \ldots, a_k) + p^2(a_2, \ldots, a_{k-1}), \end{aligned}$$

and likewise

$$p(a_2, \ldots, a_2) = p^2(a_2, \ldots, a_k) + p^2(a_2, \ldots, a_{k-1}),$$

hence,

I. *Every number whose order is even can be expressed as a quotient of sums of squares of positive integers.*

On the other hand, the continuant of a number whose order is odd has a single middle element, a_k and in that case

$$\begin{aligned} p(a_1, \ldots, a_1) &= p(a_1, \ldots, a_k)p(a_{k-1}, \ldots, a_1) \\ &\qquad + p(a_1, \ldots, a_{k-1})p(a_{k-2}, \ldots, a_1) \\ &= p(a_1, \ldots, a_k)p(a_1, \ldots, a_{k-1}) \\ &\qquad + p(a_1, \ldots, a_{k-1})p(a_1, \ldots, a_{k-2}) \\ &= \frac{p(a_1, \ldots, a_k)}{a_k}\Big[a_k p(a_1, \ldots, a_{k-1}) + p(a_1, \ldots, a_{k-2})\Big] \\ &\qquad + p(a_1, \ldots, a_{k-1})\Big[a_k p(a_1, \ldots, a_{k-1}) - p(a_1, \ldots, a_k)\Big] \\ &= \frac{1}{a_k}\Big[p^2(a_1, \ldots, a_k) - p^2(a_1, \ldots, a_{k-2})\Big]. \end{aligned}$$

since by [2]

$$a_k p(a_1, \ldots, a_{k-1}) + p(a_1, \ldots, a_{k-2}) = p(a_1, \ldots, a_k)$$

and

$$a_k p(a_1, \ldots, a_{k-1}) - p(a_1, \ldots, a_k) = p(a_1, \ldots, a_{k-2}).$$

Likewise

$$p(a_2, \ldots, a_2) = \frac{1}{a_k}\left\{p^2(a_2, \ldots, a_k) - p^2(a_1, \ldots, a_{k-2})\right\}.$$

In the quotient of $p(a_1, \ldots, a_1)$ by $p(a_2, \ldots, a_2)$, a_k divides out, and we have

II. *Every number whose order is odd can be expressed as a quotient of differences of squares of positive integers.*

If the continuant $p(a_1, \ldots, a_1)$ is of odd order, each of the other continuants [19] is of odd order. The middle element is a_k in the first, third, fifth and $2a_1$ in the second, fourth, etc. The steps that led to Theorem II., lead likewise to each of the equations:—

$$\frac{L}{M},\ L>M, \text{ order odd,}$$

$$\left.\begin{aligned}
\frac{L}{M} &= \frac{p(a_1, \ldots, a_1)}{p(a_2, \ldots, a_2)} = \frac{p^2(a_1, \ldots, a_k) - p^2(a_1, \ldots, a_{k-2})}{p^2(a_2, \ldots, a_k) - p^2(a_2, \ldots, a_{k-2})} \\
&= \frac{p_1(a_1, \ldots, a_1)}{p_1(a_2, \ldots, a_2)} = \frac{p^2(a_1, \ldots, 2a_1) - p^2(a_1, \ldots, a_3)}{p^2(a_2, \ldots, 2a_1) - p^2(a_2, \ldots, a_3)} \\
&= \frac{p_2(a_1, \ldots, a_1)}{p_2(a_2, \ldots, a_2)} = \frac{p_1^2(a_1, \ldots, a_k) - p_1^2(a_1, \ldots, a_{k-2})}{p_1^2(a_2, \ldots, a_k) - p_1^2(a_2, \ldots, a_{k-2})} \\
&\quad \cdot\quad \cdot\quad \cdot\quad \cdot\quad \cdot\quad \cdot\quad \cdot\quad \cdot\quad \cdot\quad \cdot\quad \cdot\quad \cdot \\
&\quad \cdot\quad \cdot\quad \cdot\quad \cdot\quad \cdot\quad \cdot\quad \cdot\quad \cdot\quad \cdot\quad \cdot\quad \cdot\quad \cdot \\
&= \frac{p_{2n}(a_1, \ldots, a_1)}{p_{2n}(a_2, \ldots, a_2)} = \frac{p_n^2(a_1, \ldots, a_k) - p_n^2(a_1, \ldots, a_{k-2})}{p_n^2(a_2, \ldots, a_k) - p_n^2(a_2, \ldots, a_{k-2})} \\
&= \frac{p_{2n+1}(a_1, \ldots, a_1)}{p_{2n+1}(a_2, \ldots, a_2)} = \frac{p_n^2(a_1, \ldots, 2a_1) - p_n^2(a_1, \ldots, a_3)}{p_n^2(a_2, \ldots, 2a_1) - p_n^2(a_2, \ldots, a_3)}
\end{aligned}\right\} \quad [21]$$

This result may be stated thus:

III. *Every number whose order is odd may be expressed in an unlimited number of different ways as a quotient of differences of squares of positive integers.*

If the continuant $p(a_1,\ldots,a_1)$ is of even order, the succession of continuants [19] have alternately even and odd order. The middle elements of the first, third, fifth, etc., is a_k, while the second, fourth, etc., have each $2a_1$ for the middle element. Hence we have successively

$$\frac{L}{M},\ L>M, \text{ order even,}$$

$$\left.\begin{aligned}
\frac{L}{M}=\frac{p(a_1,\ldots,a_1)}{p(a_2,\ldots,a_2)}&=\frac{p^2(a_1,\ldots,a_k)+p^2(a_1,\ldots,a_{k-1})}{p^2(a_2,\ldots,a_k)+p^2(a_2,\ldots,a_{k-1})}\\
=\frac{p_1(a_1,\ldots,a_1)}{p_1(a_2,\ldots,a_2)}&=\frac{p^2(a_1,\ldots,2a_1)-p^2(a_1,\ldots,a_3)}{p^2(a_2,\ldots,2a_1)-p^2(a_2,\ldots,a_3)}\\
=\frac{p_2(a_1,\ldots,a_1)}{p_2(a_2,\ldots,a_2)}&=\frac{p_1^2(a_1,\ldots,a_k)+p_1^2(a_1,\ldots,a_{k-1})}{p_1^2(a_2,\ldots,a_k)+p_1^2(a_2,\ldots,a_{k-1})}\\
&\cdots\cdots\cdots\cdots\cdots\cdots\\
&\cdots\cdots\cdots\cdots\cdots\cdots\\
=\frac{p_{2n}(a_1,\ldots,a_1)}{p_{2n}(a_2,\ldots,a_2)}&=\frac{p_n^2(a_1,\ldots,a_k)+p_n^2(a_1,\ldots,a_{k-1})}{p_n^2(a_2,\ldots,a_k)+p_n^2(a_2,\ldots,a_{k-1})}\\
=\frac{p_{2n-1}(a_1,\ldots,a_1)}{p_{2n-1}(a_2,\ldots,a_2)}&=\frac{p_n^2(a_1,\ldots,2a_1)-p_n^2(a_1,\ldots,a_3)}{p_n^2(a_2,\ldots,2a_1)-p_n^2(a_2,\ldots,a_3)}
\end{aligned}\right\}\quad [22]$$

This gives in the following theorem:

IV. *Every number whose order is even may be expressed in an unlimited number of different ways as a quotient of sums, as well as of differences, of squares of positive integers.*

The direct computation of these quotients for a given number is for large values of the suffix very laborious, but recurrence-formulae may be deduced by means of which, from the continuants which enter into the quotients for a given suffix, the quotients for the next higher suffix may readily be computed. Suppose that for some number of even order,

$$N=\frac{p_{2n}(a_1,\ldots,a_1)}{p_{2n}(a_2,\ldots,a_2)}=\frac{p_n^2(a_1,\ldots,a_k)+p_n^2(a_1,\ldots,a_{k-1})}{p_n^2(a_2,\ldots,a_k)+p_n^2(a_2,\ldots,a_{k-1})}$$

has been computed, then

$$N=\frac{p_{2n+1}(a_1,\ldots,a_1)}{p_{2n+1}(a_2,\ldots,a_2)}=\frac{p_n^2(a_1,\ldots,2a_1)-p_n^2(a_1,\ldots,a_3)}{p_n^2(a_2,\ldots,2a_1)-p_n^2(a_2,\ldots,a_3)}$$

may be readily calculated, for by [1]

$$\left.\begin{aligned}
p_n(a_1,\ldots,2a_1) &= p_n(a_1,\ldots,a_k)p(2a_1,\ldots,a_k)\\
&\quad+p_n(a_1,\ldots,a_{k-1})p(2a_1,\ldots,a_{k-1})\\
p_n(a_2,\ldots,2a_1) &= p_n(a_2,\ldots,a_k)p(2a_1,\ldots,a_k)\\
&\quad+p_n(a_2,\ldots,a_{k-1})p(2a_1,\ldots,a_{k-1})\\
p_n(a_1,\ldots,a_3) &= p_n(a_1,\ldots,a_k)p(a_3,\ldots,a_k)\\
&\quad+p_n(a_1,\ldots,a_{k-1})p(a_3,\ldots,a_{k-1})\\
p_n(a_2,\ldots,a_3) &= p_n(a_2,\ldots,a_k)p(a_3,\ldots,a_k)\\
&\quad+p_n(a_2,\ldots,a_{k-1})p(a_3,\ldots,a_{k-1})
\end{aligned}\right\}\quad[23]$$

The next higher quotient

$$N=\frac{p_{2n+2}(a_1,\ldots,a_1)}{p_{2n+2}(a_2,\ldots,a_2)}=\frac{p^2_{n+1}(a_1,\ldots,a_k)+p^2_{n+1}(a_1,\ldots,a_{k-1})}{p^2_{n+1}(a_2,\ldots,a_k)+p^2_{n+1}(a_2,\ldots,a_{k-1})}$$

is obtained by employing the recurrence formulae

$$\left.\begin{aligned}
p_{n+1}(a_1,\ldots,a_k) &= p_n(a_1,\ldots,2a_1)p(a_2,\ldots,a_k)\\
&\quad+p_n(a_1,\ldots,a_2)p(a_3,\ldots,a_k)\\
p_{n+1}(a_2,\ldots,a_k) &= p_n(a_2,\ldots,2a_1)p(a_2,\ldots,a_k)\\
&\quad+p_n(a_2,\ldots,a_2)p(a_3,\ldots,a_k)\\
p_{n+1}(a_1,\ldots,a_{k-1}) &= p_n(a_1,\ldots,2a_1)p(a_2,\ldots,a_{k-1})\\
&\quad+p_n(a_1,\ldots,a_2)p(a_3,\ldots,a_{k-1})\\
p_{n+1}(a_2,\ldots,a_{k-1}) &= p_n(a_2,\ldots,2a_1)p(a_2,\ldots,a_{k-1})\\
&\quad+p_n(a_2,\ldots,a_2)p(a_3,\ldots,a_{k-1})\\
p_n(a_1,\ldots,a_2) &= p_n(a_1,\ldots,a_k)p(a_2,\ldots,a_k)\\
&\quad+p_n(a_1,\ldots,a_{k-1})p(a_2,\ldots,a_{k-1})\\
p_n(a_2,\ldots,a_2) &= p_n(a_2,\ldots,a_k)p(a_2,\ldots,a_k)\\
&\quad+p_n(a_2,\ldots,a_{k-1})p(a_2,\ldots,a_{k-1})
\end{aligned}\right\}\quad[24]$$

The values of the constantly recurring continuants $p(2a_1, \ldots, a_k)$, $p(2a_1, \ldots, a_{k-1})$, $p(a_2, \ldots, a_k)$, $p(a_3, \ldots, a_{k-1})$, $p(a_2, \ldots, a_k)$, $p(a_2, \ldots, a_{k-1})$ are computed once for all, and then [23] and [24] are applied alternately in the computation of successive quotients. A different set of recurrence formulae similarly obtained can be employed in the computation of successive quotients when the number N is of odd order. For example,

$$113 = \frac{p(10,1,1,1,2,2,1,1,1,10)}{p(1,1,1,2,2,1,1,1)}$$

$p(a_1, \ldots, a_k) = 85$, $p(a_2, \ldots, a_k) = 8$,

$p(a_3, \ldots, a_k) = 5$, $p(2a_1, \ldots, a_k) = 165$,

$p(a_1, \ldots, a_{k-1}) = 32$, $p(a_2, \ldots, a_{k-1}) = 3$,

$p(a_3, \ldots, a_{k-1}) = 2$, $p(2a_1, \ldots, a_{k-1}) = 62$,

and now by successive substitution in [23] and [24] we readily obtain

$$113 = \frac{85^2+32^2}{8^2+3^2} = \frac{16\,009^2-489^2}{1\,506^2-46^2} = \frac{131\,952^2+49\,579^2}{12\,413^2+4\,664^2}$$

$$= \frac{24\,845\,978^2-758\,918^2}{2\,337\,313^2-71\,393^2} = \frac{204\,789\,589^2+76\,946\,640^2}{19\,264\,984^2+7\,238\,531^2}$$

$$= \frac{38\,560\,973\,865^2-1\,177\,841\,225^2}{3\,627\,511\,282^2-110\,301\,982^2}$$

$$= \frac{317\,833\,574\,080^2+119\,421\,234\,859^2}{29\,899\,267\,581^2+11\,234\,204\,776^2}$$

$$= \frac{59\,846\,656\,284\,458^2-1\,828\,010\,340\,118^2}{5\,629\,899\,846\,977^2-171\,964\,747\,457^2} = \text{etc.}$$

Of special interest is the inquiry under what conditions $p(a_1, \ldots, a_1)/p(a_2, \ldots, a_2)$ represents an integer. Let us suppose all the elements but one of the continuant given. Let the variable element x occupy the kth place. The problem then is, for what integral values of x will

$$\frac{p(a_1, \ldots, a_{k-1}, x, a_{k+1}, \ldots, a_{k+1}, x, a_{k-1}, \ldots, a_1)}{p(a_2, \ldots, a_{k-1}, x, a_{k+1}, \ldots, a_{k+1}, x, a_{k-1}, \ldots, a_2)} \qquad [25]$$

We expand the numerator of this quotient in terms of x by repeated application of [1] and find

$$p(a_1,\ldots,a_{k-1},x,a_{k+1},\ldots,a_{k+1},x,a_{k-1},\ldots,a_1) = Bx^2+2Gx+C,$$

where

$$B = p^2_{1,k-1}p_{k+1,k+1},\quad G = p^2_{1,k-1}p_{k+1,k+2}+p_{1,k-1}p_{1,k-2}p_{k+1,k+1},$$
$$C = 2p_{1,k-1}p_{1,k-2}p_{k+1,k+2}+p^2_{1,k-1}p_{k+2,k+2}+p^2_{1,k-2}p_{k+1,k+1}.$$

Similarly

$$p(a_2,\ldots,a_{k-1},x,a_{k+1},\ldots,a_{k+1},x,a_{k-1},\ldots,a_2) = Ax^2+2Hx+2F,$$

where

$$A = p^2_{2,k-1}p_{k+1,k+1},\quad H = p^2_{2,k-1}p_{k+1,k+2}+p_{2,k-1}p_{2,k-2}p_{k+1,k+1},$$
$$F = \tfrac{1}{2}(2p_{2,k-1}p_{2,k-2}p_{k+1,k+2}+p^2_{2,k-1}p_{k+2,k+2}+p^2_{2,k-2}p_{k+1,k+1}.$$

It follows that the integral values of [25] for integral values of x are given by the integral solutions of the equation

$$\phi = Ax^2y-Bx^2+2Hxy-2Gx+2Fy-C=0 \qquad [26]$$

For all values of $k>1$, this is a cubic form, the discussion of which can not be brought within the scope of a short paper, but for $k=1$, that is when the unknown element occupies the first place in the continuant of the numerator, we have, by virtue of [2]

$$B = p^2_{1,0}p_{2,2} = p_{2,2},\quad G = p^2_{1,0}p_{2,3}+p_{1,0}p_{1,-1}p_{2,2} = p_{2,3},$$
$$C = 2p_{1,0}p_{1,-1}p_{2,3}+p^2_{1,0}p_{3,3}+p^2_{1,-1}p_{2,2} = p_{3,3},\quad A = p^2_{2,0}p_{2,2} = 0,$$
$$H = p^2_{2,0}p_{2,3}+p_{2,0}p_{2,-1}p_{2,2} = 0,$$
$$2F = 2p_{2,0}p_{2,-1}p_{2,3}+p^2_{2,0}p_{3,3}+p^2_{2,-1}p_{2,2} = p_{2,2},$$

since

$$p_{2,1} = p_{1,0} = 1,\quad p_{2,0} = p_{1,-1} = 0,\quad p_{2,-1} = 1.$$

In this case we have therefore

$$-\phi = p_{2,2}x^2+2p_{2,3}x-p_{2,2}y+p_{3,3}=0,$$

apparently a quadratic form, but in essence linear. For since x and y are to be integers, x^2-y will be integral, and putting this equal to z our equation becomes

$$p_{2,2}z+2p_{2,3}x+p_{3,3}=0,$$

and it remains only to solve the congruence

$$2p_{2,3}x\equiv -p_{3,3}(mod\ p_{2,2}). \qquad [27]$$

If this congruence has a root x_0, which is always the case except when $2p_{2,3}$ and $p_{2,2}$ possess a divisor which is prime to $p_{3,3}$ then

$$y=\frac{p(x,a_2,\ldots,a_2,x)}{p(a_2,\ldots,a_2)}$$

will represent an integer for every value

$$x_\lambda=x_0+\lambda p_{2,2}$$

where λ is any integers which will render x_λ positive. z_λ will then be given by

$$z_\lambda=\frac{-(2p_{2,3}x_\lambda+p_{3,3})}{p_{2,2}}=-2\lambda p_{2,3}-\frac{2p_{2,3}p_{3,3}}{p_{2,2}}x_0$$

and y_λ by

$$y_\lambda=x_\lambda^2-z_\lambda.$$

The results just obtained are of importance in the theory of recurring continued fractions and may be used to construct with comparative ease a table much more extensive than the famous Canon Pellianum. For example, the integral values x which render

$$y=\frac{p(x,1,1,2,1,1,x)}{p(1,1,2,1,1)}$$

integral are given by the congruence

$$2p(1,2,1,1)x\equiv -p(1,2,1)\left[mod\ p(1,1,2,1,1)\right]$$

that is by

$$14x \equiv -4 (mod\ 12).$$

The smallest positive root of this congruence is 4, hence . integral for every value

$$x_\lambda = 4 + 12\lambda,\ \lambda = 0, 1, 2, 3, \ldots.$$

We have accordingly

λ	x_λ	x_λ^2	$-z_\lambda$	y_λ
0	4	16	5	21
1	16	256	19	275
2	28	784	33	817
3	40	1600	47	1647
4	52	2704	61	2765
5	64	4096	75	4171
6	76	5776	89	5865
7	88	7744	103	7847
8	100	10000	117	10117
9	112	12544	131	12675
10	124	15376	145	15521

and finally

$$\sqrt{y_\lambda} = (x_\lambda, \dot{1}, 1, 2, 1, 1, \dot{2x_\lambda}).$$

Volumes I and II of UNIVERSITY STUDIES are each complete in four numbers.

Index and title-page for each volume is published separately.

A list of the papers printed in the first two volumes may be had on application.

Single numbers (excepting vol. I, no. 1, and vol. II, no. 3) may be had for $1.00 each.

A few copies of volumes I and II complete in numbers are still to be had.

All communications regarding purchase or exchange should be addressed to

THE UNIVERSITY OF NEBRASKA LIBRARY

LINCOLN, NEB., U. S. A.

JACOB NORTH & CO., PRINTERS, LINCOLN

Lightning Source UK Ltd.
Milton Keynes UK
UKHW022003221118
332793UK00008B/1198/P